Lecture Notes in Earth Sciences

Lecture Notes in Earth Sciences

Edited by Somdev Bhattacharji, Gerald M. Friedman,
Horst J. Neugebauer and Adolf Seilacher

12

Stuart Turner (Ed.)

Applied Geodesy

Global Positioning System – Networks – Particle
Accelerators – Mathematical Geodesy

Springer-Verlag
Berlin Heidelberg GmbH

Editor

Stuart Turner
LEP Division, CERN
CH-1211 Geneva 23, Switzerland

Originally published as Internal Report under the Title:
Proceedings of the CERN Accelerator School of Applied Geodesy for Particle
Accelerators, Editor: S. Turner, Geneva 1987

ISBN 978-3-540-18219-1   ISBN 978-3-540-47821-8 (eBook)
DOI 10.1007/978-3-540-47821-8

2132/3140-543210

PREFACE

The CERN Accelerator School (CAS) was founded in 1983 with the aim to preserve and disseminate the knowledge accumulated at CERN (European Organization for Nuclear Research) and elsewhere on particle accelerators and storage rings. This is being achieved by means of a biennial programme of basic and advanced courses on general accelerator physics supplemented by specialized and topical courses as well as Workshops. The chapters included in this present volume are taken from one of the specialized courses, Applied Geodesy for Particle Accelerators, held at CERN in April 1986.

When construction of the first large accelerators started in the 1950's, it was necessary to use geodetic techniques to ensure precise positioning of the machines' components. Since that time the means employed have constantly evolved in line with technological progress in general, while a number of specific developments - many of them achieved at CERN - have enriched the range of available instruments. These techniques and precision instruments are used for most of the world's accelerators but can also be applied in other areas of industrial geodesy: surveying of civil engineering works and structures, aeronautics, nautical engineering, astronomical radio-interferometers, metrology of large dimensions, studies of deformation, etc.

The ever increasing dimensions of new accelerators dictates the use of the best geodetic methods in the search for the greatest precision, such as distance measurements to 10^{-7}, rigorous evaluation of the local geoid and millimetric exploitation of the Navstar satellites. At the same time, the powerful computer methods now available for solving difficult problems are also applicable at the instrument level where data collection can be automatically checked. Above all, measuring methods and calculations and their results can be integrated into data bases where the collection of technical parameters can be efficiently managed.

In order to conserve the logical presentation of the different lectures presented at the CAS school, the chapters presented here have been grouped under four main topics. The first and the fourth deal with spatial and theoretical geodesy, while the second and third are concerned with the work of applied geodesy, especially that carried out at CERN. Readers involved in these subjects will find in the following chapters, if not the complete answer to their problems, at least the beginning of solutions to them.

J. Gervaise   P.J. Bryant, Head of CAS

M. Mayoud   S. Turner, Editor

Applied Geodesy Group

CERN

LIST OF AUTHORS

BAKER, L.S.	8612 Fox Run, Potomak, USA
BEUTLER, G.	University of Berne, Switzerland
BOUCHER. C.	Inst. Géographique National, Saint-Mandé, France
BORRE, K.	Aalborg University, Aalborg Ost, Denmark
BURKI, B.	Institute for Geodesy & Photogrammetry, ETH Zurich, Switzerland
CAMPBELL, J.	Geodetic Inst. University of Bonn, FRG
CASPARY, W.F.	Univ. der Bundeswehr München, Neubiberg, FRG
COOSEMANS, W.	CERN, Geneva, Switzerland
DUFOUR, H.M.	Inst. Géographique National, Saint-Mandé, France
FISCHER, J.C.	CERN, Geneva, Switzerland
HAYOTTE, M.	CERN, Geneva, Switzerland
GERVAISE, J.,	CERN, Geneva, Switzerland
GOAD, C.C.	National Geodetic Survey, NOAA, Rockville, USA
GURTNER, G.,	Astronomical Institute, University of Berne, Switzerland
HUBLIN, M.	CERN, Geneva, Switzerland
ILIFFE, J.	University College London, United Kingdom
LASSEUR, C.	CERN, Geneva, Switzerland
MAYOUD, M.	CERN, Geneva, Switzerland
MORITZ, H.	Technical University, Graz, Austria
OLSFORS, J.	CERN, Geneva, Switzerland
QUESNEL, J.P.	CERN, Geneva, Switzerland
TROUCHE, G.	CERN, Geneva, Switzerland
UNGUENDOLI, M.	University of Bologna, Italy.
WELSCH, W.M.	Univ. der Bundeswehr München, Neubiberg, FRG
WILSON, E.J.N.	CERN, Geneva, Switzerland

CONTENTS

I. Global Positioning System and V.L.B.I.

GPS ITS DEVELOPMENT AND DEPLOYMENT

Leonard S. Baker
Retired, Director, National Geodetic Survey

ABSTRACT

The Navstar Global Positioning System (GPS) is an all-weather, space-based navigation system under development by the US Department of Defense. The military needs and requirements are well documented, but the combined utilization, savings, and deployment by the civilian users throughout the world will surpass the benefits to be reaped by DoD.

1. DEVELOPMENT

Scientists have always reached for goals that at the time seemed unobtainable. When those goals were near, new goals were envisioned to stretch our imagination, deepen our thoughts, and yes, day dream a little - "WHAT IF?".

This simplistic view is so very true of the Physical Sciences. When the first break-through was made concerning matter, it only opened one door along the way, and you immediately planned the attack on the next door. The installations, at each of the laboratories around the world, have monuments to this path of discoveries. Here at CERN there are the Proton Synchrotron, the Intersecting Storage Rings, the Super Proton Synchrotron; and now, the Large Electron Positron Collider, LEP, is under construction.

In the United States, there is renewed interest in the Super Colliding Accelerator, that may have a diameter two or three times the size of LEP. As a geodesist, who has worked on an accelerator ring, I have gained a greater appreciation of the extremely high surveying accuracies required. Each new plan for the next generation of accelerators makes me stop and wonder, "will they be able to meet the surveying accuracy requirements?". GPS will play a major role in meeting those stringent demands.

Humans have never been content to accept the norm for very long, and in 1957, when the Russians launched SPUTNIK, the gestation period of GPS began. GPS has been born and has grown quite rapidly in its short life. However, we have been treated to only a very limited preview of what GPS will be. The real GPS will not occur until the full constellation of GPS satellites are in orbit, in 1988, or 1989.

The world has changed dramatically since SPUTNIK. Everyday, each of us enjoys the benefits of the "Space Age" technology, in our work, in our play, and in our normal living. The world is moving towards being a better place to live in, not only for us, but for all humans on this planet we call earth, and GPS will contribute in so many and various ways.

GPS is an all weather, space based, navigation system, under development by the United States Department of Defence (DoD), to satisfy the requirements of the military forces. Requirements include, accurately determined positions, velocity, and time in a common reference system, anywhere on or near the Earth, on a continuous basis. The Air Force Systems Command's Space Division acts as the executive agent for the DoD in managing the GPS programme. All branches of the United States military are represented, as well as the Department of Transportation, and NATO.

The Navstar GPS navigation and time-transfer system operates on two L-band frequencies, L1 (1575.4 MHz) and L2 (1227.6 MHz). The system consists of three major segments; a space segment, satellites that transmit radio signals; a control segment, ground-based equipment to monitor the satellites and update their signals; and a user segment, equipment which passively receives and converts satellite signals into positioning and navigation information.

The GPS system has already completed Phase I, concept validation. Phase II, engineering development, is nearing completion, and a constellation of six development satellites are being maintained to support testing of both the military and civilian applications. The contract for the development of the operational satellites has been awarded and, barring any additional delays such as the shuttle disaster, operational satellites are scheduled to be placed in orbit starting this year, with a planned completion of the constellation in 1988.

The full constellation will contain 18 satellites, in six orbital planes, 60 degrees apart, and each plane will have three satellites. The satellites will operate in circular 20,200 km orbits, at an inclination of 55 degrees, with 12-hour periods. Spacing is planned to ensure that at least four satellites will be in view to the user at any time on a worldwide basis.

This is a DoD programme that could serve as a model for other programmes, anywhere. It is well planned, funded, tested, and is now being put into place. The over-riding questions that have yet to be answered are - what level of accuracy, in real time, will users other than DoD, be allotted, and - will the signals be distorted?

2. GEODETIC RECEIVERS

A number of GPS geodetic receivers are already being used to gather data, while others are undergoing tests in preparation for their employment. At the present time these include:

- MACROMETER V-1000, the first GPS geodetic receiver developed for precise surveying by Macrometrics Inc. The Aero Service Division of Western Geophysical Company of America, Houston, Texas, acquired the Macrometics Company, in March 1984.

- TEXAS INSTRUMENT TI 4100, developed by Texas Instruments, with funding from U.S. Government Agencies. This was the first dual frequency geodetic receiver to be produced utilizing both the P- and the S-code.

- GPS LAND SURVEYOR, MODEL 2002, developed by ISTAC Inc., Pasadena, California. It is an outgrowth of the SERIES technology, originally conceived and developed by MacDoran and his co-workers at the Jet Propulsion Laboratory, Pasadena, California.

- TRIMBLE 4000S, a second generation GPS instrument developed by Trimble Navigation, Ltd., Sunnyvale, California. The first GPS instrument developed by Trimble was the 4000A, a navigation instrument, developed primarily for the shipping industry.

- WM 101, a joint venture between the Wild Co., Heerbrugg, Switzerland, and Magnavox, Torrance, California. The instrument is presently undergoing testing, prior to being put into service.

- TR5S, developed by Sercel, France. These units have already collected data, but no additional information is available.

There will undoubtedly be numerous new geodetic receivers available in the near future. Many of these receivers will be second generation receivers. They will shrink in size and decrease in price as production increases to meet the demand of the multitude of users.

3. DEPLOYMENT

No one can dispute the needs, requirements, and planned use DoD has for GPS. The accomplishments will be too numerous to tabulate. However, the combined utilization, savings, and deployment by the civilian users, throughout the world, will surpass the benefits to be reaped by DoD, perhaps not in the near term, but within a very short span after the system is fully operational.

Even in the develement phase, GPS is being utilized to monitor seismic actions, both horizontal and vertical. It eliminates the time delay factor, and because the satellites are in space, ground truths can be located in regions completely outside the areas under study.

One of the major problems the scientist working at sea has always been confronted with, is accurate location while underway. Geophysical profiling is now, with the aid of computers, three-dimensional. When the profiles show a promising structure, the scientists can be confident that the coordinates are accurate, thereby ensuring that the drilling crew will be able to pierce the structure. Utilizing the GPS translocation mode, ships can now be assured of one metre accuracies, within a range of about 300 kilometres from the GPS receiver over a known location.

The most excitement GPS has generated, has been in the geodesy community. Using methods that will be discussed later by other speakers, geodesists have been able to achieve extremely high accuracies between two points. These are greater than the normal accuracies obtainable with traditional surveying equipment and approach, or sometimes exceed, the accuracies of special surveys at installations such as CERN.

When the electronic distance measuring equipment became available to the surveying and geodetic communities, it broke the bond the steel tape had had on them. Formerly, distances could only be measured if a clear line of sight was available and the angles had to be determined in order to permit the computations to be made. All that has changed. No line of sight is required, no angles are needed, only a window to the sky, a piece of space-age technology, an innovative computer program, and the radio signals from a military programme satellite.

3.1 Mapping

With the advent of satellites, mapping of the earth's surface and resources, has become a new science, a science that was only dreamed of 30 years ago. Remote sensing has been responsible for many new discoveries and has allowed new maps of the world to be produced, at least in regions where aerial photography is allowed. But the science of remote sensing has not reached the level of accuracy, or scale, required for both developed, and the developing nations. With GPS, employed in the translocation mode, and the other antenna mounted in an airplane, positions can be computed for the airplane, either in flight, or post-processed, to an accuracy of less than one metre within a range of 300 kilometres. With very little additional equipment, other than an aerial camera, the pictures taken can be used to produce maps to a scale of 1/40,000, with little or no ground control. Mapping on this scale is more than adequate for all developing nations who need to plan and establish the basic road systems, access the natural visible resources, manage the water resources, and institute major land reforms.

Most of the developing nations do not enjoy an adequate network of coordinated control, the existing system being established by methods now labelled as inadequate. The GPS instruments allow control to be established faster and less expensively than with other means. They can also be used to up-grade the existing facilities and to tie in separated areas to provide a single coordinated framework.

3.2 Air navigation

If the Precise Positioning Service (PPS) would be available to the civilian users, an accuracy of 15 metres in x, y, and z would be available for positioning aircraft in flight. That accuracy represents a distance that is less than the wing span of almost every aircraft in existance.

The present method for flight plans for commercial aircraft, is to assign a flight altitude in a designated corridor, with check points along the way. The aircraft is tracked by ground control systems that relay this information to air traffic controllers, assigned to monitor that particular flight. None of the in-flight and the ground tracking instrumentation can provide the pilot with real time locations as accurately as the PPS GPS system. Barring the problems all systems experience, a pilot, flying an airplane equiped with GPS, could take off in zero visibility, navigate to the destination, and land in near zero visibility without any outside assistance. When the coordinates, x, y, and z are established by GPS at the end of every runway, and PPS GPS is available, each of them becomes an emergency landing site.

The accuracy of the Standard Positioning Service (SPS), to be available to the world-wide community of civil users, will be 100 metres, in x, y, and z with a 95% confidence level. This is still as much accuracy as a pilot, travelling at 600 knots/hour, can use in flight, and it is only the z that is so useful to them in landing with reduced visibility.

One can speculate on the adverse side of GPS in aviation. Will inexperienced pilots, not proficient in flying by instrument, substitute the advantages of GPS for experience and attempt to fly in adverse weather conditions or after dark? Even cautious pilots become lost, perhaps due to weather fronts moving into the flight path, and fail to reach their destination even during daylight hours. GPS can add a measure of safety.

We are so prone to think only of the flying conditions of our respective countries where we have excellent facilities. Why should less fortunate people, living in distant lands without adequate air traffic control or improved airports, be subject to more hazardous flights when flying is often their only link to the outside? With GPS they will be able to take a quantum leap forward in air navigation and safety, even with the SPS, 100 metre accuracy.

3.3 Sea navigation

Shipping has used Transit satellite generated positions for many years. Transit has a 90 minute interval between fixes, therefore is not an adequate system for navigating in close quarters. When GPS is available world wide, it may spell the end of "freedom of the seas". No ship has the right to endanger the coastline of another nation, by polluting its shores with an oil spill. GPS will allow the nations of the world to assign dangerous cargo vessels to certain shipping lanes that will not cross, except at designated points. Very few ships collide when going in the same direction; the majority of collisions occur in passing or crossing situations. GPS can not alleviate the crossing situation, but can eliminate the chance of a passing collision, if shipping lanes are established, and GPS is a requirement.

3.4 Multipurpose cadastre

There is a very critical requirement for a land information system in the United States. The system is needed to improve land-conveyance and to provide for equitable taxation, information for resource management and environmental planning. We need a MULTIPURPOSE CADASTRE. We have needed it for more years than we care to admit.

The concept of the multipurpose cadastre, is a framework that supports continuous, readily available and comprehensive land-related information at the individual parcel level. The first component is a reference framework, consisting of a geodetic network. The geodetic control must be an integrated network of points on the ground, over the entire area of the parcels that are related to each other. For a true cadastre, the continental United States must have a single, densely spaced, related geodetic network. Most of the populated European countries have a densely spaced control network, that supports their cadastre. These nations have utilized, and benefited from the use of their multipurpose cadastre system for many, many years.

GPS presents to the United States, the first chance to have a truly continental integrated network of control, at an acceptable price, and within a reasonable time frame, a requirement for our cadastre. The continental network will be a part of the world network, to respond to the intercontinental needs.

3.5 Commercial application

One United States automobile manufacturer is planning to offer GPS in its 1989 models. A GPS receiver will be connected to a computer and digitized map system and will display on a small screen a continuous, detailed, map of the location of the automobile. This will be a very useful tool for travellers to a new location, and to those individuals who have difficulty finding their way. The system will allow a delivery firm to prepare a route map for the driver, increasing the efficiency of the driver and vehicle, and eliminating the, "I got lost", excuse. (How many more distractions can the human driver master?)

3.6 Rescue

GPS will become a very important element for air and sea rescue. When the search aircraft or vessel knows its exact location within a few metres, there is no doubt that the suspected area has been covered. If the aircraft, vessel, or person knows the exact location, and can relay this information, finding them becomes almost as easy as following the vehicle on the screen in the car.

3.7 Boundary locations

There are some countries that continue to have boundary disputes with their neighbours. Surprisingly enough, many of the disputes are not over the physical location, they are over the coordinates of the common turning points of the boundaries, such as the mountain peak. GPS may eliminate many of these problems by placing each country on an identical coordinate system, thereby establishing common coordinates for boundary points.

3.8 Day dreaming

If we are allowed to day dream, there are some excellent humanitarian projects where GPS could play a major role. What if we were able to build a GPS unit including a digitized map of Geneva on a small micro disk and small enough to fit into a back-pack. This could acquire the satellite signals, compute the position, and feed this information to the computer which would search the digitized map and then, verbally inform the person of his exact location.

I do not believe this is a "what if", but rather a "when" idea. When will it be available to all those people who need it now? Such a device would give them the freedom to move about with ease, with confidence, and with the assurance that they could go it alone and not be dependent upon others for guidance.

4. SUMMARY

The original GPS scientists may never hear the many thanks that surely will be voiced on behalf of their invention. They may never receive the recognition rightfully due to them, but they can take a great deal of pride in knowing that their efforts have benefited and will continue to benefit mankind, all over the earth, in so many ways, for so many years to come.

THE USE OF GPS IS ONLY LIMITED TO OUR IMAGINATION AND INGENUITY.

* * *

REFERENCES

R.W. King, E.G. Masters, C. Rizos, A. Stoltz, J. Collins; Surveying with GPS; Monograph No. 9, School of Surveying, The University of New South Wales, Kensington NSW, Australia 2033.

Need for a Multipurpose Cadastre; Panel on a Multipurpose Cadastre, Committee on Geodesy, Assembly of Mathematical and Physical Sciences, National Research Council; National Academy Press, Washington, D.C. 1980.

Positioning with GPS - 1985; Proceedings, First International Symposium on Precise Positioning with the Global Positioning System, Volume I, Rockville, Maryland, April 15-19, 1985.

GPS RECEIVER TECHNOLOGY

C. Boucher
Institut Géographique National, Saint-Mandé, France

ABSTRACT

The NAVSTAR/GPS system is already widely used for geodetic positioning, although the full capability is not yet available. Various types of receivers, using the C/A and/or P codes or codeless, are available as prototypes or commercially. This paper reviews these receivers.

1. THE GPS SYSTEM CONCEPT

The NAVSTAR/Global Positioning System (GPS) is a satellite-based system developed by the US DOD for worldwide navigation of any system moving in the Earth's environment (ground, maritime or aerospace vehicle). The capability of this system for geodesy has been illustrated many times. Although it will only be fully operational in the late eighties, a provisional constellation of satellites, available for a few hours per day, enables the geodesists not only to prove the efficiency of this system for precise positioning but also to use it for operational purposes, replacing efficiently many of the conventional procedures.

We shall review here the various receivers available for geodetic use, either prototypes, commercially available or under development. Similar surveys have already been made by J.D. Bossler and C.W. Challstrom[1], J. Hannah[2], P. Hartl et al.[3] and B. Remondi[4].

2. THE NAVSTAR SPACE SEGMENT (Fig. 1)

The GPS system uses an L-Band down-link from a cluster of several satellites in the common view of a receiver which can perform several types of measurements as described below.

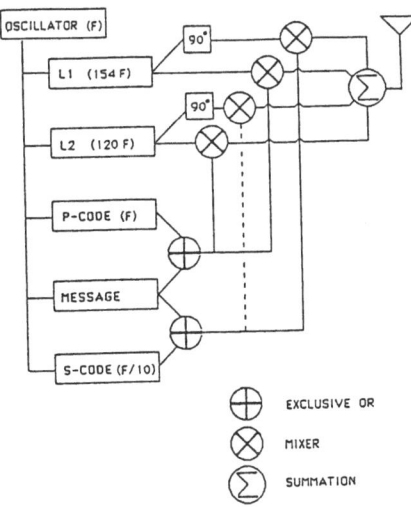

Fig. 1 GPS satellite signals

Each satellite transmits a pair of carriers at L1 = 1575.42 MHz and L2 = 1227.6 MHz. These are modulated by three levels of codes; namely C/A (also called S in the final stage of the system), P (called Y in the final stage) and message. Each code is a flow of bits, 0 or 1, represented by a ± 180 degree phase shift. The P code has a bit rate of 10.23 Mb/s, the C/A code of 1.023 Mb/s and the message code 50 b/s. The message and the P Code exist on both L1 and L2. C/A code exists only on L1 in quadrature with the P Code. For more details see Ref. 3.

3. MEASUREMENT TECHNIQUES

Three types of techniques can be conceived to perform measurements of geodetic interest:

3.1 Code correlation (Fig. 2)

This technique was adopted for normal use. The receiver produces a replica of the code (C/A or P) from a local oscillator. The code stream can be shifted in time and it is mixed with a beat signal from the received GPS signal and a pure carrier locally generated. The code is then shifted in order to obtain a maximum correlation between the received and the generated codes. The shift expresses directly the sum of the offset between satellite and receiver clocks, and the group propagation delay from the satellite to the receiver. This quantity, expressed in length through the velocity of light, is named pseudo-range.

The continuous alignment of codes cleans the received carrier on which phase or Doppler measurement now gives useful information.

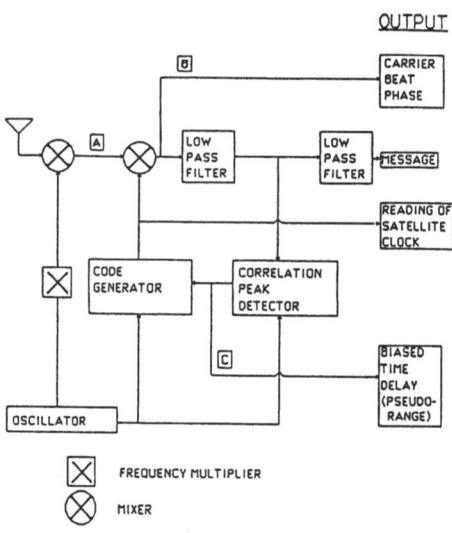

Fig. 2 GPS correlation channel

3.2 Signal squaring (Fig. 3)

The principal is to multiply the received signal by itself. In the case of the macrometer, this is done directly. As the code is in phase opposition, this technique removes it and produces a pure carrier with double frequency on which phase measurements can be performed.

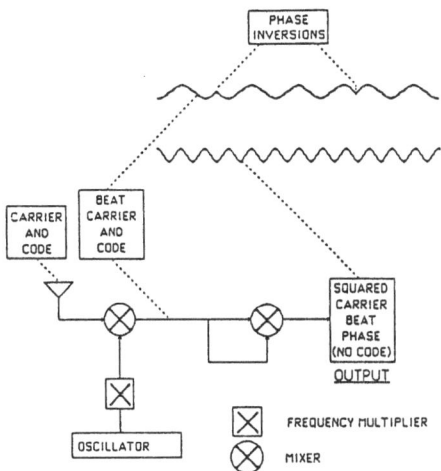

Fig. 3 GPS squaring-type channel

In the SERIES technique developed at the Jet Propulsion Laboratory, one signal is shifted in time by half the period of the code before being mixed with the other. It can be shown that epochs of code transition can be obtained precisely (see Ref. 5).

3.3 Interferometry

This last technique is identical to the one used in VLBI: each received signal is down converted and recorded. Later, signals are correlated among two or more simultaneous trackings.

To my knowledge, this has been used only for test purposes at the Haystack Observatory using a Macrometer prototype.

4. ERROR BUDGET AND MEASUREMENT MODELLING

A typical error for the pseudorange measurement is at the meter level while, for the phase, it is assumed to be millimetric (5×10^{-3} cycle). At present, the limiting factor of the GPS technique (as far as the 1 ppm application is concerned) is not the measurement in itself. The error due to the orbits, to the insufficiently accurate propagation models and to the internal clocks are larger. If the 0.1 ppm application is required all these error factors must be removed at the same time, then the problem of the measurement uncertainties becomes more crucial.

5. REVIEW OF EXISTING RECEIVERS

Several receivers are already available, either as prototypes or commercially. Table 1 lists only instruments which can be used for sub-metric geodetic positioning and which had been announced at the time of writing.

Table 1

Presently available receivers and their characteristics

Type	Manufacturer	Model	No. Chan	Frequency	Mode	No.SAT
Codeless receivers						
MAC	Macrometrics	MACROMETER 1000	6	L1	SQ and CT	6
MAC2	Western Geophys.Cy Amer. Aero Service Div.	MACROMETER II	12	L1/L2	SQ amd CT	6
SERIES	Jet Propulsion Lab.	SERIES	3	L1/L2	SD	n
SERX	Jet Propulsion Lab.	SERIES-X	3	L1/L2	SD	n
IST2002	ISTAC, Inc.	Model 2002	1(3)	L1(/L2)	SD	35
GLS1991	GEO/HYDRO	GPS Land Surveyor Model 1991	3	L1/L2	SD	n
C/A and P code						
STI5010	Stanford Telecom. Inc.	STI 5010	2	L1/L2	CC and LS	1
TI4100	Texas Instruments	TI 4100	1	L1/L2	CC and MX	4
C/A code						
4000A	Trimble	Model 4000 A	1	L1	CC and FS	4
4000S	Trimble	Model 4000 S		L1	CC	4
WM101	Wild-Magnavox	WM 101	4	L1(/L2)	CC and FS	9
TR5S	SERCEL	TR5S	5	L1	CC and CT	5
LGSS	Litton Aero Products	Litton GPS Survey Set LGSS		L1	CC	8
SEL	Standard Electrik Lorenz "low cost"		1	L1	CC and FS	4

5.1 Codeless receivers

The measurement technique is basically the signal squaring technique mentioned in Section 3.2. Several receivers are available on the market and as knowledge of the codes is not needed, they are usually double-frequency devices. On the other hand, they need to be synchronized before every experiment and the orbits need to be known from an external source.

5.2 C/A and P code receivers

The principle of these receivers is based on the code correlation. Both codes (C/A and P) must be known.

5.3 C/A code receivers

The technique is the same but only the C/A code (non-restrictive) needs to be known. As only one frequency is available the ionospheric delay must be removed by the use of models.

6. CONCLUSION

Even though the GPS system is not yet fully operational, it is widely used by the geodetic community. Many different receivers are actually available on the market. Their capabilities and measurement techniques are very different. At present, the precision of these instruments is sufficient for the 1 ppm applications. In the future, if significant improvements are realized in propogation models and in orbit determinations, this precision may be improved.

* * *

REFERENCES

1) J.D. Bossler and C.W. Challstrom, GPS instrumentation and federal policy, Proceedings of the 1st Int. Symp. on Precise Positioning with GPS, Rockville, April 15-19, 1985, NOAN (1985).

2) J. Hannah, The Global Positioning System - The positioning tool for the future, New Zealand Surveyor (1985), pp. 268-281.

3) P. Hartl, W. Scholler and K.H. Thiel, Review paper: GPS technology and methodology for geodetic applications, Proceedings of the Joint Meeting of Study groups 5B and 5C, Munich, July 1-3, 1985, Univ. der Bundeswehr Munich, Scheiltenreihe, Munich (1985).

4) B.W. Remondi, GPS Geodetic Receivers - A status update report, American Congress of Surveying and Mapping.

5) L.A. Buennagel et al., SERIES project: Final report on research and development phase, 1979 to 1983. JPL Publ. 84-16, NASA/UPL Pasadena (March 1984).

PRECISE POSITIONING WITH THE GPS

C. C. Goad
National Geodetic Survey
Charting and Geodetic Services
National Ocean Service, NOAA
Rockville, Maryland   20852
U.S.A.

ABSTRACT

A review of current measurement reduction techniques for precise
positioning using carrier phase observables from the Global
Positioning System satellites is presented. The qualities of
each are discussed. The main cause of data problems, cycle
slips, is also given along with an example, and then a possible
algorithm to correct slips over short time gaps when data from a
single station are given. It is then argued that this algorithm
will be useful for cycle slip detection in precise positioning
of a moving platform as well as in a static relative positioning
mode. Optimal estimation in the presence of unmodeled errors
is discussed.

1. INTRODUCTION

   The use of the Global Positioning System for precise surveys is now a reality. Several
government, university, and commercial agencies are using the various carrier phase measuring
receivers to obtain data which are processed interferometrically to provide base line vector
estimates which far exceed minimum first-order standards.

   The processing of these data can be separated into two generic classes depending on
the handling of cycle slips which can be caused by problems inside the tracking hardware
and by environmental causes. I call these two categories batch and interactive techniques.
The batch techniques do not require decisions to be made by an analyst during the actual
data reduction. They provide fast and accurate results but do not take advantage of the
integer nature of the ambiguities (more about ambiguities later). These batch techniques
do not yield the most precise estimates of base line vectors. Interactive techniques do
try to take advantage of the integer nature of the ambiguities, but because of the human
intervention these solutions require far more time and thus more expense. This situation
may change, however, as processing schemes and algorithms are improved. Obviously, more
automation is desired from a cost per base line point of view. Also, such improvements in
the static relative positioning processing might impact moving-platform applications as well.

The basic phase observable will be presented in Section 1. Section 2 will discuss the problem of cycle slips in phase data and how to fix them while Section 3 will address the cycle slip problem using data only from a single station. In Section 4 one technique to filter single station data will be presented with possible application to moving platforms discussed in Section 5. In Section 6, the problem of optimal estimation in the presence of unmodeled systematic errors is discussed.

2. THE PHASE OBSERVABLE

The basic carrier phase observable is the phase difference between the carrier signal transmitted at the satellite and a local receiver's oscillator. Mathematically this is expressed as

$$\phi_j^i(t_R) = \phi^i(t_t) - \phi_j(t_R) \tag{1}$$

where superscripts refer to a particular satellite; subscripts refer to a particular ground receiver; and the argument is either the transmit time, t_t, or the received time, t_R.

Equation (1) is now rewritten in a more usable form. The right side is written so that the times are expressed in terms of the received time, t_R. The distance from the satellite to receiver is given by $\rho_j^i(t_t)$. Therefore the travel time in a vacuum is given by $\rho_j^i(t_t/c)$ where c is the speed of light. Assuming that the oscillators run at a constant frequency, f, Eq. (1) is now rewritten as:

$$\phi_j^i(t_R) = \phi^i(t_R - \rho_j^i(t_t)/c) - \phi_j(t_R)$$

$$= \phi^i(t_R) - f/c \, \rho_j^i(t_t) - \phi_j(t_R) + N_j^i. \tag{2}$$

Notice that an integer N_j^i has been inserted in Eq. (2) to acknowledge that $\phi_j^i(t_1)$, the phase observable at the first measurement time, is determined modulo one cycle. If lock is maintained thereafter, the same N_j^i will appear in every phase observable between receiver j and satellite i. Some small terms have been ignored in the derivation of Eq. (2) such as the effects of relativity, ionospheric and tropospheric refraction and time tag errors. Some of these will be addressed later. Since all time arguments in Eq. (2) are implicitly or explicitly a function of the received time tag, t_R, the R designation will be dropped with the understanding that hereafter the time tag refers to the receipt time. Thus, Eq. (2) is rewritten

$$\phi_j^i(t_k) = \phi^i(t_k) - f/c \, \rho_j^i - \phi_j(t_k) + N_j^i \tag{3}$$

where k refers to the k-th sample time.

1.1 Differenced data types

Geodesists are mainly interested in using these phase observables for base line vector recovery. This information is extracted only from the term $-f/c \; \rho_j^i$ in Eq. (3) which is a function of the satellite and receiver locations. The other terms in Eq. (3) are the clock difference term, $\phi^i(t_k) - \phi_j(t_k)$, and the initial integer ambiguity, N_j^i. Selectively differencing these phase observables is one way to eliminate these non-geodetic terms and thus simplify the reduction algorithm. Goad and Remondi[1] first showed results from differencing phase observables over a common satellite to eliminate the satellite phase (or clock) term. For example, two stations, 1 and 2, at the (almost) instant, t_k, measure the phase difference between satellite 9 and the local oscillators. Using Eq. (3), this is expressed as

$$\phi_{2,1}^9(t_k) = \phi_1(t_k) - \phi_2(t_k) + f/c \; (\rho_1^9 - \rho_2^9) + N_1^9 - N_2^9. \tag{4}$$

Equation (4) is now referred to as the single difference observable between stations 1 and 2 at the k-th epoch to satellite 9. As anticipated, this observable does not contain the satellite phase value $\phi^9(t_k)$ but now contains differences in phase (or clock) values between ground receivers 1 and 2, initial ambiguity differences, and distance differences. The clock differences do not necessarily have a known relationship from one epoch to the next and thus must be estimated on an epoch-by-epoch basis. Furthermore, the receiver clock difference $\phi_1(t_k) - \phi_2(t_k)$ will be the same should another satellite participate in the difference processing.

The term involving the $\rho_1^9 - \rho_2^9$ can be calculated based on a knowledge of the station and satellite positions. $N_2^9 - N_1^9$ is the same for all epochs. Thus these two terms are "common" contributors and need not be considered on an epoch-by-epoch basis. The least-squares normal matrix possesses a structure which allows one to solve for the epoch receiver clock difference values rather easily[2]. For a given base line, no phase observables appear more than once, thus the measurements are uncorrelated and no special treatment are required. But when more than one base line are considered, the phase observables will appear more than once, and thus these measurements will, in general, be correlated. Some investigators have given results using double differences (for example, Goad and Remondi[1], Bock et al.[3]). These are obtained by taking differences of two single difference observables at a common epoch from two receivers to two different satellites. For an example, receivers 1 and 2 and satellites 4 and 9 at epoch k are chosen. Then

$$\phi_{2,1}^{4,9}(t_k) = \phi_{2,1}^4(t_k) - \phi_{2,1}^9(t_k)$$

$$= f/c \ (\rho_1^4 - \rho_2^4 - \rho_1^9 + \rho_2^9) + N_2^9 - N_1^9 - N_2^4 + N_1^4. \tag{5}$$

Now all satellite and receiver clock differences are removed. The right side contains four ranges and a linear combination of four ambiguitites. Unlike the single-difference observables, the double differences are less complicated to handle (no epoch-to-epoch clock differences to model or estimate), but the data sets are correlated. To optimally process this information one must decide whether to go with a more complex model as given by Eq. (4) for the single differences, or to choose the simpler double difference model given by Eq. (5) and then be forced to model the correlated nature of the double difference observables. Triple differences are generated by differencing two successive double differences in time. This has the effect of removing the bias and increasing the number of ranges to compute in the model to eight. Losses of lock will cause a spike in the triple-difference data rather than a step as in double-difference processing. These spikes (or outliers) are easily removed by an automatic editor or outlier detection scheme. This is an extremely desirable characteristic, especially when looking at the data initially or when the most precise results are not required. This was discussed by Goad and Remondi[1] where actual comparisons between single-, double-, and triple-differencing results were given.

In Eq. (5) the double difference is seen to possess a very desirable quality--the integer nature of the bias. Since all clock terms are removed in the differencing process, only base line and integer bias effects remain. The results of processing data from many base lines whose lengths are less than, say, 30 km are that the estimates of the biases are indeed very close to integers. This integer nature will deteriorate as the base line lengths are increased and non-cancelling effects such as ionospheric and tropospheric refraction and satellite orbit errors contaminate the solutions.

If the integers can be determined, then these biases should be forced to keep their integer values and be excluded from the set of solution parameters. This greatly strengthens the recovery of base line components. As carrier phase tracking receivers become more available, the opportunity to simultaneously collect data from several locations will become more routine. Not only is this desirable from an economical point of view, but with 3 or more receivers collecting data at the same time, vector closure is an important characteristic of the data reduction scheme when all data are processed together.

One data scheme whose least-squares weight matrix will remain uncorrelated (diagonal) no matter how many receivers are collecting data simultaneously is one which analyzes the undifferenced phases. It is quite obvious that a phase-only algorithm will consider uncorrelated data, but how does one exploit the integer nature of the biases? This question will now be answered.

1.2 The base station - base satellite concept

The base station - base satellite concept can best be presented by an example. Suppose that at the k-th epoch a sufficient number of phase measurements are collected to constitute a double difference. Again let us assume there are two stations, 1 and 2, and two satellites, 4 and 9. Then from Eq. (3), the measurements are given as follows:

$$\phi_1^4(t_k) = \phi^4(t_k) - f/c \; \rho_1^4 - \phi_1(t_k) + N_1^4$$

$$\phi_2^4(t_k) = \phi^4(t_k) - f/c \; \rho_2^4 - \phi_2(t_k) + N_2^4$$

$$\phi_1^9(t_k) = \phi^9(t_k) - f/c \; \rho_1^9 - \phi_1(t_k) + N_1^9$$

$$\phi_2^9(t_k) = \phi^9(t_k) - f/c \; \rho_2^9 - \phi_2(t_k) + N_2^9.$$

Again it should be emphasized that these four phase measurements could be used to generate one double-difference observable. Now the above relations are rewritten substituting $\bar{N}_j^1 = N_j^1 + \phi^1 - \phi_j$ in each equation. One gets

$$\phi_1^4(t_k) = - f/c \; \rho_1^4 + \bar{N}_1^4(t_k)$$

$$\phi_2^4(t_k) = - f/c \; \rho_2^4 + \bar{N}_2^4(t_k)$$

$$\phi_1^9(t_k) = - f/c \; \rho_1^9 + \bar{N}_1^9(t_k)$$

$$\phi_2^9(t_k) = - f/c \; \rho_2^9 + \bar{N}_2^9(t_k). \tag{6}$$

Arbitrarily, station 1 and satellite 4 will be chosen as the base station and base satellite, respectively. Then let us define

$$K_2^9 = [\bar{N}_2^9(t_k) - \bar{N}_2^4(t_k)] - [\bar{N}_1^9(t_k) - \bar{N}_1^4(t_k)].$$

Notice that no time argument is included in the K variable. This is intentional. After substituting for the N's, one gets

$$K_2^9 = N_2^9 - N_2^4 - N_1^9 + N_1^4. \tag{7}$$

That is, the K variable consists of the linear combination of the initial integer ambiguities. It remains the same throughout an observation session if lock is maintained. This is the essence of the algorithm. Now the four measurements are rewritten with only the fourth measurement actually changed.

$$\phi_1^4(t_k) = -f/c \ \rho_1^4 + \bar{N}_1^4(t_k)$$

$$\phi_2^4(t_k) = -f/c \ \rho_2^4 - \bar{N}_2^4(t_k)$$

$$\phi_1^9(t_k) = -f/c \ \rho_1^9 + \bar{N}_1^9(t_k)$$

$$\phi_2^9(t_k) = -f/c \ \rho_2^9 + K_2^9 + \bar{N}_2^4(t_k) + \bar{N}_1^9(t_k) - \bar{N}_1^4(t_k). \tag{8}$$

The key idea is that $\phi_2^9(t_k)$ is the only phase observable which has neither a base station subscript nor a base satellite superscript. When this occurs, the $\bar{N}$ is replaced with the K formulation. The K values are indeed the same integer biases one would obtain if double differences were generated. But this formulation allows the data to remain uncorrelated and thus simplify the overall process, no matter how many stations collect data simultaneously. Some base station and base satellite must be chosen. This selection is not at all crucial to the data reduction.

Now the technique can be stated formally. (1) Choose the base station and the base satellite. (2) When a phase measurement is encountered, if it involves either the base satellite or the base station, then the mathematical formulation free of the K's is used. Otherwise the K formulation given in the last equation of (8) is used. Epoch-to-epoch values of N must be estimated. The station positions and K variables are common contributors to all epochs. This formulation also allows for a very particular structure of the set of least-squares normal equations[2).

1.3 Ionospheric refraction

At radio frequencies the effect of the ionopshere on group velocity is to retard it inversely proportionally to the square of the frequency. A corresponding advance is created in the phase. Thus Eq. (3) must be augmented to account for this effect. The GPS satellites transmit at two frequencies, L_1 and L_2. The L_1 frequency is 1575.42 MHz which is exactly 154 times the fundamental P-code chipping rate of 10.23 MHz. The L_2 frequency is 1227.6 MHz which is exactly 120 times the P-code rate. Equation (3) is now augmented to account for the ionospheric effect,

$$\phi_j^i(t_k) = \phi^i(t_k) - f/c \ \rho_j^i - \phi_j(t_k) + N_j^i + A/f. \tag{9}$$

Equation (9) must be written twice, once each for the L_1 and L_2 frequencies. At this time we recognize that both the station and satellite L_2 phase values should be equal to the corresponding L_1 phase values scaled by the frequency ratio f_2/f_1 since they are both based on the same fundamental rate of 10.23 MHz.

$$\phi_j^i(t_k)_{L_1} = \phi^i(t_k) - \phi_j(t_k) - f_1/c \ \rho_j^i + N_j^i(L_1) + A/f_1$$

$$\phi_j^i(t_k)_{L_2} = f_2/f_1 \ [\phi^i(t_k) - \phi_j(t_k) - f_1/c \ \rho_j^i] + N_j^i(L_2) + A/f_2 \qquad (10)$$

We desire to eliminate the ionospheric effect by combining the two observations above in a linear combination as follows:

$$\phi_j^i(t_k) = \alpha_1 \ \phi_j^i(t_k)_{L_1} + \alpha_2 \ \phi_j^i(t_k)_{L_2}$$

Choosing

$$\alpha_1 = f_1^2/(f_1^2 - f_2^2)$$

$$\alpha_2 = - f_1 f_2/(f_1^2 - f_2^2) \qquad (11)$$

yields the following "corrected" observation:

$$\phi_j^i(t_k) = \phi^i(t_k) - f_1/c \ \rho_j^i - \phi_j(t_k) + \alpha_1 N_j^i(L_1) + \alpha_2 \ N_j^i(L_2). \qquad (12)$$

Equation (12) now looks exactly as the original phase equation given in (3) except the integer bias term is replaced with the linear combination of the L_1 and L_2 integer biases. Thus when ionospherically corrected data are processed, no programming changes except to linearly combine the dual frequency data into one measurement are required. Once combined, the data reduction continues without change.

2. DATA PROBLEMS

The most prevalent problem associated with GPS carrier phase observation is the occurrence of cycle slips. This occurs when the satellite signal is lost and reacquisition is required. Upon reacquiring the carrier, the integer ambiguity almost always takes on a different value which appears in the data as an abrupt jump in the carrier phase history. This jump must be accommodated in the analysis. It can be manually fixed. Additional differences can isolate the jump as is done in triple differencing, or, additional unknowns can be introduced whenever a jump occurs. As is usually the case, the technique which yields the most precise base line determination requires the most labor, and the most automatic procedure does not make use of the integer nature of the ambiguity. One possible exception will be the analysis of data over short base lines where the integer values of the biases are easily obtained. For these, cycle slips can be accommodated automatically by simply ignoring the integer cycle count. By using only the fractional part of the measurement residuals (rounded to zero), phase measurement processing with biases set to integers should proceed easily after a good estimate has been made available from some other technique such as triple differencing.

The cycle slips can occur on either the L_1 channel or the L_2 channel. They can occur at the same times or different times. The sources can be in the hardware or due to signal blockages such as trees, buildings, mountains, etc.

Large cycle slips can be identified when looking at single station data, but to date these jumps usually cannot be repaired to better than a cycle or two except when the data are combined interferometrically. The optimum solution is to fix the cycle slips on a station-by-station basis (unless the slip is due to an abnormally large gap in the data history), and a possible solution to this problem will be discussed next.

3. SINGLE SITE CYCLE FIXING

A possible technique for repairing cycle slips at a single station is based on the availability of dual frequency data. We start with Eq. (10) and note that the clock-based terms (receiver phase, satellite phase, and range term) are the same for L_1 and L_2 except for the factor f_2/f_1. The L_1 and L_2 integer ambiguities are not related, and the ionospheric terms are related by f_1/f_2. Thus when the linear combination

$$\delta_k = \phi_j^i(t_k)_{L_1} - f_1/f_2 \ \phi_j^i(t_k)_{L_2} \tag{13}$$

is formed, all the clock terms drop out and then

$$\delta_k = N_j^i(L_1) - f_1/f_2 \ N_j^i(L_2) + (1 - f_1^2/f_2^2). \tag{14}$$

Since the integers N_j should remain the same, epoch-to-epoch changes in this quantity are only due to the ionosphere. The term involving A represents 65% of the contribution to the L_1 phase measurement (with opposite sign). Figure 1 was generated from a dual frequency phase history while collecting data in Alaska in 1984. The epochs were separated by 120 seconds, and the data were collected at approximately 9:00 a.m. local time. This data set was chosen for its obvious cycle slips at epochs 6, 7, and 8. A listing of the raw phase data is given in Table 1 along with the differenced data. The Doppler contribution has been removed based on the pseudo-range determination of absolute position. What is noticed immediately is the difference in dynamic range between the Doppler corrected L_1 and L_2 phase values and differenced data. The average change from epoch to epoch of the differenced data is 0.1 cycles while it is 17 cycles for L_1 and 13 cycles for L_2 (disregarding the cycle slips). Also remember that these epoch-to-epoch differences are over a time interval of 120 s. What is proposed here is to track the difference signal and use it to detect and fix short term cycle slips. The L_1 coefficient in Eq. (13) is 1.0 and the L_2 coefficient is -1.28. Thus an isolated cycle slip would cause the δ_k to change by either 1.0 or 1.28. If the same slip occurred on both channels, the δ_k would change by 0.28, much greater than the uncertainty that one could predict. Also, if data closer in time were collected, the epoch-to-epoch changes would be reduced proportionally to the decreased interval, but the cycle slip changes would remain at the same large values. Another advantage to tracking the difference quantity is being able to follow the contribution of the ionosphere (see Bender and Larden[4]).

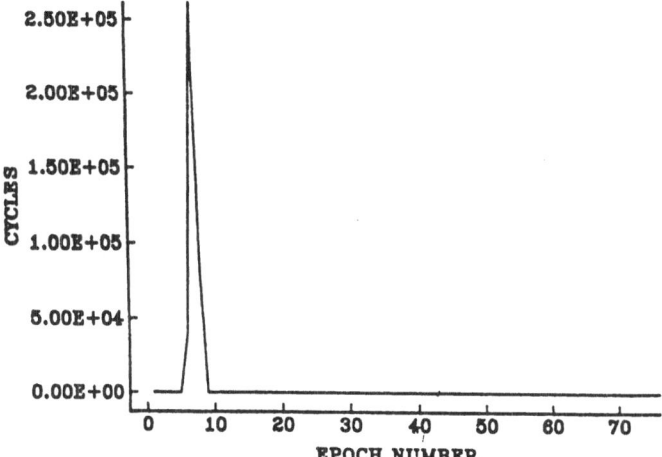

Fig. 1   Raw Phase Difference $[L_1-(F1/F2)L_2]$

4. FILTERING TECHNIQUES

A Kalman filter is easily implemented to follow the signals and the difference quantity. I have chosen to represent the state transitions of the signals as a third-order process (quadratic polynomial). This model is linear in the polynomial coefficients with state transitions computed as follows:

$$\begin{bmatrix} X(t_{n+1}) \\ \dot{X}(t_{n+1}) \\ \ddot{X}(t_{n+1}) \end{bmatrix} = \begin{bmatrix} 1 & \Delta t & 1/2\,\Delta t^2 \\ 0 & 1 & \Delta t \\ 0 & 0 & 1 \end{bmatrix} \begin{bmatrix} X'(t_n) \\ \dot{X}'(t_n) \\ \ddot{X}'(t_n) \end{bmatrix} \tag{15}$$

or

$$X(t_{n+1}) = \Phi(\Delta t)\, X'(t_n) \tag{16}$$

where the prime denotes previously updated values. The covariance of the prediction is

Table 1

L₁/L₂ phase differences and
Doppler-corrected phase values.

Epoch	Difference (cycle)	L_1 Phase (cycle)	L_2 Phase (cycle)
1	.14	.99	.66
2	.11	-17.00	-13.33
3	.11	-34.49	-26.96
4	.10	-51.47	-40.19
5	.10	-67.58	-52.74
6	39129.58	39045.94	-65.17
7	261022.78	197508.41	-49491.71
8	79421.22	79303.44	91.77
9	-.51	-134.33	-104.27
10	-.51	-149.09	-115.77
11	-.52	-163.76	-127.19
12	-.57	-178.04	-138.28
13	-.56	-192.07	-149.22
14	-.56	-206.28	-160.30
15	-.59	-220.72	-171.53
16	-.58	-235.21	-182.82
17	-.60	-249.48	-193.93
18	-.63	-264.19	-205.36
19	-.65	-279.17	-217.02
20	-.67	-294.03	-228.58
21	-.68	-309.23	-240.42
22	-.74	-324.26	-252.09
23	-.75	-339.50	-363.96
24	-.83	-355.06	-276.02
25	-.85	-371.08	-288.49
26	-.86	-387.03	-300.91
27	-.92	-403.58	-313.76
28	-.96	-420.36	-326.80
29	-1.03	-437.15	-339.83
30	-1.04	-454.38	-353.25
31	-1.17	-472.02	-366.93
32	-1.15	-489.78	-380.74
33	-1.18	-507.66	-394.65
34	-1.24	-526.18	-409.04
35	-1.25	-543.86	-422.81
36	-1.29	-561.57	-436.58
37	-1.38	-579.04	-450.12
38	-1.45	-596.62	-463.76
39	-1.54	-614.30	-477.47
40	-1.62	-631.96	-491.17
41	-1.72	-649.58	-504.82
42	-1.79	-667.15	-518.46
43	-1.90	-684.36	-531.78
44	-1.95	-701.65	-545.21
45	-2.01	-718.30	-558.15
46	-2.16	-735.47	-571.40
47	-2.27	-752.41	-584.52
48	-2.34	-769.38	-597.68
49	-2.42	-786.50	-610.97
50	-2.55	-803.56	-624.16

calculated as

$$[P_{n+1}] = \Phi(\Delta t) \ [P_n'] \ \Phi^T(\Delta t) + [Q] \tag{17}$$

where the matrix [Q] represents the uncertainty in the predictions due to errors or omissions in the transition matrix. The [Q] matrix can take on a rather dynamic behavior whereby the magnitude of the elements can be chosen to be small when smooth behavior is encountered, and large when the quadratic model cannot adequately model the state transitions. The measurements (L_1, L_2 phase; phase differences) are then used to update the predicted quantities by the standard Kalman formulae. Finally the covariance matrices are updated.

Again, the vectors X represent the values of measured carrier phase difference between the satellite transmitted phase fronts and the internal receiver oscillator values or the differenced quantities $\phi(L_1) - (f_1/f_2) \ \phi(L_2)$. The latter values are the least changing quantities (see Figs. 2, 3, and 4) and are used for deciding when lock was not maintained. If a loss-of-lock occurs, the differenced quantity and the L_2 prediction are used to obtain integer corrections to the L_1 and L_2 phase measurements. And the process continues.

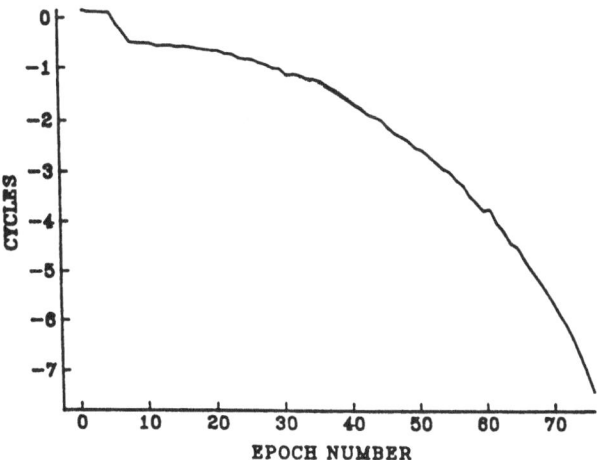

Fig. 2 Filtered Phase Difference [$L_1 - (F1/F2)L_2$]

28

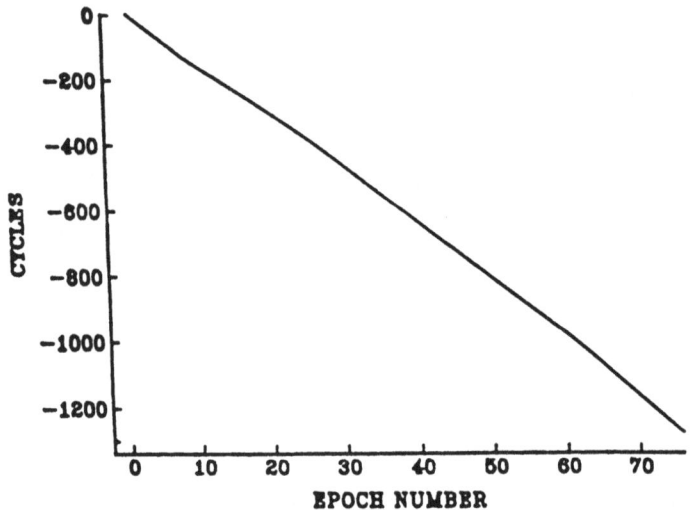

Fig. 3 Filtered Phase Data (L_1)*

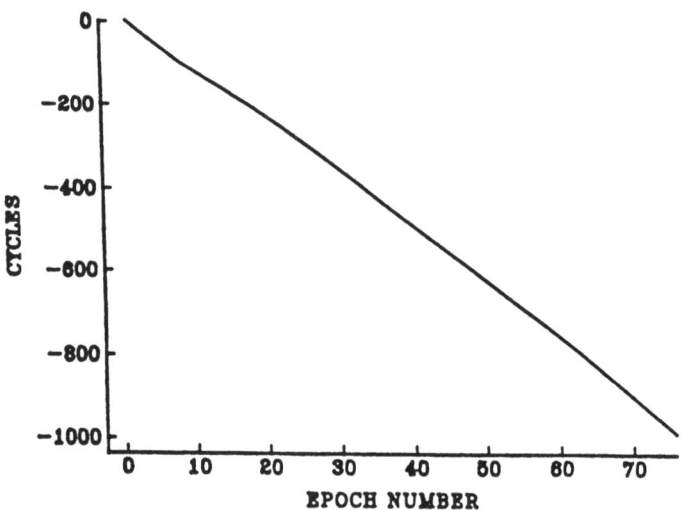

Fig. 4 Filtered Phase Data (L_2)*

* Doppler contribution has been removed.

Experience with several data sets has revealed that the corrections are good to 1 cycle or better. This alone justified preprocessing the data prior to interferometric analysis. However, should a data sampling interval much smaller than 120 seconds be chosen, then the predictor/filter performance would improve. This does not imply that the increased sampling must be used in the interferometric processing step. The single site, preprocessed data could be sampled to reduce the computational workload in subsequent data reductions. A typical sampling rate might be from 3 to 30 seconds.

5. APPLICATION TO DYNAMIC PROBLEMS

Returning to Eq. (14), it is again emphasized that the term containing the receiver to satellite range value has been removed in the differencing process. Obviously, if the receiver was in motion (say in an airplane), the term would not appear and thus the same gradually changing signal as given in Fig. 2 would be expected. However, the ability to predict the L_2 phase values will be degraded by the arbitrary motion of the platform. But still, this is one very important quantity that can be used to measure whether or not lock has been maintained.

6. UNMODELED ERRORS

No matter which technique one chooses, it is clear that unmodeled errors are always present and will contaminate the data reductions to some (hopefully small) level. When combining data from more than two stations, it is also clear that cancellation of these unmodeled errors, such as environmental effects, will be more complete for the closer stations. Expressed mathematically we say that the measurement errors are correlated based on station separation. Given same gain matrix, K, the linear estimate can be written

$$\hat{x} = K\,y$$

where x is the vector of estimated parameter values, y is the vector of phase measurements. The key idea here is to realize that the vector y is composed of more than just parameters x and noise e. Let us express y as

$$y = A\,x + B\,z + e$$

where the design matrices A and B map the effects of the parameters to be estimated, x, and the unadjusted parameters (but which have error), z respectively. Usually $K = (A^T W A)^{-1} A^T W$ where the W matrix is the inverse of the "noise" covariance matrix. But in this case, the properly modeled covariance matrix consists of both noise and contribution of the unadjusted errors. The proper weight matrix should be modeled as

$$W^{-1} = B \, cov(z) \, B^T + cov(e)$$

where $Cov(e) = E(ee^T)$,

$Cov(z) = E(zz^T)$, and the mean values of z and e are assumed to be zero.

This is almost identical to the approach used in collocation[5]. By properly modeling the correlation between phase measurements, the estimation of base line components and the integer ambiguities could accommodate systematic error contribution. Put a different way, modeling correlations based on station separation may keep errors present on long base lines from contaminating integer bias recovery on short base lines when simultaneous solutions with many stations' data are attempted. More rigorous error statistics should also be available from the data reduction.

Most often residual errors due to environmental sources such as the troposphere or ionosphere are difficult, if not impossible, to model in the matrix B. Thus, some rule is usually used and is expressed mathematically as a function of distance between stations. In this regard, any rule that is used should not destroy the positive-definite structure of the weight matrix, W[6]. This is identical to the standard collocation implementation.

REFERENCES

1) C. C. Goad, B. W. Remondi, Initial relative positioning results using the Global Positioning System. Bull Geoid., 58, 193-210, (1984).

2) C. C. Goad, Precise relative positioning using GPS carrier phase measurements in a nondifference mode, in Goad, C. C.(ed) Proceedings of the First International Symposium on Precise Positioning with the Global Positioning System, Volume 1, National Geodetic Survey, Rockville, Maryland, pp 347-356, (1985).

3) Y. Bock, C. C. Counselman, S. A. Gourevitch, R. W. King, A. R. Paradis, Geodetic Accuracy of the Macrometer[TM] Model V-1000. Bull. Geod., 58, 211-221, (1984).

4) P. L. Bender, D. R. Larden, GPS Carrier Phase Ambiguity Resolution Over Long Base lines. Proceedings of the First International Symposium on Precise Positioning with Global Positioning System, Rockville, Maryland, April 15-19, (1985).

5) H. Moritz, Physical Geodesy, W. H. Freeman and Company, San Francisco, 364 pp., (1967).

6) S. K. Jordan, Self-consistent statistical models for the gravity anomaly, vertical deflections, and undulation of the geoid, J. Geophys. Res., 77 (20), 3660-3670, (1972).

GPS ORBIT DETERMINATION USING THE DOUBLE DIFFERENCE PHASE OBSERVABLE

G. Beutler
Astronomical Institute, University of Berne, Switzerland

ABSTRACT

After discussing orbital precision requirements for GPS space
vehicles a review of simple but correct orbit determination (and
orbital biases elimination-) procedures is given. The 1984 Alaska
GPS experiment and the 1985 High Precision Baseline Test (HPBL-Test)
are presented as typical examples where orbit determination
techniques are mandatory. The results indicate that we may expect
accuracies of 1 part in 10^7 in the near future for GPS-derived regional and continental networks -- provided orbit determination
techniques are combined with the network processing mode.

1. INTRODUCTION

In this article we summarize theory and applications of GPS orbit determination facilities developed at the Astronomical Institute of the University of Berne, Switzerland. Most
of the material presented here may be found in Refs. 1-3.
Although only partially deployed, the Global Positioning System (GPS) is already being exploited for high precision geodetic surveys. The Macrometer[R] V-1000, a first generation
interferometric receiver using only the L1 carrier phase of the GPS signal, has clearly
demonstrated its capability in differential positioning. In 1984 the Alaska GPS Experiment
Ref. 2, in 1985 the March HPBL-test (High Precision Baseline-Test, see e.g. Ref. 3, clearly
demonstrated the potential of measuring networks with GPS on a regional or even continental
scale with an accuracy of a few centimeters - provided the orbits are known or are being estimated with an accuracy of 1-2 meters.

These orbital precision requirements suggest that an extended worldwide tracking
network equipped with high-precision receivers will be required. Such a network is not now
available to civilian users nor will one be available in the immediate future. Plans for
several regional civilian tracking networks are presently being developed and/or imple-

mented, and these may provide acceptable orbits over certain regions. However, the geodesist using GPS for large-scale, high-precision surveys may not be able to assume that the GPS orbits are known to sufficiently high accuracy. He may have to estimate so-called orbital biases along with the parameters in which he is actually interested, i.e., the relative coordinates of the receivers.

Methods to estimate such orbital biases have been developed and widely used in the processing of Transit Doppler data. With the exception of some of the so-called short arc procedures, these algorithms do not describe the orbits by physical parameters. For example, one technique is to parallely shift and rotate the orbit. What usually results is non-physical in the sense that the resulting orbit is not a particular solution of the equations of motion of the satellite.

Here we attempt to demonstrate that there is no need for such non-physical methods in the determination of GPS satellite orbits. As a matter of fact, it is quite simple to model the orbital biases for these satellites to any precision required in a purely physical way. Because of the orbital characteristics of GPS satellites (almost circular orbits with semi-major axes of about 26,500 km), we may assume the earth's gravity field to be known. As a rigorous modelling of the gravitational attraction due to the sun and the moon is not a serious problem, the only significant external force on the satellites which is not adequately known a priori is radiation pressure.

The subsequent discussion is divided into three sections. Orbital precision requirements for geodetic applications are discussed in Section 2. Then the necessary theory is developed in Section 3. In Section 4 we present applications.

2. PRECISION REQUIREMENTS

The accuracy of orbits needed to obtain baseline estimates of a certain accuracy depends mainly on the length of the baseline (see Ref. 4, Eq. (84)):

$$\frac{db}{b} \doteq \frac{dr}{\rho} \tag{1}$$

where b   is the length of the baseline
    ρ   is the range (receiver to satellite)
    dr   is the orbit error
    db   is the induced baseline error.

For GPS satellites we have approximately

$$\rho = 25000 \text{ km} . \tag{2}$$

Accepting, for example,

$$db = 1 \ cm \hspace{10cm} (3)$$

as a maximum for the baseline error introduced by the orbit, we obtain the values in Table 1 for the maximum orbit error, dr, allowed.

Table 1

Maximum permissible orbit error, dr, of 1 cm
for an accuracy in a baseline of length b.

b (km)	dr (m)
0.1	2500.0
1.0	250.0
10.0	25.0
100.0	2.5
1000.0	.25

Rather than assuming a specific value for a baseline error independent of baseline length, we may wish to talk about a relative error expressed as parts per million (ppm) of the baseline length. Table 2 gives the values of dr for a number of relative baseline errors.

Table 2

Maximum permissible orbit error, dr,
for a certain relative accuracy in the baseline

db/b (ppm)	dr (m)
5	125.0
1	25.0
0.5	12.5
0.1	2.5

Summarizing Tables 1 and 2:
  a) The orbital accuracy required for surveying depends highly on the length of the base lines to be measured.

(b) for "local surveys" (diameter of surveyed region smaller than 100 km), an orbital accuracy of 5 m to 10 m will be sufficient in most cases.

(c) for regional or even continental surveys, orbital accuracies of 1 m to 4 m will be required.

3. PRINCIPLES OF ORBIT DETERMINATION

The orbit of every satellite is a particular solution of a system of second-order differential equations:

$$\vec{r}^{(2)} = \vec{f}(t;\vec{r},\vec{r}^{(1)}, p_1,p_2,\ldots,p_n) \tag{4}$$

where $\vec{r} = \vec{r}(t)$ is the position of the satellite in a nonrotating (with respect to inertial space), geocentric coordinate system.

$\vec{r}^{(i)}$, i=1,2, are the first and second time derivatives of $\vec{r}(t)$

p_i, i=1,2,...,n are parameters defining the forces acting on the satellite ("dynamical" parameters, e.g. parameters describing radiation pressure).

To define an orbit uniquely (one particular solution of Eq. (4)), additional information has to be supplied. Normally the problem is formulated as an initial value problem:

$$\vec{r}(t_0) = \vec{r}_0(k_1,k_2,\ldots,k_6)$$
$$\vec{r}^{(1)}(t_0) = \vec{r}_0^{(1)}(k_1,k_2,\ldots,k_6) \tag{5}$$

where k_i, i=1,2,...,6 are six parameters uniquely specifying the vectors on the right-hand sides of Eq. (5). Possible choices for these parameters are:

Components of vectors $\vec{r}_0$, $\vec{r}_0^{(1)}$
Osculating orbital elements at time t_0.

If we know the right-hand sides of Eq. (5) (parameters k_i, i=1,2,...,6) and the dynamical parameters p_i, i=1,2,...,n, the orbit of a satellite is uniquely defined. We therefore may state:

(a)   Orbit determination in its usual, more restricted, sense is defined as the problem of determining the six parameters k_i, i=1,2,...,6 defining the initial values on the right-hand sides of Eqs. (5).

(b)   Orbit determination in its most general sense is the problem of determining the six parameters k_i, i=1,2,...,6 defining the initial values or boundary values and the dynamical parameters p_i, $i=1,2,\ldots,n$.

For the application we have in mind here (modelling the orbits of GPS satellites) most of the dynamical parameters in Eq. (4) may be assumed to be known (coefficients of the earth's gravity field, gravitational force of the sun and the moon). For utmost accuracy, however, we will have to estimate some of those parameters defining radiation pressure.

To solve an orbit determination problem, at some point one needs observations. An observation may be defined as a value of a function of satellite positions, ground positions, and nuisance parameters, like clock offsets or clock drifts of satellite and receiver clocks. More specifically, the following GPS observables have been instrumented in presently available receivers:

- pseudoranges using the C/A-code
- pseudoranges using the P-code
- Doppler measurements
- phase measurements of carriers.

The carrier phase measurements are potentially the most powerful measurements as they can be made with the greatest precision. We therefore assume that we will be dealing with measurements of this kind for orbit determination problems.

Every orbit determination is actually an orbit improvement process using observations such as those given above. These observations are nonlinear functions of the satellite position $\vec{r}(t_i)$ (and possibly of the velocity $\vec{r}^{(1)}(t_i)$) at observation time t_i. $\vec{r}(t_i)$ in turn is a nonlinear function of the parameters of the orbit. The linearization of the orbit determination problem is therefore done in two steps: (a) the observation has to be approximated as a linear function of $\vec{r}(t_i)$, which is straight forward; and (b) $\vec{r}(t_i)$ has to be represented by a linear function of the unknown parameters k_i (and possibly some of the p_i). This is done in the following way. Since Eq. (4) is nonlinear, we must linearize it and determine our orbit iteratively, where in each iteration step we assume we have a <u>known</u> approximate orbit $\vec{r}_a(t)$ at our disposal.

$$\vec{r}_a^{(2)} = \vec{f}(t;\vec{r}_a,\vec{r}_a^{(1)},p_1,p_2,\ldots,p_n)$$

$$\vec{r}_a(t_0) = \vec{r}_{a0}(k_{a1},k_{a2},\ldots,k_{a6})$$

$$\vec{r}_a^{(1)}(t_0) = \vec{r}_{a0}^{(1)}(k_{a1},k_{a2},\ldots,k_{a6}) \quad , \qquad\qquad (6)$$

where the k_{ai}, $i=1,1,\ldots,6$ are approximate values of the unknown parameters.

The true, initially unknown, orbit is now assumed to be a linear(ized) function of the parameters k_i, $i=1,2,\ldots,6$:

$$\vec{r}(t) = \vec{r}_a(t) + \sum_{i=1}^{6} \vec{z}_i(t)(k_i - k_{ai}) \tag{7}$$

where

$$\vec{z}_i(t) = \left. \frac{\partial \vec{r}_a}{\partial k_i} \right|_{k=k_a} \quad i=1,2,\ldots,6$$

If we deal with the more general problem, additional terms involving the dynamical parameters appear on the right-hand side of Eq. (7):

$$\vec{r}(t) = \vec{r}_a(t) + \sum_{i=1}^{6} \vec{z}_i(t)(k_i - k_{ai}) + \sum_{i=1}^{n} \vec{z}_i^*(t)(p_i - p_{ai}) \tag{8}$$

where

$$\vec{z}_i^*(t) = \left. \frac{\partial \vec{r}_a}{\partial p_i} \right|_{p=p_a} \quad , \quad i=1,2,\ldots,n \tag{9}$$

The functions $\vec{z}_i(t)$, $\vec{z}_i^*(t)$ are solutions of an initial (or boundary) value problem, which follow from the primary problem (6) by taking the (total) derivatives of all the equations in (6) with respect to k_i, $i=2,3,\ldots,6$ and p_i, $i=1,2,\ldots,n$. The resulting set of differential equations are usually called the systems of variational equations. It is elementary to show (see Ref. 1) that each of the $\vec{z}_i$ or $\vec{z}_i^*$ is a particular solution of a linear second order differential equation system.

Let us summarize: In every iteration step of the orbit improvement process, we have to solve one system of nonlinear differential equations of type (6), and, for each orbit parameter we have to estimate (for each of the $\vec{z}_i$, $\vec{z}_i^*$) one linear differential equation system.

In our GPS software system we solve the nonlinear systems (Eqs. (4), (5) with the technique of numerical integration, the partial derivations $\vec{z}_i$, $i=1,2,\ldots,6$ are computed approximately: (a) the k_i, $i=1,2,\ldots,6$ are chosen to be the osculating elements pertaining to the initial epoch t_o; (b) in formula (9) the approximate orbit $\vec{r}_a$ is replaced by the Kepler orbit defined by these osculating elements. It is then possible to compute the partials analytically using the well known formulae of the two-body problem.

Let us conclude this section by briefly reviewing our modelling of the force field acting on GPS-satellites. (In Ref. 2 the influences of the constituents of the force field on the GPS orbits are discussed in some detail):

- The earth's gravity field is expressed by spherical harmonics, where we may select the coefficients of either the GEM-10 or the GRIM-3L1 model. (Usually we cut off the development after degree and order 8).

- Point mass attractions from sun and moon are taken into account.
- The radiation pressure model we use at present is very simple from the point of view of application: Only two parameters define the model, only two parameters may be estimated from observation data. In Fig. 1 we define three unit vectors: $\vec{e}_1$ is the antenna axis, $\vec{e}_2$ is the solar panels axis and $\vec{e}_0$ is the unit vector pointing to the sun.

Fig. 1 Model of GPS Space Vehicle

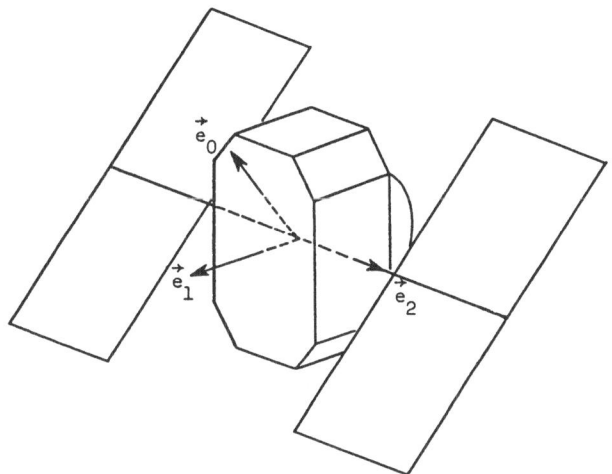

Legend

$\vec{e}_1$: Antenna axis
$\vec{e}_2$: Axis of solar panels
$\vec{e}_0$: Unit vector to sun

We model the acceleration $\vec{a}$ excerted by radiation pressure as

$$\vec{a} := -p_0 \cdot \vec{e}_0 + p_2 \cdot \vec{e}_2 \tag{10}$$

The first term is called direct solar radiation pressure (it would be the only one if the space vehicles were spherically symmetric or if the surfaces were totally absorbing the radiation), the second is usually referred to as "y-bias". Several effects may contribute to this latter term. Misalignment of the $\vec{e}_2$-axis ($\vec{e}_2$ not perpendicular to $\vec{e}_0$) certainly contributes to this term. Fliegel et al. (see Ref. 5) discuss thermal radiation of the space vehicle as another candidate.

Parameters p_0 and p_2 may be estimated from observation data. This means that the partial derivatives of the orbit with respect to these parameters have to be computed too (see Eqs. (9)). At present we do that by numerical integration.

Very often it is useful to introduce a priori knowledge for some parameters, into the adjustment. If we have e.g. precise ephemerides at our disposal as a priori orbit source, it would be a pity not to use the fact, that these orbits are good to - let us say - 10 meters along track and 2 meters in the other directions. In our parameter estimation program we take a priori information for parameters into account by adding an artificial observation equation for the (improvement of) parameter under consideration. To this observation equation a weight proportional to the inverse of the (estimated) a priori variance of the parameter is assigned.

4. APPLICATIONS

The results for two campaigns are summarized here:

(a) 1984 Alaska GPS Campaign (see Ref. 6).
(b) March 1985 High Precision Baseline Test (HPBL-Test) (Ref. 3).

In the first case we are looking at a typical regional network ($\sim$ 2500 km × 1000 km) where orbit improvement had to be done as a byproduct of the analysis, the main interest lying on the estimation of site coordinate. It became clear that network-processing mode and the estimation of "orbital biases" were the two keys to the success of this particular analysis. The situation was quite different for the second analysis. Here orbit estimation was the main topic, the quality of receiver coordinates "only" served to prove the quality of the orbits estimated from GPS observations from four fiducial sites whose coordinates were known from VLBI observations.

4.1 The 1984 Alaska GPS Campaign

During the summer of 1984, the U.S. National Geodetic Survey (NGS) and the Jet Propulsion Laboratory operated Mobile VLBI systems in Alaska and Canada as part of a series of measurements by the National Aeronautics and Space Administration's Crustal Dynamics Project. Given the great expense and complex logistics of moving mobile VLBI systems to and around such a distant location, NGS decided to measure the same baselines by GPS in the hopes that as GPS technology and techniques mature and a history of GPS/VLBI comparisons emerge, these measurements may be performed solely by the less expensive GPS systems. The sites occupied by both the mobile VLBI and GPS systems are shown in Fig. 2, the approximate distance between the sites may be found in Table 3.

During this GPS/VLBI intercomparison, five TI-4100 GPS receivers were available. The original GPS observing plan was to occupy Nome, Fairbanks, Sourdough, Yakataga, and

Whitehorse for 2 consecutive days. The second scheduled session kept the same receivers at Nome and Fairbanks and moved the other receivers to Sandpoint, Kodiak and Yellowknife for 2 additional observing days. The principal observing session on each day consisted of two satellite scenarios. The first began in early afternoon local time, included SV's 6, 8, 9 and 11 and lasted almost 3 hours. This was immediately followed by the second scenario which included SV's 4, 9, 11, and 13 and lasted about 1 hour. A second observing session was scheduled for each day approximately 12 hours after the primary one, but it turned out that these sessions did not produce useful results.

The first observing occupation scheme went as planned and produced data that contained only a few cycle slips. The second occupation scheme had to be severely modified as it lasted for the next 2 weeks due to reciever failures in Alaska. Coincidentally these data are heavily burdened with cycle slips.

Fig. 2   The 1984-Alaska-GPS-Campaign

1   Yakataga
2   Fairbanks
3   Kodiak
4   Nome
5   Sandpoint
6   Sourdough
7   Yellowknife
8   Whitehorse

500 km

Table 3

Approximate distances between stations in kilometers

Station	Kodiak	Nome	Sand	Sour	White	Yaka	Yellow
Fair	849	848	1285	276	789	603	1631
Kodiak		1025	557	671	1044	632	2135
Nome			1006	1004	1591	1276	2460
Sand				1179	1598	1188	2681
Sour					591	329	1575
White						414	1105
Yaka							1510

Fair:	Fairbanks	White :	Whitehorse
Sand:	Sandpoint	Yaka :	Yakataga
Sour:	Sourdough	Yellow :	Yellowknife

After the pre-processing step (see Ref. 6) roughly 20'000 double difference observations (8000 from scheme 1, 12000 from scheme 2) could be processed by the parameter estimation program. In the program run combining the two observation schemes, thus giving the entire Alaska network with eight sites as main result, a total of 330 parameters had to be estimated of which 180 were orbit parameters. Since we had precise ephemerides at our disposal, very small variances could be associated with the artificial observations constraining the orbit parameters (1σ errors of 15 m along track and 3 m in the other directions were used, see Ref. 6 for more information).

Consistency and integrity of both, the GPS and VLBI solutions could be demonstrated by a 7-parameter Helmert transformation between the two final networks. Table 4 gives residuals and transformation parameters of this transformation.

4.2 The March 1985 High-Precision Baseline Test (HPBL-Test)

This campaign took place from March 29 to April 5 in the United States of America.

In March/April 1985 space vehicles 4, 6, 8, 9, 11, 12 and 13 could be observed. The daily observation period started approximately at 3 a.m. and lasted till 11 a.m. Nine TI-4100 receivers, three AFGL dual frequency instruments and two SERIES-X receivers were distributed over nine sites (see Fig. 3 and Table 5), among them Haystack, Fort Davis, Richmond and Mojave which we used as fiducial points (coordinates known from VLBI-surveys). (For Hat Creek and Owens Valley VLBI results are available too.)

Table 4

Residuals of a Helmert Transformation between the GPS Solution for the 1984 Alaska GPS
Campaign and the VLBI-Solution transformed into WGS-72 Observations Schemes 1 and 2

Station	VLBI - GPS (m)		
	x	y	z
YAKATAGA	0.146	-0.027	-0.147
FAIRBANKS	0.023	0.098	-0.106
KODIAK	-0.301	0.039	-0.038
NOME	-0.085	0.083	0.115
SANDPOINT	0.132	-0.153	0.253
SOURDOUGH	0.157	0.030	-0.173
YELLOWKNIFE	-0.112	-0.106	0.237
WHITEHORSE	0.041	0.036	-0.140

RMS of Transformation   =   0.16 m
Rotation around x-axis =   0.18   +- 0.03 arc sec
Rotation around y-axis =   -0.01   +- 0.03 arc sec
Rotation around z-axis =   0.04   +- 0.01 arc sec
Scale factor   =   -0.36   +- 0.07 mm/km

Fig. 3 Receivers and sites in the 1985 HBPL test

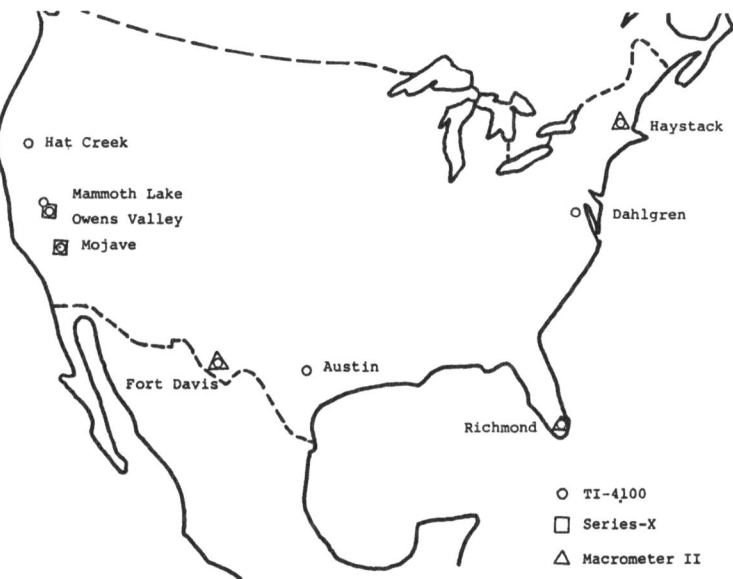

Table 5

Distances in km between sites of the March 1985 HPBL-Test

	BIGP	FTDA	HAYS	HATC	MAMM	MOJA	NSWC	RICH
AUST	2035	598	2678	2417	2100	1866	2081	1778
BIGP		1508	3930	484	71	245	3557	3742
FTDA			3136	1933	1577	1314	2586	2363
HAYS				4033	3959	3904	669	2045
HATC					416	729	3731	4066
MAMM						313	3597	3800
MOJA							3498	2046
NSWC								1442

Station names:
AUST Austin ARL
BIGP Owens Valley Big Pine
FTDA Fort Davis
HAYS Haystack Obs., Westford

HATC Hat Creek, CA
MAMM Mammoth Lake, Casa 1956, Ca
MOJA Mojave, Barstow, Ca
NSWC Dahlgren, Va
RICH Richmond, Perrine F

Table 6

Observation schedule for TI-4100 (TI) and AFGL (MM) receivers

Day	AUST	BIGP	FTDA	HAYS	HATC	MAMM	MOJA	NSWC	RICH
Mar 29	TI	TI	MM				TI	TI	TI,MM
Mar 30	TI	TI	TI,MM	TI	TI	TI	TI	TI	TI,MM
Mar 31		TI	TI,MM	TI,MM	TI	TI	TI	TI	TI,MM
Apr 01	TI	TI	TI,MM	MM	TI	TI	TI	TI	MM
Apr 02	TI	TI	TI,MM	MM	TI	TI	TI	TI	TI,MM
Apr 03	TI	TI	MM	MM	TI	TI	TI	TI	TI,MM
Apr 04	TI	TI	TI	TI,MM	TI	TI	TI	TI	TI,MM
Apr 05	TI	TI	TI,MM	TI,MM	TI	TI	TI	TI	TI,MM

Station names: see Table 5

Because we did not have a complete set of SERIES-X data at our disposal, only Macrometer and TI-observations were used for the investigation. The observation schedule for the

two instrument types are given in Table 6. The TI-receivers recorded data each 30 seconds, (which leads to impressive numbers of observations (5000 to 12000 double differences could be used per day), the Macrometer recorded data at a much lower rate (one epoch per 6 minutes).

The present analysis was severly handicapped by a number of circumstances, the most important being that

(a) No surface weather data had been available to us when we processed the campaign.

(b) The GESAR software implemented in the TI-4100 receivers was not working properly. (The consequence was that one clock offset had to be estimated for each TI-receiver for each session.)

In order to study repeatability of the estimated parameters the campaign was artificially divided into two parts:

Part A: March 29 - April 1
Part B: April 2 - April 5

Either all observations of part A or all observations of part B were processed.
Tests were made with 1 day arcs, 2 day arcs, 4 day arcs. "Only" broadcast ephemerides were available as a priori information. Each arc was characterized by six osculating elements and by two radiation pressure parameters (see Eq. (10)). For the latter parameters the following a priori values were used for all satellites

$$p_0 = 0.94 \cdot 10^{-7} \ m/s^2$$
$$p_2 = 0 \qquad m/s^2 .$$

$$(11)$$

The perigee was constrained by

$$\sigma(\omega) = 0\overset{..}{.}1 .$$

$$(12)$$

No constraints have been imposed on the other elements.

Results:

(a) Radiation Pressure Results

It soon became evident that for short arcs (1 day) it is not necessary (and does not make sense) to improve the a priori values for p_0, p_2 given in (11).   For 2-day arcs only p_0 could be estimated with a satisfactory reliability. Only for 4-day arcs the y-bias (p_2) gave a significant contribution.

The estimates for p_0 were quite consistent: The results obtained in parts A and B of the campaign differ from each other by only 1 % - 2 %, a similar statement holds for the different arc-lengths used. The situation is somewhat less satisfactory for p_2, the y-bias parameter (see Ref. 3).

(b) <u>Osculating Elements</u>

Two facts are worth being mentioned:

(1)   During the campaign, the broadcast ephemerides were of a surprisingly high quality: the positions derived from our a posteriori orbit estimates differed only for 10 - 25 m from the braodcast predictions. (The exception was satellite 4 where differences of up to 100 m occured.)

(2)   Fig. 4 illustrates the "state of the art" of our orbit modelling capabilities: There a two-day arc (April 2, 3, solution B21) is compared with a four day arc (April 2-5), solution B42). In solution B21 for each arc seven parameters were estimated (six osculating elements and p_0), in solution B42 8 parameters (six elements and p_0, p_2). As one may see the differences between the two solutions are of the order of a few meters, which indicates that the orbits should be good to a few parts in 10^{-7} (see Section 2).

Fig. 4

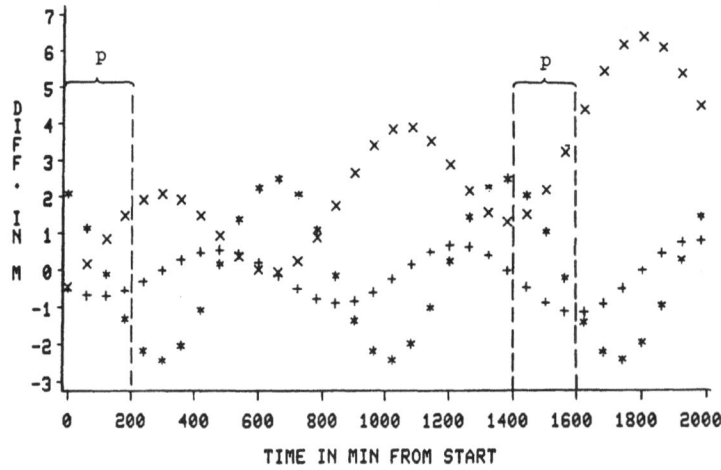

SOLUTION B21 − SOLUTION B42

SVN= 8   DATE(START)= 1985 4 2 3.0
DIFF. RADIAL(+),ALONG TRACK(X),OUT OF PLANE(*)

p: observation period

Table 7

Differences in the estimated station coordinates with respect to
solution B21 (only stations other than fiducial stations)

Station	Difference	d(latitude) cm	d(longitude) cm	d(height) cm
Hat Creek	A10 - B21	-2.4	0.3	3.3
	A21 - B21	-4.9	-8.0	6.1
	A42 - B21	-1.6	3.9	7.7
	B10 - B21	-2.4	0.3	3.3
	B42 - B21	0.1	2.3	-0.6
Owens Valley	A10 - B21	1.2	-1.7	3.6
	A21 - B21	-1.2	-5.1	4.6
	A42 - B21	0.8	0.6	4.4
	B10 - B21	1.1	-1.4	3.6
	B42 - B21	0.0	1.3	-0.4
Mammoth Lake	A10 - B21	-1.3	-0.4	-2.3
	A21 - B21	-1.7	-4.8	-0.8
	A42 - B21	-1.2	2.2	-0.3
	B10 - B21	-1.3	-0.4	-2.3
	B42 - B21	0.0	1.6	-0.5
Austin	A10 - B21	2.1	5.6	5.8
	A21 - B21	2.3	5.3	8.2
	A42 - B21	4.6	-0.6	12.0
	B10 - B21	-2.1	5.6	5.8
	B42 - B21	0.2	-2.0	2.5
Dahl- gren (NSWC)	A10 - B21	4.0	0.4	0.5
	A21 - B21	2.4	0.5	5.4
	A42 - B21	4.9	-1.5	15.6
	B10 - B21	4.0	-0.4	0.5
	B42 - B21	-3.1	7.0	8.7

Where: Aik: - Solution using all observations from March 29 to April 1.

   - i = arc length in days.

   - k = number of radiation pressure parameters estimated.

   Bik: - Solution using all observations from April 2 to April 5.

   - i = arc length in days.

   - k = number of radiation pressure parameters estimated.

(c) Coordinates

An independent check of the quality of the estimated orbits is possible by comparing the coordinates (as obtained in the various solutions) for the stations Hat Creek, Owens Valley, Mammoth Lake, Austin and Dahlgren, because these reciever locations were estimated without constraints in all program runs.

If our orbits would be good to - let us say - 1 ppm, we would expect differences of up to 4 meters in the different solutions, since we have baselines of up to 4000 km. The results in Table 7 show differences that are almost two orders of magnitude smaller!

Table 7 also shows, that the differences between the three different orbit types (1-, 2-, 4-day arcs, different radiation pressure models) cause only coordinate changes on the sub-decimeter level, which indicates adequate modelling.

We also see that the repeatability "A-solutions minus B-solutions" is of the same order of magnitude. (Remember that solutions A and B are indpendent in the sense that they have no observations in common and that the orbits are completely independent from each other.)

It is interesting (but not surprising), that the largest differences occur in height - hopefully surface weather data and water vapour radiometer data will improve the situation somewhat.

*   *   *

REFERENCES

1) G. Beutler, D.A. Davidson, R. Langley, R. Santerre, P. Vanicek, D.E. Wells, Some Theoretical and Practical Aspects of Geodetic Positioning with the Global Positioning System using Carrier Phase Difference Observations, Dept. of Surveying Engineering Technical Report No. 109, University of New Brunswick, Canada (1984).

2) G. Beutler, W. Gurtner, I. Bauersima, R. Langley, Proc. First Int. Symp. on Precise Positioning with GPS, Vol. 1, 99-111, (1985).

3) G. Beutler, W. Gurtner, M. Rothacher, T. Schildknecht, Determination of GPS Orbits using Double Difference Carrier Phase Observations from Regional Networks, Fourth Int. Geodetic Symp. on Satellite Positioning, Austin, Texas (1986).

4) I. Bauersima, Navstar/Global Positioning System (GPS) II, Mitteilungen der Satellitenbeobachtungsstation Zimmerwald, No. 10, University of Berne (1983).

5) H.F. Fliegel, W.A. Feess, W.C. Layton, N.W. Rhodus, Proc. First Int. Symp. on Precise Positioning with GPS, Vol. 1, 113-119 (1985).

6) W. Gurtner, GPS papers presented by the Astronomical Institute of the University of Berne in the Year 1985, Mitteilungen der Satellitenbeobachtungsstation Zimmerwald, No. 10, University of Berne (1983).

ACCURACY PROBLEMS WHEN COMBINING TERRESTRIAL AND SATELLITE OBSERVATIONS

W.M. Welsch
Institut für Geodäsie, Universität der Bundeswehr München, D-8014 Neubiberg

ABSTRACT

The combination of terrestrial and satellite aided network observations requires hybrid models of two kinds: a hybrid functional
model and a stochastic model being able to process the heterogeneous observations appropriately. An appropriate combination of
stochastic information means to adjoin proper weights to the various types of observations on one hand and on the other hand to separate the relevant internal accuracy from that accuracy which is
affected by a number of network external influences. The first
task can be overcome by the technique of estimating variance components of the individual observation groups, the second one can be
handled by applying S-transformations to covariance matrices. The
paper deals with problems of the hybrid stochastic model. Examples
are given.

1. INTRODUCTION

The combination of classically performed terrestrial and satellite aided geodetic
network observations rank among the many tasks being attended with the introduction of the
new observation techniques. Doing so, several aspects are important. One of them is the
functional connection of the heterogeneous observations. This is the problem of knotting
together different reference systems whose interrelations are not clearly defined. As a
rule the task is mastered by similarity transformations. It has often been treated, as an
extensive bibliography proves.

The stochastic hybrid model, however, has been hardly ever dealt with. Therefore,
the following representation intends to make a contribution, first by analyzing the accuracies of individually adjusted terrestrial and satellite networks. These accuracies are
of two kinds: accuracies which are influenced by numerous network external effects, and
internal accuracies which are relevant for relative accuracy statements within the network.
The combination of those internal and external accuracies in a hybrid stochastic model has
to be treated with care. This is one problem. The other problem is harmonizing the
stochastic model by adjoining proper weights to the individual and heterogeneous groups of
terrestrial and satellite aided observations.

Therefore, when comparing or combining terrestrial and satellite aided network observations the first task is to get a deeper insight into the accuracies of the individual
and heterogeneous observations.

2. THE ACCURACY OF SATELLITE AIDED NETWORK OBSERVATIONS

First information about the accuracies of satellite aided network observations can be taken from the various data processing software packages. The covariance matrices provided are, however, affected with all kinds of network external effects such as instrumental and propagation medium influences, ephemeris errors etc. In many cases, especially in cases of small networks and short observation periods with multi-station instrumentations, those effects exercise a systematic influence by shifting, rotating and scaling the network - both the coordinates and the elements of the covariance matrix C_x . In case only the mutual or relative positions of the network points are of interest, this "external" covariance matrix C_x does not provide an appropiate measure of the relative or internal accuracies of the point positions. To achieve such a measure the external effects have to be split off. According to the "Internal Error Theory of Networks" (Ref. 1; Appendix A) this can be done by translations and rotations and a change of scale common to all network points. That means the covariance matrix C_x of the network points provided by those data processing software packages and representing externally affected accuracies has to undergo a similarity transformation (S-transformation) in order to reveal the "real" internal accuracy C_x^i of the network point determination.

The diagonal block structure of the covariance matrix of the GPS measurements of the INN VALLEY Network (Appendix C) shows the following characteristics:

$$ {}_s C_x : \quad s_x = \pm 1.5 \text{ cm}, \quad s_y = \pm 1.3 \text{ cm}, \quad s_h = \pm 2.0 \text{ cm} \tag{1} $$

(subscript ${}_s \bullet$ for satellite). These values contain external effects. After their removal by S-transformation the internal accuracies are

$$ {}_s C_x^i : \quad s_x = \pm 1.3 \text{ cm}, \quad s_y = \pm 1.2 \text{ cm}, \quad s_h = \pm 1.6 \text{ cm} \quad . \tag{2} $$

3. THE ACCURACY OF TERRESTRIAL NETWORK OBSERVATION RESULTS

If, for instance, horizontal directions, spatial distances and zenith angles are introduced into a free three-dimensional network adjustment, freedom of the geodetic datum exists only with respect to 3 translations and the rotation around the h-axis. The remaining 3 datum dispositions are established by the scale of distances and the orientation of the zenith angles (plumbline or normal of the ellipsoid). Due to these dispositions of 3 datum constraints which are network external effects the network is somehow restrained from being absolutely free. The restraints are reflected in the covariance matrix of the adjusted coordinates, too. Thus the covariance matrix provided by a "free" network adjustment of this kind provides external accuracies which are, in the case of the terrestrial INN VALLEY Network,

$$_T C_x \; : \quad s_x = \pm\, 0.4 \text{ cm}, \quad s_y = \pm\, 0.4 \text{ cm}, \quad s_h = \pm\, 3.6 \text{ cm} \tag{3}$$

(subscript $_T\bullet$ for terrestrial). After S-transformation the corresponding results of the internal accuracies are

$$_T C_x^i \; : \quad s_x = \pm\, 0.4 \text{ cm}, \quad s_y = \pm\, 0.4 \text{ cm}, \quad s_h = \pm\, 2.1 \text{ cm} \quad . \tag{4}$$

As expected, the S-transformation influences especially the accuracy in heights by rotations around the x- and y-axes since strong datum effects are filtered off.

4. COMPARISON OF SATELLITE AND TERRESTRIAL OBSERVATION RESULTS

The first possibility to achieve assertions on accuracies of satellite aided network observations in comparison to terrestrial network observations is the simple comparison of specific results obtained from the individual adjustments of the terrestrial and satellite observations, respectively.

The coordinates of those adjustments can be compared with each other after a 7-parameter similarity transformation which splits off network external effects caused for instance by a different geodetic datum.

$$_{T-S} dx \; = \; _T x - m R_S x \tag{5}$$

$$_{T-S} dx^T = \; |dx \; dy \; dh|$$
coordinate differences to be investigated after transformation

$_T x$ centre of gravity related terrestrial coordinates

$_S x$ centre of gravity related GPS coordinates

m   scale factor

R   rotation matrix .

Applied to the example of the INN VALLEY Network the following results have been obtained (Ref. 2):

The coordinate differences in dx range from -2.4 cm to +3.0 cm, in dy from -3.2 cm to +2.8 cm, in height dh from -12.8 cm to +12.6 cm. The r.m.s. error in position is $s_p = \pm\, 2.6$ cm , while the r.m.s. error in height is $s_h = \pm\, 8.0$ cm . s_p is an excellent result, it corresponds quite nicely to what can be derived from Eq. 1. The uncertainty in height s_h can be questioned, it is much higher than the r.m.s. value $s_h = \pm\, 2.0$ cm of the GPS derived heights and also higher than $s_h = \pm\, 3.6$ cm of the terrestrial heights. The conclusion may be drawn that the GPS heights cannot be burdened with the discrepancy. It is most likely that the refraction influenced terrestrial heights are much weaker and not so consistent; therefore, they cannot compete with the satellite aided height observations.

Most suitable for comparison purposes are also those elements which are by themselves independent of network external influences and geodetic datum invariant, i.e. spatial distances.

The differences

$$_{T-S}d\delta = {}_T\delta - {}_S\delta \tag{6}$$

of the terrestrial and satellite network distances are, with only one exception, positive, ranging from +0.2 cm to +9.8 cm; related to the length of the distances a scale factor of m_S = +2.47 ppm can be derived. The differences of the scaled distances range from -2.6 cm to +6.6 cm having a relative accuracy of ± 2.0 ppm.

The above-mentioned statements are the usual ones when terrestrial and satellite observations are compared. However, the rigorous combination of both the observation types yields accuracy statements which go much further.

The first step is some considerations on the functional and stochastic hybrid network models.

5. THE COMBINATION OF TERRESTRIAL AND SATELLITE AIDED NETWORK OBSERVATIONS - THE FUNCTIONAL MODEL

The observation equations of a combined or hybrid adjustment model have to be formulated with respect to a common reference system. For the following models the "satellite" coordinates, i.e. the network point coordinates determined by satellite aided observations, and their covariance matrix were transformed onto the terrestrial coordinate system refering to the national geodetic datum.

5.1 Gauss-Markov model

A relatively simple model regards the satellite aided observations as direct coordinate observations which are added to the observation equations of the terrestrial measurements. Therefore, the combined system of observation equations is

$$
\begin{aligned}
{}_T\ell + {}_T v &= Ax \,, \quad {}_T P_1 = {}_T C_1^{-1} \\
{}_S\ell + {}_S v &= Ix \,, \quad {}_S P_x = {}_S C_x^{-1}
\end{aligned}
\quad , \quad
C_1 = \begin{vmatrix} {}_T C_1 & 0 \\ 0 & {}_S C_x \end{vmatrix}
\tag{7}
$$

($_T\ell$ terrestrial observations, $_S\ell$ satellite coordinates, $_T P_1$, $_T C_1$ and $_S P_x$, $_S C_x$ weight and covariance matrices respectively of the observations).

The vector x of unknowns contains, among others, the point coordinates x, y, h common to both types of networks.

In case the satellite coordinates are used as approximate coordinates, $_S\ell = 0$. Then the normal equations are

$$(A^T \ _T P_1 A + \ _S P_x)x = A^T \ _T P_1 \ _T \ell \ , \tag{8}$$

where the satellite coordinates are only represented by their weight matrix $_S P_x$. The cofactor matrix of the adjusted coordinates

$$Q_x = (A^T \ _T P_1 A + \ _S P_x)^+ = (_T P_x + \ _S P_x)^+ = \ _{T+S} Q_x \tag{9}$$

demonstrates that the weights of the terrestrial and satellite coordinates are directly added. The mutual interaction will depend on the size and on the "nature" of the two partners. With point coordinates observed by satellite aided techniques, Q_x is regular in general.

5.2 Gauss-Helmert model

Both the sets of terrestrial and satellite coordinates are considered observations within a system of condition equations with unknowns. The unknowns are the transformation parameters p (Eq. A.1). The system of equations is

$$_T x + \ _T v - (_S x + \ _S v) + G \cdot p = 0 \tag{10a}$$

or

$$Bv + G \cdot p + w = 0 \tag{10b}$$

with

$$B = \begin{vmatrix} 1 & 0 & 0 & -1 & 0 & 0 & 0 \\ 0 & 1 & 0 & 0 & -1 & 0 & 0 & \cdots \\ 0 & 0 & 1 & 0 & 0 & -1 & 0 \\ & & & \cdot & \cdot & \cdot & \end{vmatrix}$$

$$v^T = \begin{vmatrix} _T v_x & _T v_y & _T v_h & _S v_x & _S v_y & _S v_h & \cdots \end{vmatrix}$$

$$w^T = \begin{vmatrix} _T x - \ _S x & _T y - \ _S y & _T h - \ _S h & \cdots \end{vmatrix} \ .$$

The weight matrix of the observations results from the individually adjusted terrestrial and satellite networks in

$$P_x = \begin{vmatrix} A^T {}_T P_1 A & 0 \\ 0 & {}_S P_x \end{vmatrix} = \begin{vmatrix} {}_T P_x & 0 \\ 0 & {}_S P_x \end{vmatrix} . \tag{11}$$

The solution is according to the well-known algorithms. Attention has to be paid to the inversion of P_x which is possibly positive semi-definit.

If one transforms model Eq. 10 owing to

$$B\upsilon = \bar{\upsilon} \tag{12}$$

into the Gauss-Markov model

$$w + \bar{\upsilon} + G \cdot p = 0 \quad , \tag{13}$$

the weight matrix $\bar{P}_x$ results from

$$\bar{Q}_x = {}_T Q_x + {}_S Q_x \quad , \tag{14}$$

if the Q-matrices are the inverses of the corresponding P-matrices.

6. THE COMBINATION OF TERRESTRIAL AND SATELLITE AIDED NETWORK OBSERVATIONS — THE STOCHASTIC MODEL

There are two major problems involved in the proper combination of the stochastic information of terrestrial and satellite aided network observation.

6.1 The combination of heterogeneous observations

The stochastic information contained in geodetic observations can often not be utilized in its entirety in "usual" adjustments especially in cases of heterogeneous observation material since with the unit variance s_0^2 only one single quantity of the stochastic model is estimated on a global scale. If, however, some knowledge about the stochastic structure of the observations is pre-given, the estimation of the variance components s_i^2 of individual groups of observations is made possible. This a posteriori estimation technique (Appendix B) is suitable to adjoin proper weights to the various observation groups.

Due to the linear structure of the stochastic model in Eq. 7

$$C_1 = \sigma_0^2 Q_1 = \sigma_T^2 {}_T Q_1 + \sigma_S^2 {}_S Q_x$$

$$\sigma_0^2 \boxed{Q_1} = \sigma_T^2 \begin{vmatrix} {}_T Q_1 & 0 \\ 0 & 0 \end{vmatrix} + \sigma_S^2 \begin{vmatrix} 0 & 0 \\ 0 & {}_S Q_x \end{vmatrix} \quad , \tag{15}$$

the unit variance σ_0^2 as well as the variance components σ_T^2 and σ_S^2 can be estimated (Eq. B.16). Even a refinement is possible if one splits $_T Q_1$ into its components

$$\sigma_T^2 \, _T Q_1 = \sigma_{dir}^2 \, Q_{dir} + \sigma_{dis}^2 \, Q_{dis} + \sigma_{zen}^2 \, Q_{zen} \tag{16}$$

(dir directions, dis distances, zen zenith angles)
and estimates the respective variance components.

For the combination of the terrestrial observations with the internally highly accurate GPS measurements of the INN VALLEY Network there were 6 separate groups of observations established for the VCE: horizontal directions, zenith angles, 3 groups of spatial distances and the GPS observations. By this means the interaction of the satellite measurements and the individual groups of terrestrial observations can be studied thoroughly (Ref. 3).

Applying VCE to the Gauss-Markov model (Eq. 13) the linear structure (Eq. 14) to be considered is

$$\bar{C}_1 = \sigma_0^2 \bar{Q}_1 = \sigma_0^2 \bar{Q}_x = \sigma_T^2 \, _T Q_x + \sigma_S^2 \, _S Q_x$$

$$\sigma_0^2 \boxed{\bar{Q}_x} = \sigma_T^2 \boxed{_T Q_x} + \sigma_S^2 \boxed{_S Q_x} \quad . \tag{17}$$

The estimation procedure is according to Eq. B.13.

6.2 Combination of internal and external stochastic information

There are several possibilities of combining the stochastic information in terms of internal and external accuracies.

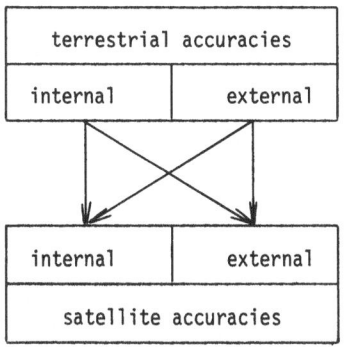

Combination of stochastic information

The combination of only internal accuracies provides, after proper weighting by VCE, internal accuracies within the combined network. An analogous conclusion can be drawn if exclusively external accuracies are combined. It can be questioned, of course, whether such a combination is reasonable in this case - are apples and oranges assembled? The same question arises if internal accuracies are combined with external ones. However, after S-transformation in all cases the same internal accuracy of the combined network can be revealed. This is illustrated in the following (see Appendix A):

$$_T Q^i_x = H _T Q_x H \quad , \quad _S Q^i_x = H _S Q_x H \tag{18}$$

$$_{T+S} Q^i_x = H _T Q_x H + H _S Q_x H \tag{19}$$

$$= H(_T Q_x + _S Q_x)H \quad .$$

Equation 19 shows that the internal covariance matrix of the combined network can be achieved either by the sum of the internal covariance matrices of the terrestrial and satellite networks or by the S-transformed sum of the external covariance matrices. Due to the idempotence of H the same is valid if an external covariance matrix is combined with an internal one.

Applied to Eq. 9

$$_{T+S} Q^i_x = H _{T+S} Q_x H = (H(_T P_x + _S P_x)H)^+ = (H _T P_x H + H _S P_x H)^+ \quad , \tag{20}$$

and applied to Eq. 14

$$\bar{Q}^i_x = H \bar{Q}_x H = H(_T Q_x + _S Q_x)H = H _T Q_x H + H _S Q_x H \tag{21}$$

the same statement holds (Eq. A.11).

7. THE COMBINATION OF TERRESTRIAL AND SATELLITE AIDED NETWORK OBSERVATIONS
   - RESULTS

The INN VALLEY Network serves again as an example for numerical experiments (Appendix C).

7.1 The accuracy of coordinate observations

As the first measure the accuracy of the satellite aided coordinate observations as obtained from a hybrid network adjustment is investigated.

Introducing the external GPS covariance matrix $_S C_x$ (Tab. 1, row 2) leads to coordinate accuracies of the combined network which cover the terrestrial accuracies (Tab. 1, row 1) totally. After applying VCE the a priori weights of the GPS coordinate observations turn out to be by a factor 4.5 less accurate than before (Tab. 1, row 3 vs. 2). The accuracy situation of adjusted coordinates of the combined network (Tab. 1, row 4) which represents at the same time the accuracy of the adjusted coordinate observations is governed by the external accuracy of the GPS coordinates (Eq.1). For further processing, see next paragraph.

Table 1

Accuracies of coordinate observations

row	status	accuracies of coordinates		
		s_x [cm]	s_y [cm]	s_h [cm]
terrestrial network				
1		0.4	0.4	3.6
combined terrestrial and GPS ($_S C_x$) network				
2	1	1.5	1.3	2.0
3	2	6.7	5.8	8.9
4	3	1.7	1.5	3.4
combined terrestrial and GPS ($_S C_x^i$) network				
5	1	1.3	1.2	1.6
6	2	5.2	4.8	6.4
7	3	0.4	0.4	3.1

Status: 1 = a priori input standard deviations of coordinate observations;
2 = a priori output standard deviations after VCE;
3 = standard deviations of the adjusted coordinate observations after VCE; they represent also the standard deviations of the adjusted netpoint coordinates.
All standard deviations are r.m.s. values.

Using the internal GPS covariance matrix (Eq. 2; Tab. 1, row 5) one obtains after VCE the internal observational accuracies of the GPS measurements (Tab. 1, row 6) which lead to standard deviations of adjusted coordinate observations and coordinates (Tab. 1, row 7) being the same as the terrestrial ones. An exception is the standard deviation s_h demonstrating the good influence of GPS measurements to an increased (external) accuracy of heights in this combined network.

7.2 Accuracy of adjusted coordinates

The results of the combined terrestrial and GPS-network are:

$_S C_x$ (Eq. 1) and $_T C_x$ (Eq. 3) lead to $_{S+T} C_x$ after VCE (Tab. 1, row 4):

$$_{S+T} C_x : \quad s_x = \pm 1.7 \text{ cm}, \quad s_y = \pm 1.5 \text{ cm}, \quad s_h = \pm 3.4 \text{ cm} . \tag{22}$$

$_S C_x^i$ (Eq. 2) and $_T C_x^i$ (Eq. 4) lead to $_{S+T} C_x^i$ after VCE:

$$_{S+T} C_x^i : \quad s_x = \pm 0.4 \text{ cm}, \quad s_y = \pm 0.4 \text{ cm}, \quad s_h = \pm 1.6 \text{ cm} . \tag{23}$$

S-transformation of Eq. 22 leads to Eq. 23 as does S-transformation of any other combination of external and internal accuracies, for instance of the results as given in
Tab. 1, row 7. Note that the internal accuracy of heights has been improved by the introduction of GPS-measurements.

A more detailed investigation of the behaviour of the accuracies of the original and
adjusted terrestrial observations and the coordinate observations is given by Ref. 3.

8. CHANGES IN POSITIONS AND HEIGHTS DUE TO THE NETWORK COMBINATION

The coordinates based on satellite aided observations were transformed onto the adjusted coordinates of the terrestrial network prior to the network combination. The terrestrial network coordinates were serving as approximate coordinates in all cases of hybrid adjustments. However, in contrast to the terrestrial adjustment with a datum definition based on the national geodetic datum, the satellite coordinates and their covariance
matrices determine the geodetic reference system within the combined adjustment models.

If one considers the mean values of coordinate unknowns resulting from the various
combinations of stochastic information, one can realize that systematic changes in positions and heights result from the application of the external covariance matrices. This
effect is due to the external influences which govern the external covariance matrices; in
contrast, the use of the internal covariance matrices leaves on an average the hybrid network coordinates in the position of the terrestrial network.

This effect is the clearest with the height coordinates. Therefore, the behaviour of
the changes in heights will be discussed in the following.

The observed GPS heights differ from the terrestrial heights between -12.9 cm ...
+12.3 cm (mean ± 0 cm) before the combination. After combination with $_S C_x$ (Eq. 1) and
VCE, height changes occur from -7.4 cm ... +3.6 cm (mean -2.5 cm). Particularly interesting are the results of the combination with the internal GPS covariance matrix $_S C_x^i$
(Eq. 2): they reach from -11.0 cm ... +6.5 cm (mean ± 0 cm).

Considering the essential results of the combination of terrestrial with GPS heights, the distinct influence of the GPS heights to the height determination of the network points becomes clear. Especially the extreme values -11.0 cm and +6.5 cm are highly significant improvements of two, terrestrially only weakly determined heights.

9.   CONCLUSIONS

An appropriate combination of the stochastic information obtained from terrestrial and satellite aided network observation requires

- a hybrid functional and stochastic model of adjustment
- variance component estimation techniques
- S-transformation of covariance matrices.

Some results are

- various combinations of external and internal covariance matrices can be applied;

- after variance component estimation an identical internal covariance matrix of the adjusted coordinates of the hybrid network can be obtained from different combinations by S-transformations;

- the covariance matrices of satellite aided coordinate observations as obtained from various standard software packages are strongly affected by external influences;

- for comparison of accuracies only internal covariance matrices (or invariant functions) are suitable;

- GPS observations can contribute to an improvement of the internal accuracy of even very precise local terrestrial networks.

Appendix A

S-TRANSFORMATION

The S-transformation parameters p are 3 translations, 3 rotations and 1 scale factor for a three-dimensional network

$$p^T = |t_x\ t_y\ t_z\ \varepsilon_x\ \varepsilon_y\ \varepsilon_z\ m| \quad, \tag{A.1}$$

the coefficient matrix G is for n points

$$G^T = \begin{vmatrix} 1 & 0 & 0 & 1 & 0 & 0 & \cdots & 1 & 0 & 0 \\ 0 & 1 & 0 & 0 & 1 & 0 & \cdots & 0 & 1 & 0 \\ 0 & 0 & 1 & 0 & 0 & 1 & \cdots & 0 & 0 & 1 \\ 0 & -z_1 & y_1 & 0 & -z_2 & y_2 & \cdots & 0 & -z_n & y_n \\ z_1 & 0 & -x_1 & z_2 & 0 & -x_2 & \cdots & z_n & 0 & -x_n \\ -y_1 & x_1 & 0 & -y_2 & x_2 & 0 & \cdots & -y_n & x_n & 0 \\ x_1 & y_1 & z_1 & x_2 & y_2 & z_2 & \cdots & x_n & y_n & z_n \end{vmatrix}. \tag{A.2}$$

The transformation of a C_x covariance matrix is carried out with

$$H = (I - G(G^T G)^{-1} G^T) \tag{A.3}$$

according to Ref. 1 to

$$C_x^i = H C_x H \ . \tag{A.4}$$

The matrices G and H have the following important properties:

$$r\{G\} = d \tag{A.5}$$

($r\{\cdot\}$ rank), if the number of transformation parameters applied is d.

$$r\{G^T G\} = d \ , \tag{A.6}$$

so that $(G^T G)$ is regular and $(G^T G)^{-1}$ exists.

$$\bar{G} = G(G^T G)^{-1} G^T = \bar{G}^T \tag{A.7}$$

is symmetric and idempotent

$$\bar{\bar{G}}\bar{G} = \bar{G} , \quad r\{\bar{G}\} = tr\{\bar{G}\} = d \tag{A.8}$$

($tr\{\cdot\}$ trace) .

The transformation matrix H is symmetric and idempotent

$$H = H^T , \quad HH = H , \quad r\{H\} = tr\{H\} = tr\{I\} - tr\{\bar{G}\} = 3n-d , \tag{A.9}$$

and the Moore-Penrose generalized inverse H^+ is the matrix itself

$$H^+ = H . \tag{A.10}$$

From the idempotence (Eq. A.9) it follows that a S-transformation can only once be applied to a coveriance matrix C_x ; a second S-transformation of the same matrix would not change it.

The rank defect of the resulting matrix C_x^i is d , if $r\{H\} = 3n-d \leq r\{C_x\}$; $d = 7$ at the most in case of a three-dimensional network. On this condition, with P_x as the inverse of C_x and with Eq. A.10 also

$$C_x^i = HC_xH = (HP_xH)^+ \tag{A.11}$$

holds (Ref. 4).

Appendix B
_____

VARIANCE COMPONENT ESTIMATION

In the following only a short review is given. For detailed studies see for instance Ref. 5 - 7.

The stochastic part of the Gauss-Markov model

$$\ell + v \; = \; Ax \; , \quad C_1 \tag{B.1}$$

ℓ vector of observations $(n{\times}1)$
v vector of residuals $(n{\times}1)$
A design matrix $(n{\times}u)$
x vector of unknowns $(u{\times}1)$

Consists of the covariance matrix C_1 of the observations which can be divided in two

$$C_1 \; = \; \sigma_0^2 Q_1 \tag{B.2}$$

σ_0^2 theoretical unit variance
$Q_1 = P_1^{-1}$ cofactor matrix of the observations .

Q_1 is considered pre-given, σ_0^2 is after the adjustment (a posteriori) unbiasedly estimated by the empirical unit variance s_0^2 , i.e.

$$s_0^2 \; = \; \frac{v^T P_1 v}{r} \; , \tag{B.3}$$

r redundancy of the observations

$$r \; = \; \mathrm{tr}\{P_1 Q^{vv}\} \; = \; n - u \tag{B.4}$$

$(\mathrm{tr}\{\cdot\}$ trace operator$)$,

$$Q^{vv} \; = \; Q_1 - A(A^T P_1 A)^+ A^T \; = \; Q_1 - A Q_x A^T \tag{B.5}$$

Q^{vv} cofactor matrix of the residuals $(n{\times}n)$,

$$v \; = \; (A Q_x A^T P_1 - I)\ell \; = \; - Q^{vv} P_1 \ell \tag{B.6}$$

and

$$v^T P_1 v \; = \; \ell^T P_1 Q^{vv} P_1 Q^{vv} P_1 \ell \; = \; s_0^2 \, \mathrm{tr}\{P_1 Q^{vv} P_1 Q^{vv}\} \quad . \tag{B.7}$$

Frequently observations of different types have to be adjusted in one hybrid model, i.e. horizontal directions, spatial distances, zenith angles drop in the adjustment model of a terrestrial network, or even terrestrial and satellite aided network observations have to be combined. In such a case the original Gauss-Markov model is split into several parts corresponding to the c different groups of observations

$$\begin{vmatrix} \ell_1 \\ \vdots \\ \ell_c \end{vmatrix} + \begin{vmatrix} v_1 \\ \vdots \\ v_c \end{vmatrix} = \begin{vmatrix} A_1 \\ \vdots \\ A_c \end{vmatrix} \cdot x \quad . \tag{B.8}$$

Also the stochastic model falls into groups

$$C_1 \; = \; \sigma_0^2 Q_1 \; = \; \sigma_1^2 Q_1 + \ldots + \sigma_c^2 Q_c \; = \; \sigma_1^2 P_1^{-1} + \ldots + \sigma_c^2 P_c^{-1} \quad . \tag{B.9}$$

While the model was homogeneous, only the estimation of the unit variance σ_0^2 was of interest. In the inhomogeneous case, however, the wish may exist to estimate the variance components $\sigma_1^2 \ldots \sigma_c^2$ as well.

This can be achieved by dissecting Eqs. B.5 and B.7 according to

$$Q^{vv} \; = \; \begin{vmatrix} Q_{11}^{vv} & \cdots & Q_{1c}^{vv} \\ \vdots & \ddots & \vdots \\ Q_{c1}^{vv} & \cdots & Q_{cc}^{vv} \end{vmatrix} \tag{B.10}$$

with

$$Q_{ii}^{vv} \; = \; Q_i - A_i Q_x A_i^T \qquad i = 1 \ldots c$$

$$Q_{ij}^{vv} \; = \; \quad - A_i Q_x A_j^T \qquad i,j = 1 \ldots c \; , \; i \neq j$$

and

$$v^T P_1 v \; = \; \sum_{i=1}^{c} v_i^T P_i v_i \tag{B.11}$$

with

$$v_i^T P_i v_i \; = \; \sum_{j=1}^{c} \mathrm{tr}\{P_j Q_{ji}^{vv} P_i Q_{ij}^{vv}\} s_j^2 \tag{B.12}$$

$$= \; \sum_{j=1}^{c} t_{ij} s_j^2 \quad .$$

After composing to one system of equations

$$
\begin{vmatrix} v_1^T P_1 v_1 \\ \vdots \\ v_c P_c v_c \end{vmatrix} = \begin{vmatrix} t_{11} & \cdots & t_{1c} \\ \vdots & & \vdots \\ t_{c1} & & t_{cc} \end{vmatrix} \cdot \begin{vmatrix} s_1^2 \\ \vdots \\ s_c^2 \end{vmatrix} \tag{B.13}
$$

the variance components s_j^2 can be solved for by inversion. The solution has to be found iteratively until all components converge $s_i^2 \rightarrow s_j^2$. The final estimation result is un-biased.

Eq. B.13 is the general formula which has to be applied if the linear structure of the Q_1 matrix is "overlapping", i.e.

$$
\sigma_0^2 \boxed{Q_1} = \sigma_1^2 \boxed{Q_1} + \sigma_2^2 \boxed{Q_2} \tag{B.14}
$$

or "sequential"

$$
\sigma_0^2 \boxed{Q_1} = \sigma_1^2 \boxed{\begin{array}{c|c} Q_1 & \\ \hline & 0 \end{array}} + \sigma_2^2 \boxed{\begin{array}{c|c} 0 & \\ \hline & Q_2 \end{array}} \ . \tag{B.15}
$$

In this case the equation system B.13 can be simplified. The estimation formula is

$$
v_i^T P_i v_i = s_i^2 \ \mathrm{tr}\{P_i Q_{ii}^{vv}\} \tag{B.16}
$$

for each variance component s_i^2 independent of the others. The estimation process is iterative, too.

One very important fact of the estimation of variance components is its independence of the geodetic datum or other network external effects.

Appendix   C

THE INN VALLEY NETWORK

Terrestrial observations

The INN VALLEY Network of the Institute of Geodesy of the Bundeswehr University Munich covers a space of 25 km × 15 km × 1.3 km. All points but one are monumented by concrete pillars. The terrestrial coordinates were determined by 48 spatial distances (three classes of accuracies), 44 horizontal directions, 44 zenith distances and 8 × 2 components of the deflections of the vertical.

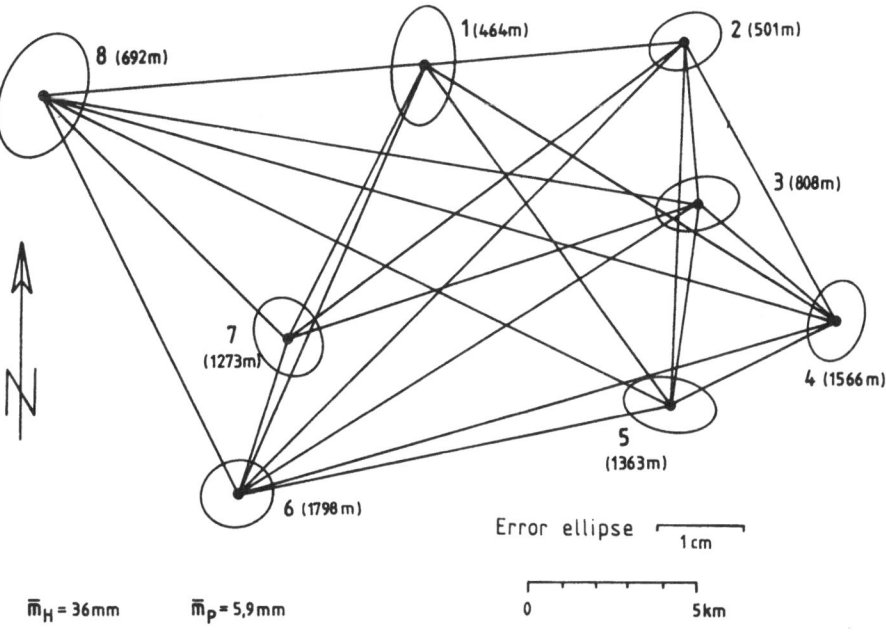

Fig. 1   The INN VALLEY Network: terrestrial observations

GPS observations

In November 1984 coordinate differences of the INN VALLEY Network points were deter-mined by Macrometer V 1000 measurements (Ref. 2).

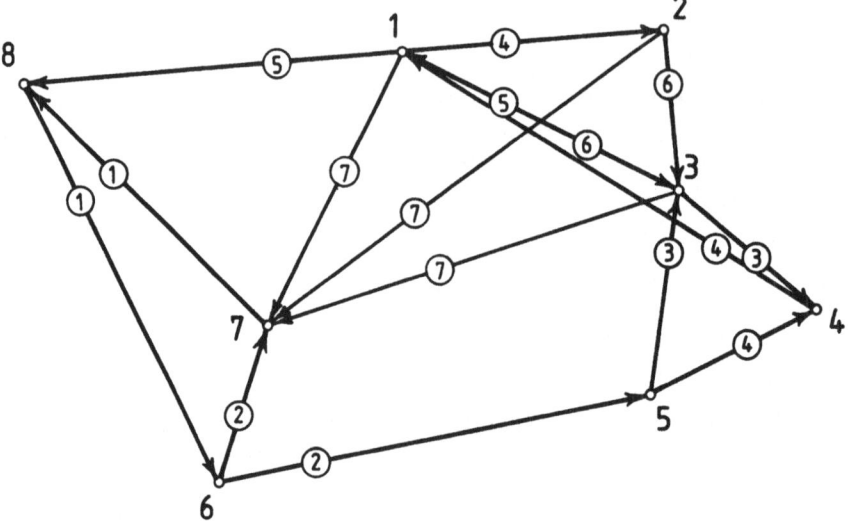

Fig. 2   The INN VALLEY Network: Macrometer measurements
(ⓘ   simultaneous observations)

*   *   *

REFERENCES

1) P. Meissl, Zusammenfassung und Ausbau der Inneren Fehlertheorie eines Punkthaufens, in: K. Rinner, K. Killian, P. Meissl, Beiträge zur Theorie der geodätischen Netze im Raum, Deutsche Geodätische Kommission, Reihe A, Heft 61, München 1969

2) H. Heister, A. Schödlbauer, W. Welsch, Macrometer Measurements 1984 in the INN VALLEY Network, in: First International Symposium on Precise Positioning with the Global Positioning System, proceedings, 567-578, Rockville 1985

3) W. Welsch, W. Oswald, The hybrid adjustment of terrestrial and satellite aided network observations - investigation of accuracies, XVIII. International FIG Congress, invited paper 503.1, Toronto 1986

4) A. Albert, Regression and the Moore-Penrose Pseudoinverse, Academic Press, New York and London 1972

5) E. Grafarend, A. Kleusberg, B. Schaffrin, An introduction to the variance-covariance-component estimation of Helmert-Type, Zeitschrift für Vermessungswesen 105 (1980)4, 161-180

6) H. Pelzer, Grundlagen der mathematischen Statistik und der Ausgleichungsrechnung, in: H. Pelzer (ed.), Geodätische Netze in Landes- und Ingenieurvermessung II, Konrad-Wittwer-Verlag, Stuttgart 1980, 3-120

7) W. Welsch, Grundlagen, Gebrauchsformeln und Anwendungsbeispiele der Schätzung von Varianz- und Kovarianzkomponenten, Vermessung, Photogrammetrie und Kulturtechnik 82 (1984)9, 296-301

VERY LONG BASELINE INTERFEROMETRY

J. Campbell
Geodetic Institute University of Bonn, Fed. Rep. of Germany

ABSTRACT

This contribution reviews the technique of Very Long Baseline Interferometry as a tool for high precision measurements of relative point positions and spatial baseline
orientation. The geodetic and geophysical applications of
these measurements are discussed in relation to the objectives of global and regional programs of Earth dynamics
research. Following a description of the method and the
instrumentation, the systematic errors limiting the accuracy of VLBI baseline vector determinations are discussed
and the different approaches of error elimination are indicated. Some of the most interesting results recently obtained by the different groups involved in geodetic VLBI
are shown.

1. INTRODUCTION

Radio interferometry originated as an astronomical observation method
with the aim of increasing as much as possible the low angular resolving
power of single telescopes. In its simplest form a radio interferometer consists of two antennas separated by a given distance D, the baseline, and
connected via the receivers and a phase stable electrical connection to a
phase meter. The response of such a system to the observed radio source
carries information on its angular size and structure, which are of a prime
interest to astronomers. For the study of extremely compact sources, however,
the resolution of the connected element radio interferometers (CERI) proved
insufficient and some means to increase the baseline length, which is proportional to the resolving power, had to be found. A major breakthrough was
made in 1967, when independent stations both in Canada and in the U.S.A.
were able to produce interference fringes by using very precise oscillators
and tape recorders, thus eliminating the need for a cable connection (BROTEN
et al. 1967[1]); BARE et al. 1967[2]). First experiments that were explicitly
aimed at achieving geodetic accuracy took place in 1969 on the 845-km baseline between the Haystack Observatory, Massachusetts and the 43-m antenna
of the National Radio Astronomy Observatory in Greenbank, West Virginia
(HINTEREGGER et al. 1972[3]). In order to reach the high group delay resolution
of ±1 ns attained in the above experiments, a bandwidth synthesis technique
had been developed (ROGERS 1970[4]). With this technique it became possible to

use tape recording equipment with a limited bandwidth and sample the broad-
band receiver window at a set of widely spaced frequencies. Both the introduction of independent high stability frequency standards and the bandwidth
synthesis technique have contributed to transform radio interferometry into
a geodetic measuring system of unprecedented accuracy, now commonly known
as Very Long Baseline Interferometry (VLBI).

The following description of the VLBI technique, its potential and its
achievements, is primarily oriented towards the geodetic applications with
emphasis on the achievement of high group delay accuracy, while omitting
the astronomical source imaging techniques.

2.   FUNDAMENTALS OF THE VLBI-TECHNIQUE

The most striking difference between Very Long Baseline Interferometry
and other space techniques is, that the interferometric observables are obtained a posteriori by the alignment in a processor of the two identical
signal streams received at different times at the two telescopes. Thus the
telescopes plus the processor could be seen to embody one instrument and the
baseline separating the telescope sites may be (and often is) named an instrumental calibration constant.

Fig. 1

Functional diagram of a VLBI - system

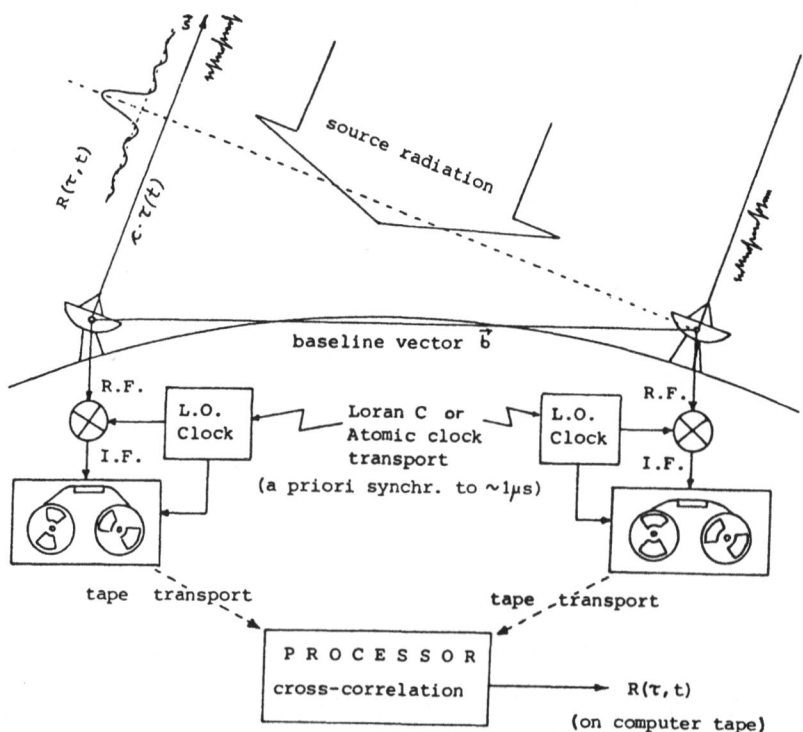

The main elements of a VLBI-system are shown in Fig. 1: The radio signals coming from a given source are observed simultaneously at two or more stations at a preselected frequency in the GHz-region, converted to baseband (video frequency) and recorded (usually in a digitized form) on high data-rate tape. Before registration the signal streams are provided with precise time information derived from the local frequency standards. Later, when the tapes are brought together in the playback processor, this permits a coherent correlation of the approximately aligned signal streams for a certain interval. Because the actual time lag is still unknown, the correlation is done at a number of different delays seperated by $\Delta\tau = 1/2B_s$ where B_s is the synthesized bandwidth of the observed channels.

During correlation the data stream from one station is delayed quasi-continuously in such a way that the changing geometric delay τ_9 is almost completely compensated for. This gives rise to a rather low residual fringe frequency, the phase of which slowly varies on the scale of a few turns per minute.

The output of the correlator is usually described by the complex cross-correlation function $R(\tau,t)$ which translates the response of an interferometer system to a radio source, see e. g. (MORAN 1976[5])).

RECEIVED SIGNALS

$$\text{Station 1}: x_1(t) = \int_{-\infty}^{+\infty} X_1(\omega)\, e^{j(\omega - \omega_0)t}\, d\omega$$

$$\text{Station 2}: x_2(t-\tau) = \int_{-\infty}^{+\infty} X_2(\omega)\, e^{j[(\omega - \omega_0)\, t - \omega\tau]}\, d\omega$$

CROSSCORRELATION FUNCTION $\quad R(t,\tau_i) =$

a) computed: $\dfrac{1}{2T} \displaystyle\int_{-T}^{+T} x_1(t)\, x_2(t - \tau_i)\, dt$

b) model: $K\,\dfrac{\sin \pi B_s (\tau - \tau_i)}{\pi B_s (\tau - \tau_i)}\, e^{j[\omega_f t - \pi B_s (\tau - \tau_i)]}$

$\underbrace{\phantom{\text{amplitude term}}}\qquad \underbrace{\phantom{\text{phase term}}}$

$\qquad$ amplitude term $\qquad\qquad$ phase term

$\quad$ K = constant proportional to SNR, T = integration time

$\quad B_s$ = maximum spanned bandwidth, ω_f = fringe frequency,

$\quad \tau$ = true delay, τ_i = discrete correlation lags

The crosscorrelation function amplitude has its first zero at $\pi B_s(\tau-\tau_i) = \pi$, hence $\Delta\tau = 1/B_s$, which is the halfwidth of the main peak.

Fig. 2

Response of a Mk II - VLBI - System to a point source

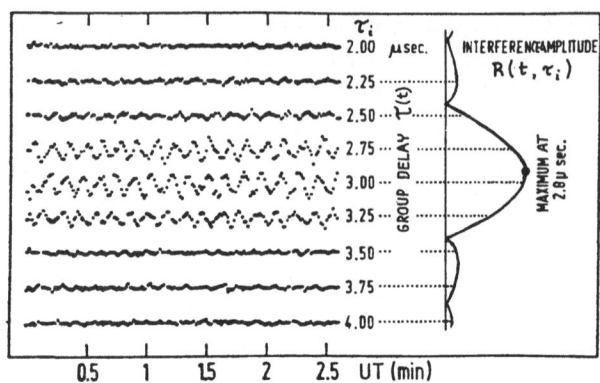

In Fig. 2 the aspect of a typical fringe pattern as produced by a single 2 MHz-channel interferometer observing a point source is shown. Here the delay spacing is 1/2 · 2 MHz = 0.25 µs.

The analysis of the fringe pattern yields a wealth of astronomical and geodetic information. As many as four - albeit partially dependent - observables can be derived from the crosscorrelation function:

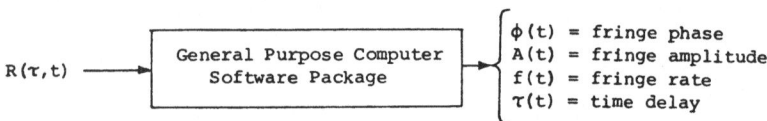

The actual travelling time τ of a particular wavefront between the antennas at two sites can be expressed in two ways, i. e. as the delay of a wave group formed by the wide-band signals (group delay) and as the phase difference of a given monochromatic constituent of the signal stream (phase delay). The group delay is defined as the derivative of phase versus frequency in the band and can be estimated unambiguously by finding the maximum of the crosscorrelation function, which for an ideal square bandpass assumes the form of a sinx/x function (see amplitude term above). The delay observable yields a full baseline solution and therefore plays the most important role in geodetic VLBI.

The <u>fringe phase</u> and the <u>fringe rate</u> (fringe frequency) are obtained from the sine and cosine parts of the crosscorrelation function. Due to the close relationship of φ and f, these observables are determined simultaneously, either from an ordinary sine wave adjustment or using the fourier transform into the frequency domain, where the maximum of S(f) is estimated. These methods, which allow to establish the function φ(t) over a certain interval of time (usually the duration of an uninterrupted source scan) are often referred to as "phase tracking" methods (THOMAS 1972[6]). In local interferometry (CERI) this method is the common way of calibrating the short baselines and measuring source positions. Due to large phase fluctuations on longer baselines the phase observable can be used in VLBI only under special conditions.

The <u>fringe rate</u> is unambiguous, but it is insensitive to the z-component of the baseline. So, compared to the delay observable it plays a less important role in geodetic baseline determinations.

The <u>fringe amplitude</u> is of essential importance for source structure mapping.

Figure 3 shows a typical example of the output of the Mk III fringe analysis software package developed by A.R. WHITNEY (1976)[7].

Fig. 3

Multichannel delay resolution function (Mk III fringe analysis)

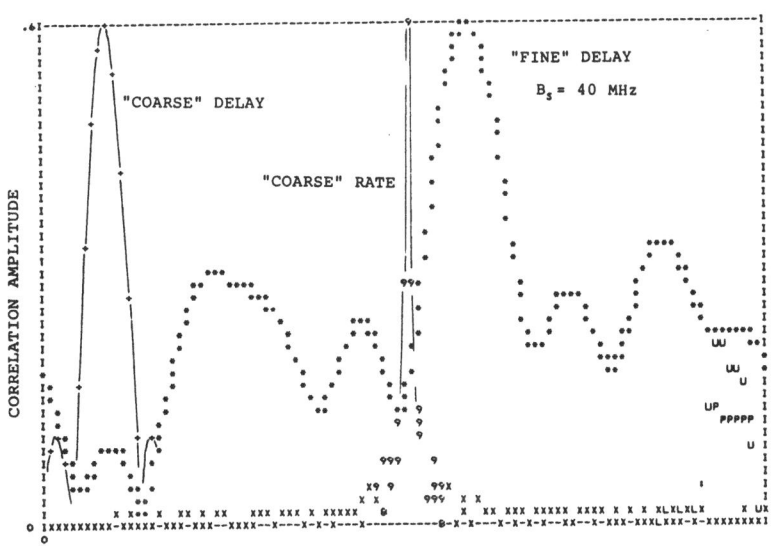

The accuracy with which the time delay τ is estimated from the cross-correlation function depends chiefly on the halfwidth of the main peak, which is given by Δτ (see above).

There are different possible methods of delay estimation (WHITNEY et al. 1976[7]), all of which yield a precision σ_τ of roughly 1 % of Δτ, depending of course on the SNR (signal to noise ratio) and the available integration time per observation.

The signal-to-noise ratio after averaging a total of 2 B_0T correlated signal samples can be expressed by

$$SNR = \frac{S}{2k} \cdot \frac{\sqrt{A_1 \cdot A_2}}{T_S} \cdot \sqrt{2\ B_0 T} \ ,$$

where k is Boltzmann's constant.

This expression shows that, apart from the fundamental relationship Δτ = $1/B_s$, the delay accuracy is proportional to the flux density S of a point source, the geometric mean of the antenna apertures A_1 and A_2, the square root of the recorded bandwidth B_0 and integration time T, and inversely proportional to the geometric mean of the system noise temperatures at both stations. If the ratio λ/D becomes very small, i. e. at high observing frequencies and on long baselines, many of the compact sources are resolved. This results in a marked decrease of the correlated flux density with the effect that sources which show strong fringes on baselines of a few hundred km become very weak on intercontinental baselines (see e. g. KELLERMANN et al. 1971[8]). Therefore in the latter case a high system sensitivity is of particular importance.

Efforts to improve the sensitivity of a Very-Long-Baseline interferometer have been concentrated on the sampling rate $2B_0$. With the newly developed Mk III system it is possible to record a data stream of 112 Megabit per second on 28 tracks of 2 MHz bandwidth each (CLARK 1979[9]). This has produced an increase of the sensitivity by a factor of 5.3 over the commonly used Mk II system.

The available coherent integration time T depends on the stability of the frequency standards and on the state of the atmosphere (troposhere and ionosphere). Due to the latter the ultimate stability that is achievable is limited to about 10^{-15} over time scales of 10^2 - 10^4 seconds. Hydrogen maser frequency standards guarantee a stability of 10^{-14} over the same periods of time which is acceptable for geodetic applications. Using the above expression for the SNR the instrumental phase error

$$\sigma_\phi = (SNR)^{-1}$$

and the group delay error

$$\sigma_\tau = (2\pi \text{ SNR } B_s)^{-1}$$

can be computed. To illustrate these expressions, a typical example for a VLBI system consisting of two 20m antennas equipped with Mk III data acquisition terminals is given:

X - band (8.4 GHz) : 8 channels 2 MHz each; T = 300 sec

Correlated flux of radio sources: 1 Jansky

Θ_A = 0.057° (two antennas of 20m diameter and 50 % eff.)

Θ_R = 160° K (uncooled paramps)

SNR = 18.2 $\sigma_\tau = \pm 3°$

With spanned bandwidth = 360 MHz:

σ_τ = 0.025 ns   ($\hat{=}$0.7 cm)

This extremely high instrumental precision is, of course, curtailed to a certain degree by systematic instrumental and environmental error sources as will be discussed later in this section.

The fundamental observation equation relating the <u>time delay</u> with the baseline and source vectors may be written as

$$\tau(t) = -\frac{1}{c}\, \vec{b} \cdot \vec{s}(t)$$

where

$$\vec{b} \cdot \vec{s}(t) = b_x\cos\delta\cosh + b_y\cos\delta\sinh + b_z\sin\delta$$

with the baseline components b_x, b_y, b_z

the radio source position α, δ

and the Greenwich hour angle of the source

h = GST - α

The baseline vector components are referred to the instantaneous Earth rotation axis. The unit vector of the source points to the apparent position at the time of observation (see Fig. 4).

At this stage there are 3 + 2 · n fundamental parameters to be determined in a least-squares fit: the three baseline components b_x, b_y, b_z and the coordinates α, δ of n observed sources.

In order to make theory consistent with observation, the above model has to be supplemented with terms allowing for a number of physical and instrumental effects listed below. Some of these effects can be predicted to a high degree of accuracy while others have to be parameterized or supplemented with additional measurements. In Fig. 5 an example of a geodetic VLBI software system is shown.

Fig. 4
Geometric VLBI Model

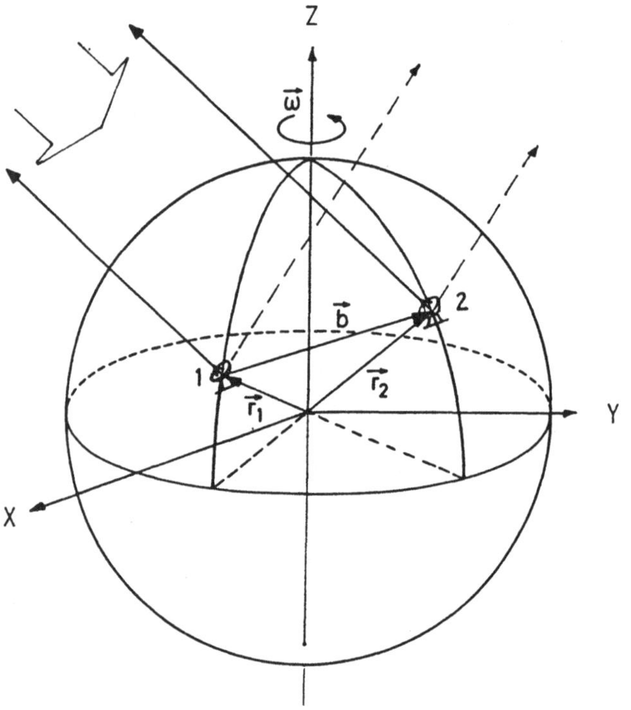

Model refinements:
  a) Effects of precession and nutation

Precession and nutation are motions of the Earth's axis with respect to the celestial system represented by the positions of the observed compact radio sources. These motions are caused by external forces, the gravitational attraction of the members of the solar system, acting upon the non-spherical, inhomogeneous and visco-elastic body of the Earth. The main coefficients of the standard model have been determined empirically by long astronomical observing series and therefore are not error-free. In turn, any misalignment of the Earth's axis due to these errors will be detected by the highly sensitive VLBI observations. Significant corrections to the precession constant and to some of the nutation coefficients have been derived from the analysis of only a few years of VLBI data (HERRING et al. 1985[10]).

Fig. 5

Flow diagram of a geodetic VLBI software system

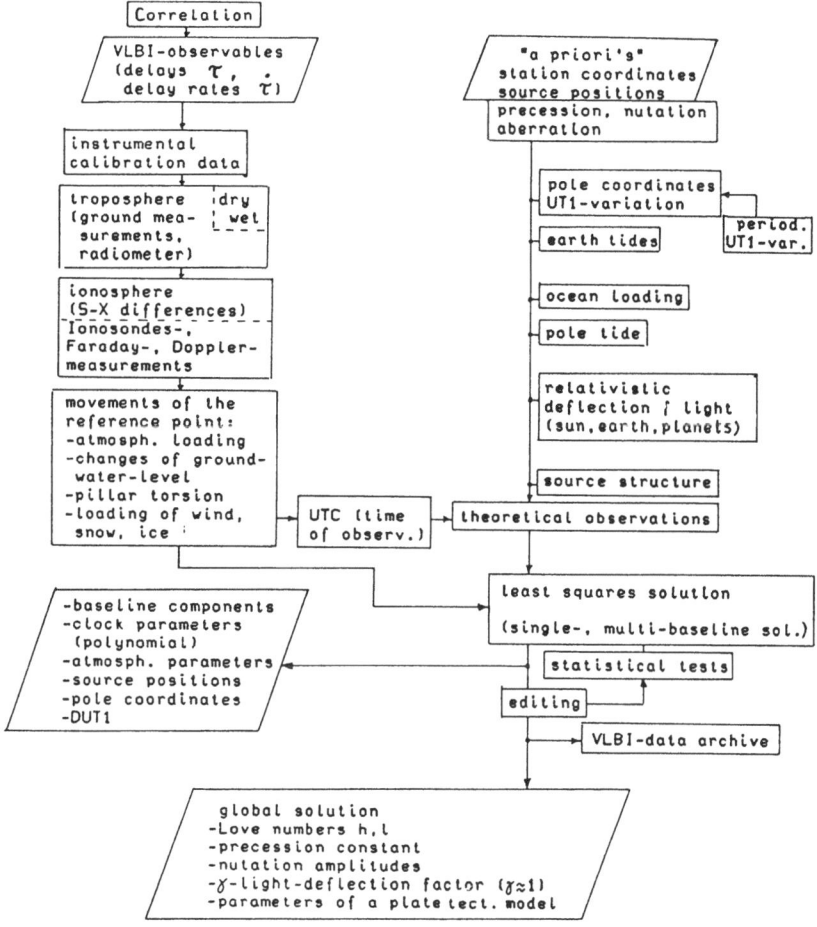

b) Relativistic effects (space-time geometry)

- special relativity. Due to the relative motion of the VLBI
  antennas during observation the finite speed of light causes
  significant additional delays which can be precisely modelled
  (diurnal aberration etc.).

- general relativity. The effect of gravity on the propagation
  of electromagnetic waves is considerable (near the sun the
  delay may be many orders of magnitude larger than the
  measurement error). VLBI observations have been used to
  verify Einstein's theory to an accuracy of about 0.1 %.

c) Instrumental effects

- local oscillator instabilities. Usually a clock model is introduced that accounts for the unknown epoch difference and a
  differential drift rate (two parameters). Unmodelled clock
  offsets and rates may lead to large systematic errors
  especially in the z-component of the baseline.
- change in the electronic circuitry and cable delays. While a
  constant delay is absorbed by the clock offset parameter, any
  delay changes have to be monitored by a phase and delay calibration system.
- deformations of the telescope structure, displacements of the
  reference points. These effects can be checked by models supplied
  with local geodetic measurements.

d) Environmental effects

The effect of the atmosphere on VLBI-observations is considered
the most serious problem, because at widely separated stations the
look-angles of the telescopes aimed at the same source differ greatly as well as the meteorological conditions themselves. The
ionosphere, which is a highly dispersive medium for radiation in
the radio frequency band, can be dealt with by using two different
observing frequencies. Moreover, its influence diminishes considerably with increasing frequency. At 8 GHz the extra-zenital path due
to the ionosphere reduces to about 2 dm. The remaining effect, however, is very unstable; therefore the dual frequency method using
a second frequency at 2.3 GHz is applied. The neutral atmosphere,
essentially the troposphere, presents the same problems in VLBI
as in Doppler and radar satellite observations. Its influence on
radio signals adds up to an extra-zenital path of 2.0 - 2.5 m.
The contribution of the dry air is rather stable and can be
described by suitable models. The wet component, although much
smaller, changes rapidly and has to be monitored during the observations. The most promising method appears to be the radiometer technique, which consists of measuring the micro-wave thermal
emission from water vapor near 22 GHz in the line-of-sight.

e) Geophysical effects

- polar motion and UT1-variations. These are effects that change
  the components of the baseline but not its length. To a certain
  degree of accuracy (~ 1 m) these variations can be obtained from
  the regularly published bulletins of the BIH to apply "corrections" to the observables (or to the model). In recent years the
  observations have proven to be more accurate than the corrections.
  Therefore the problem has been inverted, the changes showing in

the baseline components being used as precise polar motion and
Earth rotation data.
- Earth tides. The tides of the solid Earth are mainly semidiurnal
  and diurnal effects of a few decimeters in amplitude that change
  the baseline components as well as the length. The solid Earth
  tides are predictable to a few cm, but it is also possible to
  derive significant estimates of Love's numbers from VLBI observations. For stations close to the shores an additional correction due to the ocean tides has recently been included in the
  model.
- crustal motions. The theory of plate tectonics which stipulates
  that the Earth's crust is formed by a mosaic of separate plates
  that are in motion relative to each other has now been universally accepted. Predictions derived from geophysical evidence
  yield motions of a few centimeters per year. These small displacements constitute one of the main interests of a continued
  series of VLBI experiments aimed at the detection of the present
  rates of motion.

f) Radio-astrometric effects

- Precise source positions are usually determined simultaneously
  with the baseline components. Even so, special astrometric campaigns are carried out in order to establish consistent radio-
  source catalogues for fundamental astronomy, optical counter-
  part identification and for the support of geodetic programs.
  The accuracy of some thirty compact radio sources, most of which
  are located in the northern hemisphere, is now at a level of
  2 - 3 milliarcseconds as determined from VLBI-experiments.
  Work is in progress to improve the source coverage over the
  sky and to provide a firm connection to the existing optical
  reference system.
- Most of the observed compact sources tend to show structure at
  the level of a few milliarcseconds. This effect, in particular
  any changes in the structure, poses a limit on the accuracy of
  the radio-reference system. Permanent monitoring of the structure, which can be done by using the geodetic VLBI data, will
  help to reduce the contribution of these errors in the future.

The total geodetic error budget on baselines of about 1000 km sums up to
about 1 cm in length and 3.5 cm in the individual components (see Fig. 6).
Intercontinental baselines of about 6000 km show errors of 2 - 3 cm in
length and 10 cm in the components. These extremely high accuracies
($\sim$ 0.01 ppm) form the basis of the increasing efforts invested in the use
of the VLBI-technique for the determination of such subtle effects as variations in the Earth's rotation and crustal motion.

Fig. 6

Error Budget of Geodetic VLBI Accuracy

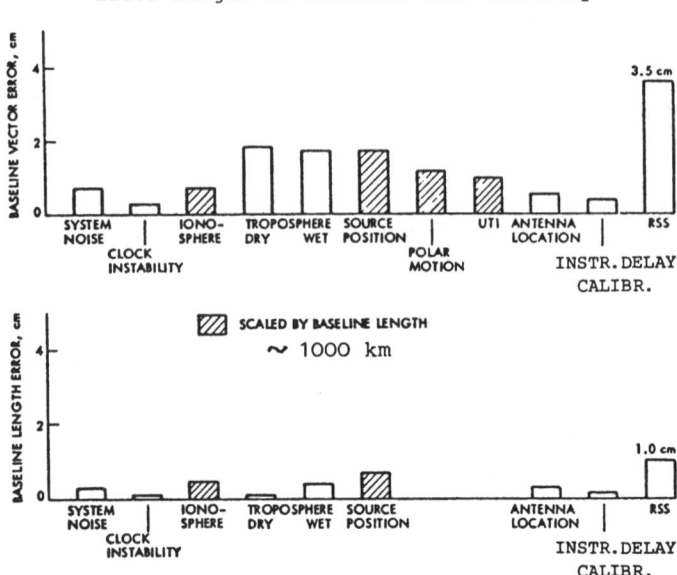

2. SCIENTIFIC INTEREST; PROGRAMS AND RESULTS

A detailed description of the geophysical applications of VLBI has been presented as early as 1969 at a conference held in London, Canada on "Earthquake displacement fields and the rotation of the Earth" (SHAPIRO, KNIGHT 1970[11]). In subsequent years virtually all of the goals could be realized or at least reach the stage of initial successful verification. In the following paragraph the scientific goals and projects are summarised and some of the recent results of the ongoing campaigns are shown.

Geodetic and geophysical interest in VLBI is based fundamentally on the use of an inertial reference frame of highly compact extragalactic radio sources. With the VLBI-technique it is possible to measure very accurately the baseline vectors and their changes in time between distant points on the Earth's crust. Therefore the primary objectives to use VLBI observations are:

1. the realization of a unified global reference system in order to satisfy the needs of global geodetic and navigational systems;
2. the monitoring of polar motion and changes in the Earth's rotation rate to enable a better understanding of the kinematics and dynamics of the Earth-Moon system and the structure of the Earth's interior;
3. the estimation of the Earth's elasticity parameters from directly measured tidal deformations;
4. the determination of plate motion and plate stability to improve the understanding of global plate tectonics;

5. the investigation of regional crustal movements in order to deduce the
building up of strain and provide input to earthquake prediction programs.

Other important activities are:

6. the determination of improved values for the precession and nutation
constants, and

7. the precise verification of the effects of general relativity.

The accuracy requirements necessary to attain these goals are ± 5 to
± 10 cm in each baseline component for items 1 and 2 (± 0.1 ms for UT1)
and ± 1 to ± 3 cm in baseline length repeatability for items 3 through 5.
By now these accuracies have been demonstrated in hundreds of VLBI campaigns
on baselines connecting nearly all major continents of the globe.

Corresponding to the international nature of the anticipated goals
several programs of multilateral cooperation have been initiated, among
which the following most important projects should be mentioned:

1. NASA Crustal Dynamics Project (CDP).
This project is part of a US Federal program involving several government
agencies for the application of space technology to crustal dynamics and
earthquake research. Major cooperative arrangements have been made with
European and other countries extending the project to a global research
program (NASA 1983[12]).

The VLBI part of the CDP comprises regular experiments (10 - 20 a year)
of one to three days duration between the major geodetic VLBI facilities
in the US, Europe and in and around the northern Pacific. In addition, so-
called bursts of observations using the mobile VLBI units are carried out
each year in the tectonically active zones of California and Alaska. These
campaigns are aimed at the creation of a detailed picture of the local
crustal motion pattern to assist the investigation of earthquake mechanisms.

The global station distribution is shown in Fig. 7. Some of the stations
are still in the process of being fully equipped for Mk III VLBI.

The CDP in its present form is planned to extend through 1988. Follow-up
programs will be set up in order to insure the repetition of the measurements at regular intervals of one or two years.

2. Project IRIS (International Radio Interferometric Surveying).
The goal of project IRIS is to conduct joint activities between the NGS-
(US National Geodetic Survey) POLARIS-Network and other international
stations that are dedicated to full-time geodetic work, such as the VLBI-
station of Wettzell, for the regular monitoring of Polar Motion and UT1.

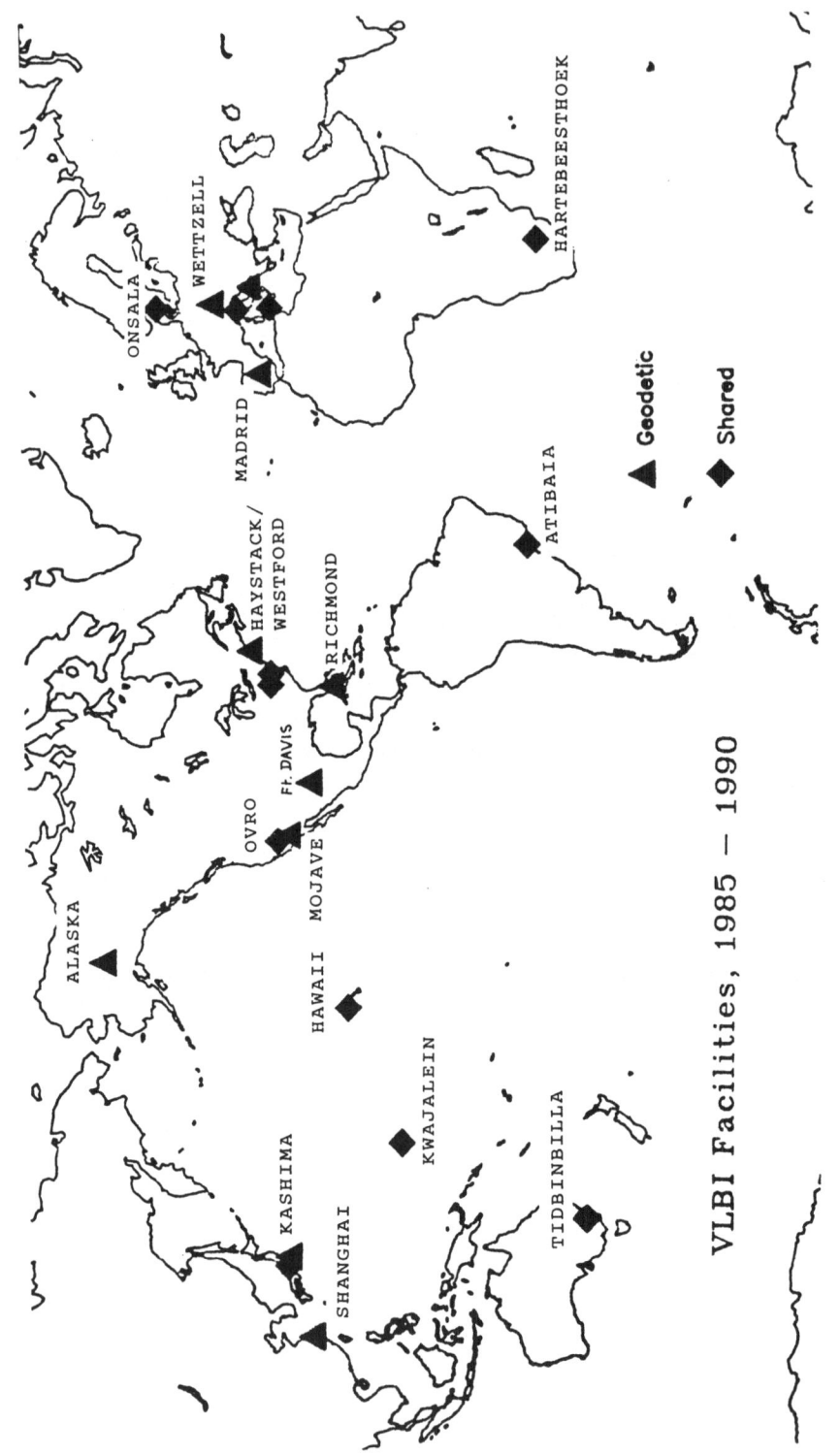

VLBI Facilities, 1985 – 1990

ONSALA

WETTZELL

HARTEBEESTHOEK

MADRID

ATIBAIA

Geodetic

Shared

HAYSTACK/
WESTFORD

RICHMOND

Ft. DAVIS

OVRO

ALASKA

HAWAII

MOJAVE

KWAJALEIN

KASHIMA

TIDBINBILLA

SHANGHAI

Fig. 7 Global VLBI network

The IRIS activities consist of a series of VLBI observing sessions of
24 hours duration at five-day intervals. The IRIS observations normally
involve three stations in the United States (the Westford telescope in
Massachusetts, the George R. Agassiz Station (Ft. Davis) in Texas and the
Richmond Observatory in Florida) and the 20 m radiotelescope of the Wettzell Observatory in the Federal Republic of Germany. One session per
month also includes the Onsala Space Observatory in Sweden (see Fig. 8).

Fig. 8
IRIS earth rotation network

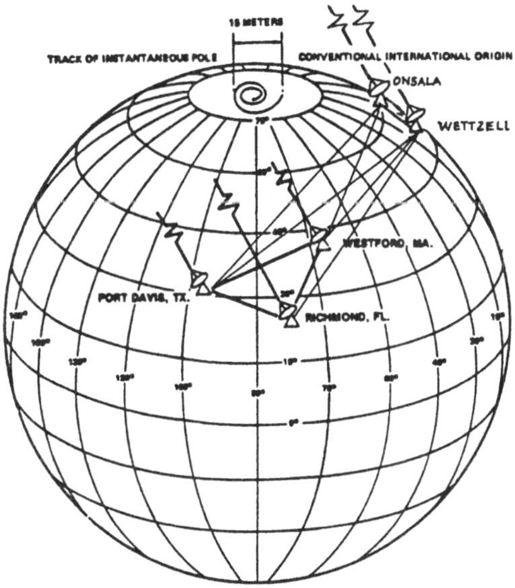

The National Geodetic Survey (NGS)/Rockville, Md. regularly analyses all
of the IRIS data to obtain polar motion and UT1 time series, which are
published monthly in the IRIS Bulletins A. Thus, since Jan. 5, 1984, the
date that the Wettzell Observatory began regular operations, the IRIS
system has been routinely providing the x and y components of polar motion
with an accuracy of 1 to 2 marcsecs, and UT1 with an accuracy of 0.04 to
0.1 msecs (CARTER, ROBERTSON, MACKAY, 1985[13]).

Figures 9 and 10 show the remarkably smooth trace of the instantaneous
pole of rotation from Sept. 26, 1980 (the time of the MERIT preliminary
campaign) until March 15, 1986. In particular, a significant improvement
of accuracy can be seen from the moment that the full IRIS network with
the station of Wettzell has become operational in January 1984 (Fig. 10).

In addition to the regular 5-day observations a special campaign of
daily observations on the baseline Westford - Wettzell is being performed

Figs. 9 and 10
Merit and POLARIS/IRIS pole position
Sept. 1980 - April 1984

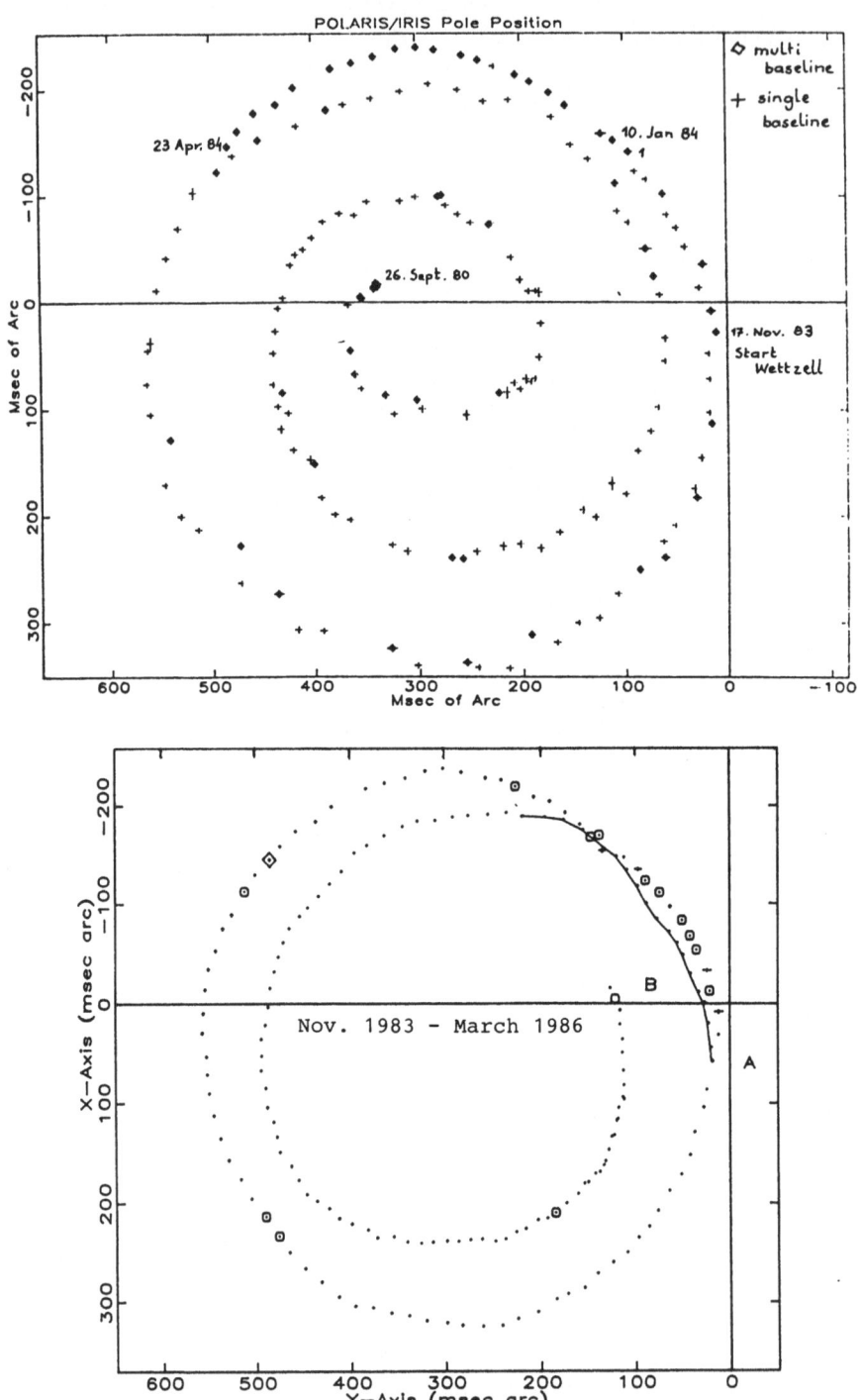

with the aim of looking at the short period UT1-variations. A frequency analysis on 90 days of daily VLBI observations already permitted the detection
of the predicted 13.6 day and 9.1 day periods caused by tidal deformation of
the polar moment of inertia of the Earth (Fig. 11). Both the amplitudes (0.9
and 0.2 msec) and phases are in good agreement with the theoretical values.
These results represent a first purely empirical confirmation of theoretical-
ly derived short period tidal oscillations in the Earth's rotational speed,
which until now had been hidden in the noise and the poor temporal resolution of the BIH data (ROBERTSON et al. 1985[14], CAMPBELL, SCHUH 1986[15]).

The NASA-CDP and the IRIS VLBI campaigns have been providing baseline
length results since the late seventies when the Mk III VLBI system became
fully operational. As the length of a baseline vector is independent from
changes in its orientation, the observed baseline length series are free from
errors in the Earth rotation parameters and can be treated as an independent
result of the campaign. However, as has been shown in the error budget (Fig.
6), other factors such as source positions (including the associated precession and nutation models) and atmospheric effects may still cause problems
when long series are to be interpreted. At present only two baselines across
the Atlantic (Onsala - Haystack and Onsala - Westford) cover a long enough
period (4 years) to be able to make first estimations of a meaningful relative motion. Both indepently observed series seem to indicate the same annual
drift rate of 1 - 2 cm per year between the North-American and the Eurasian
plate (Fig. 12). Before any firm conclusions can be drawn the further evolution of the much more frequently observed baseline Westford - Wettzell should
be awaited. The continental baseline Westford - GRAS (Ft. Davis, Texas) also
shows an appreciable change, which could confirm theories of an elastic
behaviour of the plates.

While the North-Atlantic section of the CDP and IRIS networks has been
observed on a routine basis, great efforts are made to include new stations,
in particular Shanghai, China, Hartebeesthoek, South Africa and Atibaia,
Brazil (Fig. 7). The stations of Kashima, Japan and Kauai, Hawaii have been
active since the beginning of 1984. On this baseline, where large motions
are expected, a contraction of 5 cm per year has been derived from the first
two epochs. The length variation on another Pacific baseline, Kashima -
Kwajalein, of -8 cm in one year is the largest so far observed. The agreement of these observations with the predicted rates derived from geophysical
evidence is very good (KONDO, HEKI, TAKAHASHI, 1986[16]).

Thus in recent years the VLBI technique has lived up to its expecta-
tions. By now large amounts of data have proven conclusively that the anticipated accuracy levels can be effectively realized in routine observing
campaigns.

Fig. 11
Short period UT1 variations
(obtained form 90 days of daily VLBI observations)

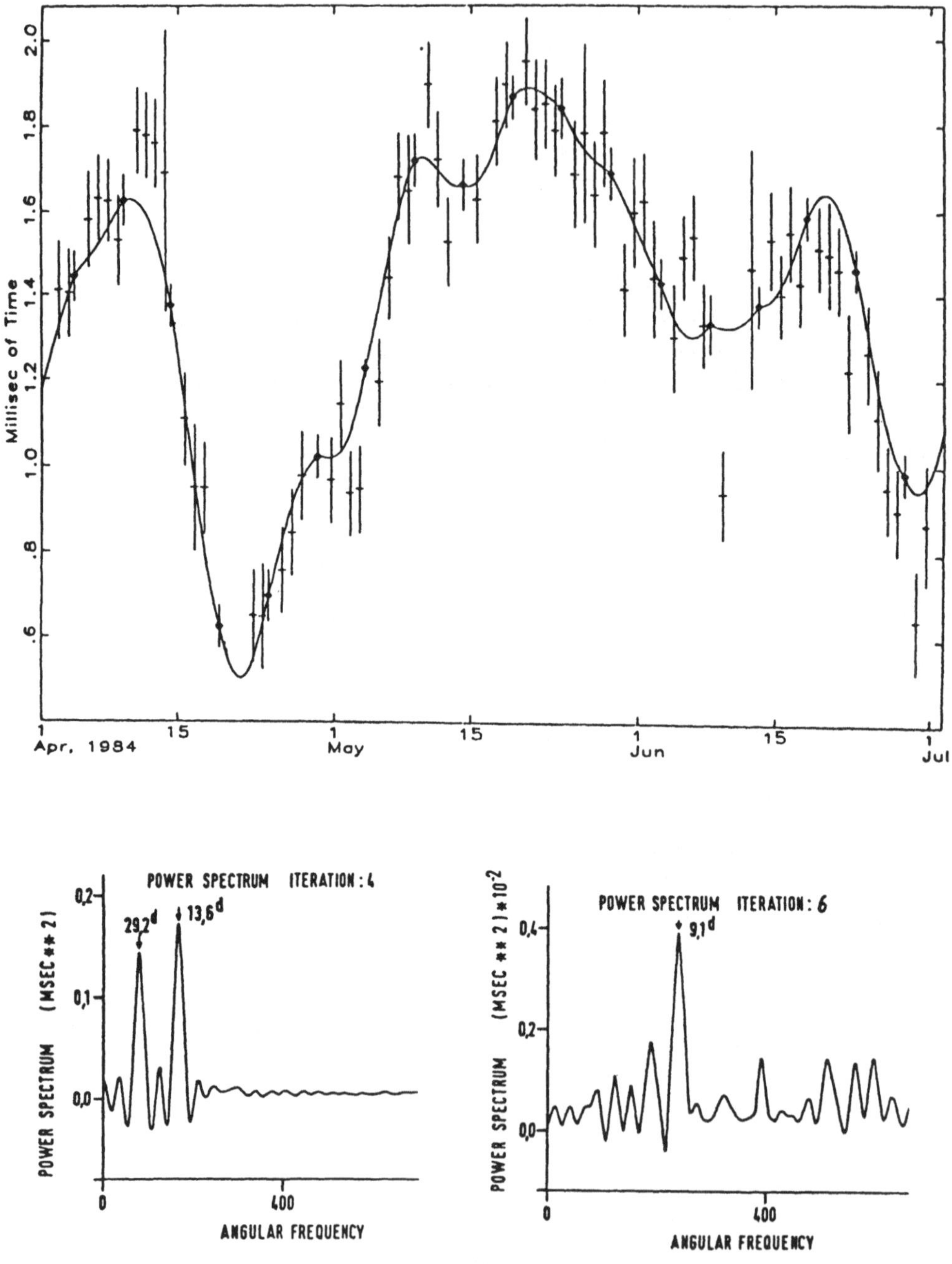

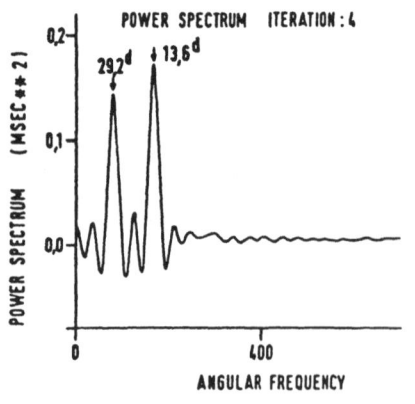

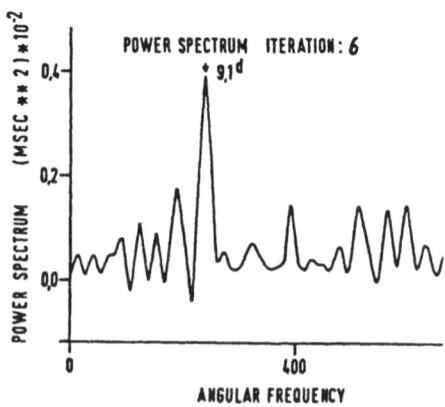

Fig. 12
Transcontinental and intercontinental baseline evolution 1981 - 1985

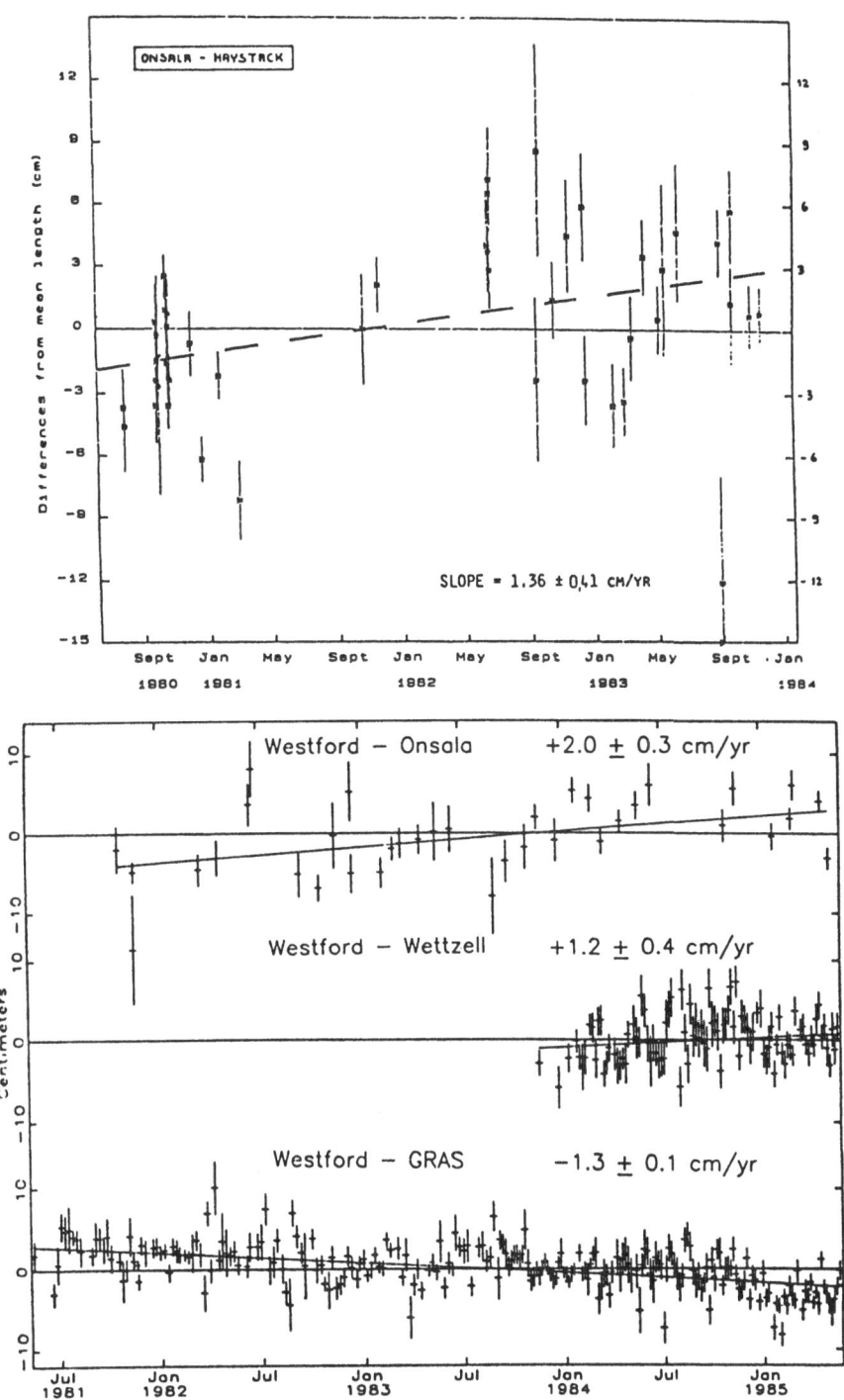

REFERENCES

1) BROTEN, N. W. et al.: Long Baseline Interferometry: A New Technique. Science, Vol. 156, p. 1592 - 1593, 1967.

2) BARE, C. et al.: Interferometer Experiments with Independent Local Oscillators. Science, Vol. 157, p. 189 - 191, 1967.

3) HINTEREGGER et al.: Precision Geodesy via Radio Interferometry. Science, Vol. 178, p. 396 - 398, 1972.

4) ROGERS, A. E. E.: Very Long Baseline Interferometry with Large Effective Bandwidth for Phase Delay Measurements. Radio Science, Vol. 5, p. 1239 - 1248, 1970.

5) MORAN, J. M.: Very Long Baseline Interferometric Observations and Data Reduction. In: Methods of Experimental Physics, Vol. 12, Part C, Radio Observations, Academic Press, New York, 1976.

6) THOMAS, J. B.: An Analysis of Long Baseline Radio Interferometry. In: The Deep Space Network Progress Report, Technical Report, 32 - 1526, Vol. VII, VIII, XVI, Jet Propulsion Laboratory, Pasadena, Calif., 1972.

7) WHITNEY, A. R. et al.: A Very-Long-Baseline Interferometer for Geodetic Applications. Radio Science, Vol. 11, p. 421 - 432, 1976.

8) KELLERMANN et al.: High-Resolution Observations of Compact Radio Sources at 6 and 18 cm. Astrophys. Journ., Vol. 169, p. 1-24, 1971.

9) CLARK, T. A.: Mark III System Overview. Proc. Radio Interferometry Techniques for Geodesy. Cambridge, Mass., June 19-21, 1979, NASA Conf. Publ. 2115.

10) HERRING, T. A., GWINN, C. R., SHAPIRO, I. I.: Geodesy by Radio Interferometry: Studies of the forced Nutations of the Earth. Journ. of Geophys. Research, 1985.

11) SHAPIRO, I. I., KNIGHT, C. A.: Geophysical applications of long baseline radio interferometry. In: Earthquake Displacement Fields and the Rotation of the Earth, edited by L. Mansinha, D. E. Smylie, and A. E. Beck, p. 284, Springer, New York, 1970.

12) NASA: The NASA Geodynamics Program: An Overview, NASA Geodynamics Program Office, Technical Paper 2147, Jan. 1983.

13) CARTER, W. E., ROBERTSON, D. S., MACKAY, J. R.: Geodetic Radio Interferometric Surveying: Applications and Results. Journ. of Geophys. Research, Vol. 90, p. 4577 - 4587, 1985.

14) ROBERTSON, D. S., CARTER, W. E., CAMPBELL, J., SCHUH, H.: Daily Earth Rotation determinations from IRIS very long baseline interferometry. Nature, Vol. 316, p. 424 - 427, 1985.

15) CAMPBELL, J., SCHUH, H.: Short-period Variations of Earth Rotation
                    determined by VLBI. Proc. Tenth. Int. Symp. on Earth
                    Tides, Sept. 1985, Madrid, (in press) 1986.

16) KONDO, T., HEKI, K., TAKAHASHI, Y.: Pacific Plate Motion detected by
                    the VLBI experiments conducted in 1984-1985. Proc.
                    Symp. on Application of Space Techniques to Astronomy
                    and Geophysics, p. 98 - 107, University of Tokio
                    Press, 1986.

17) CLARK, T. A. et al.: Precision Geodesy using the Mark-III Very-Long-
                    Baseline Interferometer System. IEEE Transactions on
                    Geoscience and Remote Sensing, Vol. GE-23, No. 4,
                    p. 438 - 449, 1985.

II. Surface Geodetic Networks and Underground Geodesy

THE LEP TRILATERATION NETWORK

J. Gervaise and J. Olsfors
CERN, Geneva, Switzerland

ABSTRACT

The installation and the alignment of a particle accelerator
require a very high accuracy. For the new CERN project, the
Large Electron Positron ring (LEP) of 27 km circumference, a
reliable control network has been created. Measurements have
been made with the two-colour Terrameter, giving an accuracy of
10^{-7}, and precision levelling. Computations took into
account a very fine assessment of the geoid, through collocation
on a large mass model and astro-geodetic measurements with a
zenithal camera. Resulting local coordinates were good to
$\sigma \leq 1.5$ mm for planimetry and $\sigma < 5$ mm for altimetry
above a local ellipsoid. In December 1984, seven points of this
network were observed using GPS (Global Positioning System)
satellite receivers. The measurements were carried out by
GEO-HYDRO Inc. with three Macrometer V-1000 Surveyor instruments.
Six independent baselines of this GPS network were measured twice
during three consecutive days. Although the observation constellation was rather poor, a common evaluation of all observations
(with the Bernese GPS Software Package) gave rms-errors of
roughly 1 - 2 mm per coordinate, showing an excellent internal
consistency of the GPS solution. The comparison with the
terrestrial solution was established through a Helmert transformation. The overall rms error of the transformation is 4 mm.

1. INTRODUCTION

    LEP (Large Electron-Positron Collider) is intended to develop more exactly the
extensive new synthesis of electro-weak interaction born out of the discoveries of the W
and Zo bosons. In such a machine, electrons and their antiparticles, positrons, are
accelerated simultaneously. Whilst turning in opposite directions, they collide head-on
and are annihilated and transformed into energy which, in turn, results in new particles
and antiparticles. LEP is a 27 km-long underground ring located in the Pays de Gex,
France, and partially in Switzerland, with eight experimental areas. The existing CERN
machines will act as injectors to LEP, which represents an important economy. This
implies that the LEP geodetic network should be adapted conveniently to the previous
networks (SPS).

Although geodesy, topography and cartography may be but a small part of the preliminary studies for a large civil engineering project, they are an indispensable part, and the work of the geodesist must precede that of the other engineers. This is all the more true for the construction of a 27 km particle collider, a project requiring the highest precision. In search for greater accuracy, the geodesist must look beyond the causes and effects that are already appreciated to other influences, albeit small ones, at the very limits of measurement.

The complexity of the geodesy has grown with each new accelerator, notably due to certain influences which become more significant with the increasing size of accelerators. The influence of the earth's curvature is a first example of this ever-increasing complexity. For LEP there is another influence to consider, that of the deviation of the vertical, which produces effects that were not foreseen at the start of the project.

Due to the curvature of the earth, the normal height of the centre (P_o) of the Proton Synchrotron (PS) is 0.8 mm above the horizontal plane of the synchrotron. Because the Linac is aligned in the horizontal plane of the injection point, the origin of the levelling had to be 0.3 mm above the height of this injection point. For the Intersecting Storage Rings (ISR) transfer tunnels the corrections reached 9.5 mm for TT1 and 15.8 mm for TT2. The ISR being virtually circular, as is also the PS, there were no height corrections around the ring itself. The corrections for the Super Proton Synchrotron (SPS) transfer tunnels were made in like manner. However, due to the existence of straight sections of 256 m, the machine can no longer be considered circular. To keep the accelerator in the plane, the height corrections in these straight sections reached 3.5 mm. The geometry of LEP has been calculated in the CERN tridimensional system (x, y, z) but, as z is not measurable, all z-coordinates have been converted to normal heights (h) above the reference ellipsoid.

The surrounding topography gives rise to distortions of the equipotential surfaces. The extent of these distortions has been calculated by theory and confirmed by measurements of the deviation of the vertical. This has allowed the definition of the theoretical altitudes (h) of all the points of LEP. The deviation of the vertical, caused by the non-homogeneity of the surrounding topography, necessitates a correction for the determination of points at the bottom of the pits in relation to the geodetic points on the surface. For a 140 m pit at the foot of the Jura, this correction is 6 mm.

The deviation of the vertical also affects the gyroscope measurements which provide the orientation for the excavation of the LEP tunnel.

The precision required during the various steps of the construction of LEP necessitates, as was the case for the SPS, a surface geodetic network of very high standard. This is why, as soon as the project was approved, the decision was taken to equip the Applied Geodesy Group of CERN with an LDM 2 Terrameter. This two-wavelength electromagnetic distance measurement device arrived at CERN in Autumn 1982.

2. **PRINCIPLE OF MEASUREMENT WITH INSTRUMENTS USING TWO DIFFERENT WAVELENGTHS IN THE VISIBLE SPECTRUM**

The use of two-colour electromagnetic distance measuring instruments will modify considerably the approach to the problems of very high accuracy geodesy.

Rather than consider the time dt taken by the light to cover a length element dl, we use the proportional quantity c.dt, called the optical path, the length covered in the vacuum during a time dt. The ratio of the speed of an electromagnetic wave in a vacuum to that in matter is known as the absolute index of refraction and given by n = c/v. This index is actually frequency-dependent. The dependence of n on the wavelength of light is a well-known effect called dispersion. The optical path becomes : n . v . dt = n . dl. Thus, the optical path L, going from A to B, is defined by :

$$L_{(AB)} = \int_A^B n \; dl \qquad (1)$$

The principle of two-colour electromagnetic distance measuring instruments was proposed by Prilepin in 1957 and Bender & Owens in 1965, in order to make the distance measurements less dependent on the refractive index of air. This index is essentially variable for long paths through the atmosphere and impossible to measure accurately with one-frequency EDM instruments. Along the same path, simultaneous measurements of the optical path with different frequencies allow the desired correction for the refractive index to be largely determined.

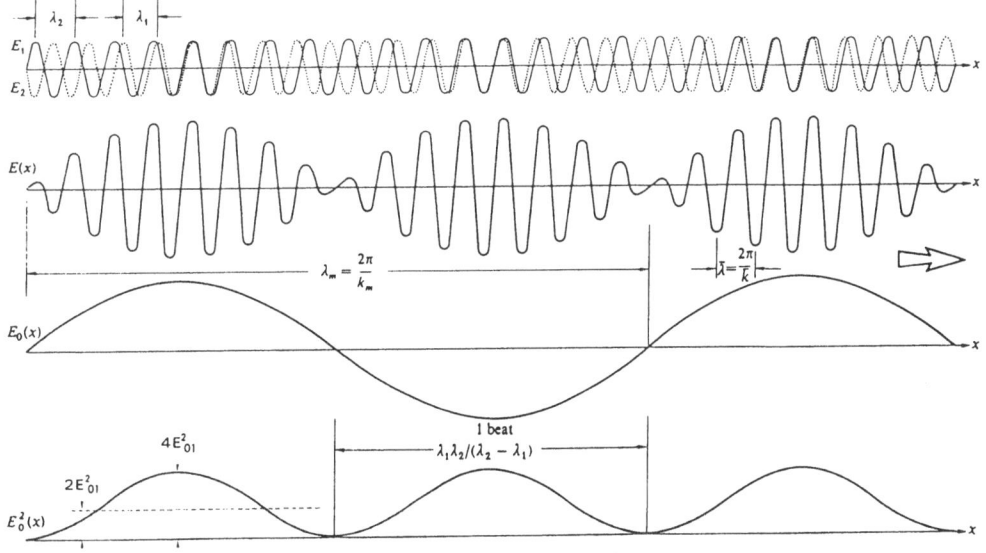

Fig. 1   Definition of group velocity u

The two-colour instruments measure two different optical paths due to the dispersion of the refractive index in relation to the wavelength. We assume a geometrical path length L of 15 km between light source and reflector, i.e. 30 km outgoing and incoming. The one-way optical length will be L + S, where S is the additional contribution due to the atmosphere. Light being modulated, L + S is inversely proportional to the group velocity u. To study the addition of waves of slightly different frequency, it is sufficient to consider two waves λ_1 and λ_2 travelling at velocities v_1 and v_2. The total disturbance arising from the combination of these waves having equal amplitude and zero epoch angles may be regarded as a travelling wave of frequency f_m with modulated amplitude E_0 (x,t) (Fig. 1).

The irradiance is proportionnal to E_0^2 (x,t) and oscillates about a value of $2 E_0^2$. The disturbance E (x,t) consists of a high frequency (ω) carrier wave, amplitude modulated by a cosine function. The rate at which the modulation envelope advances is known as the <u>group velocity u</u>. In a non-dispersive media, in which the <u>phase velocity v</u> is independent of wavelength, u = v. Specifically, in a vacuum, v = u = c. For optical media in regions of normal dispersion, the refractive index increases with frequency inasmuch as u = v (dv/dλ); u is smaller than v. Clearly, one should define also a group index of refraction : n_g = c/u which must be carefully distinguished from n. So, S can be defined by :

$$S = \int_0^L (n_g - 1)\, dl. \qquad (2)$$

For red light alone, S is about 400 cm. If the red line of a helium-neon laser (632.8 nm) and the blue line of a helium-cadmium laser (441.6 nm) are chosen, the extra optical paths S_B and S_R for the blue and red respectively will differ by about 10%, giving a difference $\Delta S = S_B - S_R$ of about 40 cm. By convention, the dispersive power of the atmosphere is a number A without dimension :

$$A = (n_g^B - n_g^R) / (n_g^R - 1), \qquad (3)$$

its inverse 1/A is called the constringency of the atmosphere.

One can thus write the difference of the optical path :

$$\Delta S = \int_0^L A\, (n_g^R - 1)\, dl. \qquad (4)$$

A being independent of the atmospheric density along the path and only weakly dependent on the atmospheric composition, we can replace it with a good approximation by its average value 〈 A 〉 and take it outside the integral sign :

$$\Delta S = \langle A \rangle \int_0^L (n_g^R - 1)\, dl = \langle A \rangle\, S_R, \qquad (5)$$

then $\qquad\qquad\qquad\qquad S_R = S / \langle A \rangle$ $\qquad\qquad\qquad\qquad$ (6)

and $\qquad\qquad\qquad\qquad L = L_R - S_R = L_R - (L_B - L_R) / \langle A \rangle,$ $\qquad\qquad$ (7)

which shows that the true distance is obtained by measuring only the optical lengths L_B and L_R over the same path thus removing the effect of the atmosphere from the optical measurement.

We call refractivity the value $N = (n - 1)\, 10^6$ and N_{go} the group refractivity in standard atmospheric conditions, temperature being $15^{\circ}C$ and atmospheric pressure being 760 mm of mercury.

We can calculate $\langle A \rangle$ for the wavelengths used in the Terrameter – 632.8 nm in the red, 441.6 nm in the blue. The Barrell & Sears formula is particularly convenient for visible wavelengths :

$$N_{go}\, 10^6 = 287.604 + 3\,(1.6288/\lambda^2) + 5\,(0.0136/\lambda^4), \qquad\qquad (8)$$

where λ is the wavelength of the radiation in nm. This allows the constringency to be calculated : $1/A = 21.12$.

If we use Edlen's equation with the Cauchy form for visible and near infra-red light,

$$N_{go}\, 10^6 = 272.600 + 4.608/\lambda^2 + 0.066/\lambda^4. \qquad\qquad (9)$$

In this case, $1/A = 21.16$.

In the Terrameter, the following simplified formula assumes a 0.03% CO_2 atmosphere, with P_w the partial pressure of water vapour, and P_s the partial pressure of dry air. P is atmospheric pressure in millibars; T is temperature in degrees Kelvin.

$$N_{go} = (n-1) \cdot 10^8 = (C_1\, D_s + C_2\, D_w) \cdot 10^8 \qquad\qquad (10)$$

$$D_s = P_s/T \left[1 + P_s\,(57.9 \cdot 10^{-8} - 9.325 \cdot 10^{-4}/T + 0.25844)/T^2 \right]$$

$$D_w = P_w/T \left[1 + P_w\,(1 + 3.7 \cdot 10^{-4} \cdot P_w)\,(- 2.37321 \times 10^{-3} + 2.23366/T - 710.792/T^2 + 7.75141 \cdot 10^4/T^3) \right]$$

$$C_1 = \text{(He-Cd) } 80.874255 \cdot 10^{-6} \qquad \text{(He-Ne) } 84.735399 \cdot 10^{-6}$$

$$C_2 = \text{(He-Cd) } 69.094806 \cdot 10^{-6} \qquad \text{(He-Ne) } 73.700935 \cdot 10^{-6}$$

The value of P_w can be determined from the relative humidity RH and the saturated water vapour pressure; i.e. :

$$P_w = RH \cdot P \text{ saturated.} \qquad (11)$$

In this case, $1/A = 20.86$.

The measurement of ΔS gives the atmospheric correction to the optical path length. To measure ΔS, it is necessary to modulate the polarisation of the red and blue light simultaneously at frequencies of 3 GHz. The signal-to-noise ratio allows a great accuracy to be obtained and also the reduction of systematic errors. ΔS is determined with a precision equal to a small fraction of the modulation wavelength under good atmospheric conditions.

3. LDM 2 TERRAMETER (Terratechnology Corporation)

The first prototype, presented in June 1969 at Boulder, Colorado, at the "International Symposium on Electromagnetic Distance Measurements", is now produced in a limited series. Unlike conventional EDM instruments which approximate the refractive index of air along the measurement path by sample measurements of temperature, air pressure and water vapour, the Terrameter makes a direct and precise measurement of the refrative index by using simultaneous distance measurements at two optical wavelengths, one in the red part of the spectrum and the other in the blue. The instrument calculates correction terms from the optical path length difference between the two wavelengths and computes the corrected base line distance. This automatically eliminates the first-order effects of temperature, air pressure and water vapour.

The basic principle of operation is the same as that of the Fizeau velocity of light experiment. In that experiment, light was returned from a distant retro-reflector to the photodetector only if there existed a proper temporal and spatial relationship between the outgoing and incoming light. Light returned to the detector only if the transit time of the light was exactly equal to the integral number of modulation periods. The principle of the Fizeau cog-wheel was adapted first by Michelson and then by Bergstrand in 1950. A KD*P (potassium dideuterium phosphate) crystal cell is used in the Terrameter to modulate the polarization of light by a linear electro-optical effect, known as Pockel's effect, inasmuch as the induced birefringence is proportional to the applied electrical field and, therefore, the applied voltage. Pockel's effect exists only in certain crystals which lack a centre of symmetry. The operating principle for such a device is that the birefringence is varied electrically by means of a controlled applied electric field. The retardance can be altered as desired, thereby changing the state of polarization of the incident linear wave. In this way, the system functions as a polarization modulator. A Pockel cell is simply an appropriate non-symmetric, oriented, single crystal immersed in a controllable electric field.

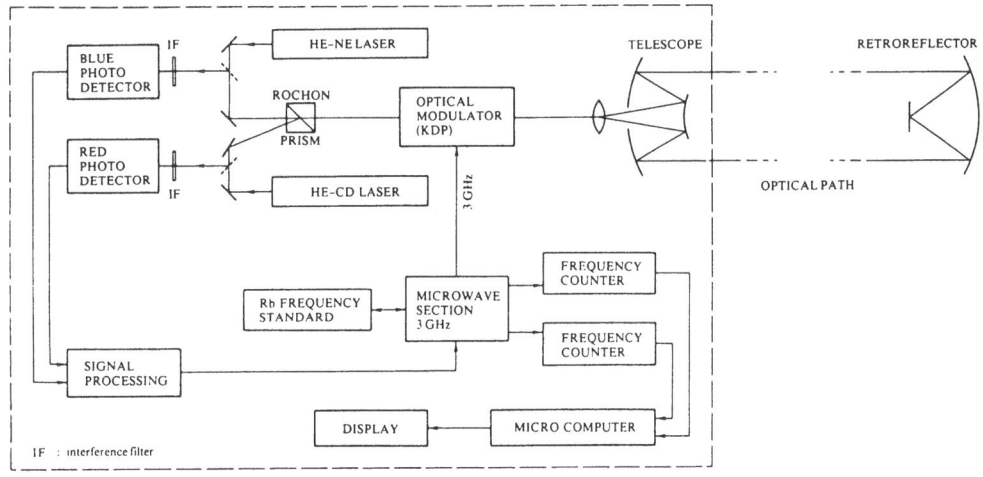

Fig. 2   Scheme of the LDM2 Terrameter

In the Terrameter (Fig. 2), red and blue light from the He-Ne and He-Cd lasers enters a Rochon prism at the proper angle and polarization to make the outgoing beams co-linear. The light passes through a microwave modulator that modulates the ellipticity of the polarized light at 3 GHz. The light, transmitted by a Cassegrainian telescope, traverses the path being measured and is returned by the retro-reflector. The beam is received by the same optics used for transmission, and passes through the modulator a second time, where the ellipticity of the polarization is increased or decreased depending upon the phase of the modulator excitation. The beams emerging from the prism are separated by colour and directed to the photodetectors.

The analogue outputs of the photodetectors are processed and used to control one VCO (voltage controlled oscillator) for each colour. The VCO output can be varied between 4 and 5 MHz, depending on input frequency giving a 2995.5 MHz signal which after amplification to 4 Watts drives the optical modulator.

Because there is only one modulator for the two colours, it must be switched alternatively between the red and blue frequencies.

In order to adjust the modulating frequency so that a minimum of light is received at the photodetectors, each VCO is switched between two frequencies, one below and the other above the centre frequency (Fig. 3).

As can be seen on the figure, the minima are much better defined than the maxima. The frequency difference between the minimas is :

$$f_c - f_{c'} = v/2D, \qquad\qquad (12)$$

where v is the speed of light in the air, and D is the one-way distance. The amount of VCO frequency shift Δf (determined by the "Approximate distance" set by the operator on the control panel) is about one quarter of $f_c - f_c'$.

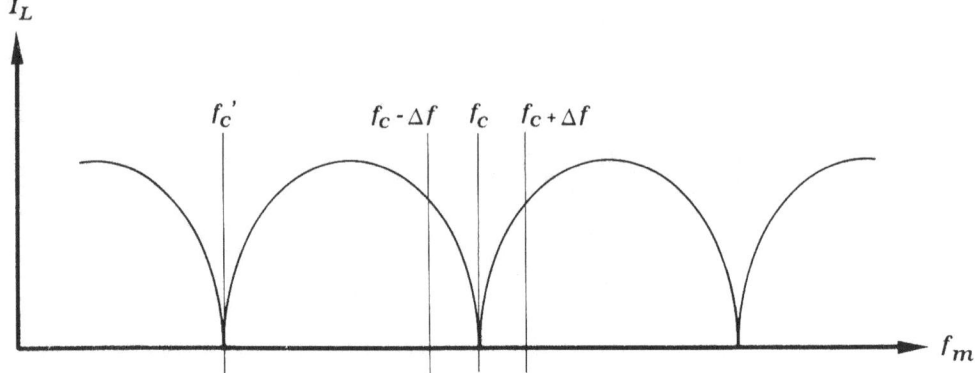

Fig. 3   Received light intensity vs. modulator frequency

The internal frequencies are measured and these values used in the microcomputer to calculate the true distance. Since A is a weak function of the meteorological parameters, rough end-point values of temperature, pressure and relative humidity are entered using thumbwheels on the instrument panel (Fig. 4).

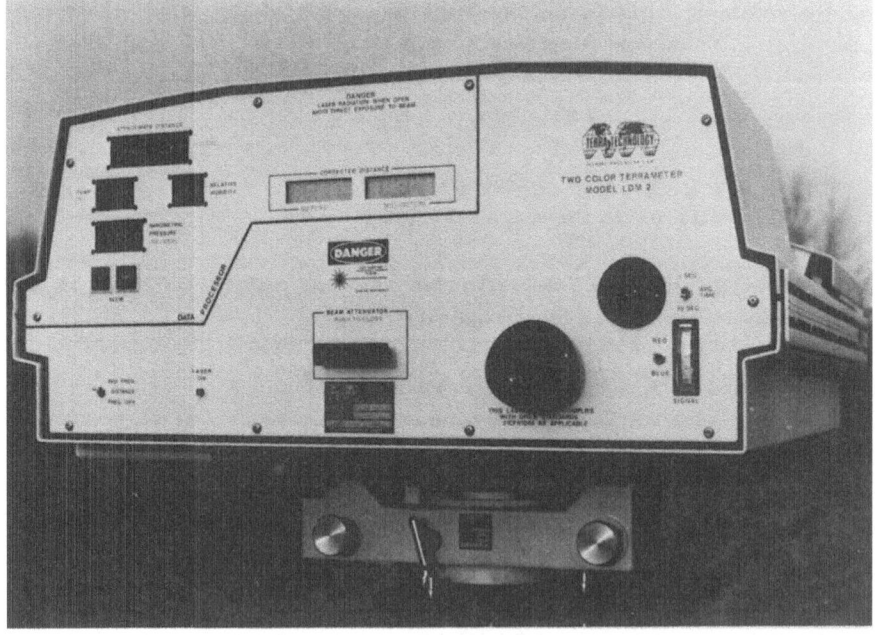

Fig. 4   Back view of the Terrameter

With these data the microcomputer determines the true distance, which is digitally displayed on a liquid crystal readout on the back panel of the instrument. A hard copy of all input data, internal frequencies and true distance may be obtained every ten seconds by connecting a terminal or a printer to the Terrameter's computer interface. A rubidium frequency standard is used as the reference for counters that measure the internal frequencies that are used for the distance calculations. This standard is several orders of magnitude more accurate than ever required and no subsequent recalibration is necessary. The repeatability is $2 \cdot 10^{-10}$ after 10 minutes of warm-up, with a long term stability of $4 \cdot 10^{-11}$.

The laser beams from the instrument are directed to a distant retro-reflector (Fig. 5). An aligning telescope is used to acquire the reflected beam, and the operator can then adjust the fine pointing of the instrument by viewing the frosted glass aperture mounted on the rear.

Fig. 5   Retroreflector

The Terrameter's range is from 350 m to 15 km. The outgoing power of the He-Ne laser is 5 mW and that of the He-Cd laser is 10 mW. The polarisation is modulated at 3 GHz, i.e a modulation wavelength of 10 cm. The beam dispersion in the atmosphere is limited to $2 \cdot 10^{-4}$ radians.

The frequencies are servo-controlled in order that the return signal arrives in opposite phase to the outgoing signal. Thus the number N of wavelengths on the outward and return path is an integer plus a half. The red and blue frequencies are adjusted such that N is equal for the two colours. The ratio of the two indices is equal to the ratio of the two frequencies. Due to the dispersion, it is possible to calculate n_B and n_R indices, all the other parameters being known except N. It is thus necessary to determine an approximate value of the distance to better than half a wavelength (0.05 m). The instrument is automatic, apart from the introduction of the approximate distance and meteorological values. However, due to the elaborate optical and electronic systems, it needs maintenance and internal adjustments which require the presence of an electronics engineer throughout the observation period. Also the bulkiness and weight inhibit easy use in all field geodetic measurements. It has been used principally to measure the deformations of the earth's crust, the stability of large construction sites and landslips. It can, with difficulty, be adapted to the measurement of particular geodetic networks such as those of CERN.

4. MEASUREMENT OF THE LEP GEODETIC NETWORK WITH THE TERRAMETER

The first critical operation was to determine the instrumental constant of the device. The Applied Geodesy Group has at its disposal a 500-m outdoors baseline which has a pillar with forced centring at every 50 m. With the new model of the Distinvar, the absolute accuracy over a distance of 50 m is 0.03 mm, which means, on the 500 m of the base, a precision of 0.1 mm (relative precision of $2 \cdot 10^{-7}$). It has been necessary to increase the length of this base by building two more pillars, one located 478 m to the north and the second 549 m to the south. Figure 6 shows all the measurements carried out with the Terrameter. Values measured with the Distinvar between pillars 2 and 12 have been included in the adjustment, the results of which have given a constant for the instrument-reflector set of 0.0275 m.

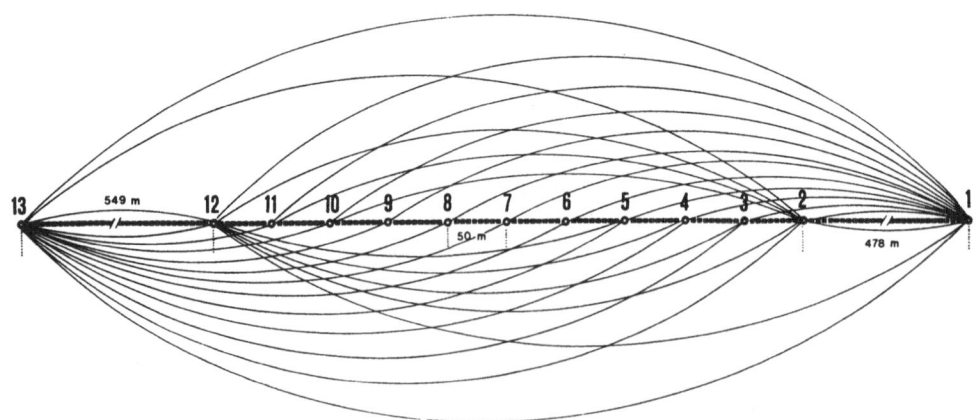

Fig. 6   Determination of the constant of the Terrameter

The constant of the instrument-reflector set having been determined with the maximum accuracy, it was possible to measure a geodetic figure, the nucleus of the LEP network, in order to carry out the acceptance tests of the Terrameter.

Figure 7 shows the location of the geodetic points which surround the LEP ring. Measurements were made from all points except the pillar of point 232 located on the water tower of Collex-Bossy. The wooded, hilly and rather dense urbanization of the area meant that not all pillars could be positioned at ground level.

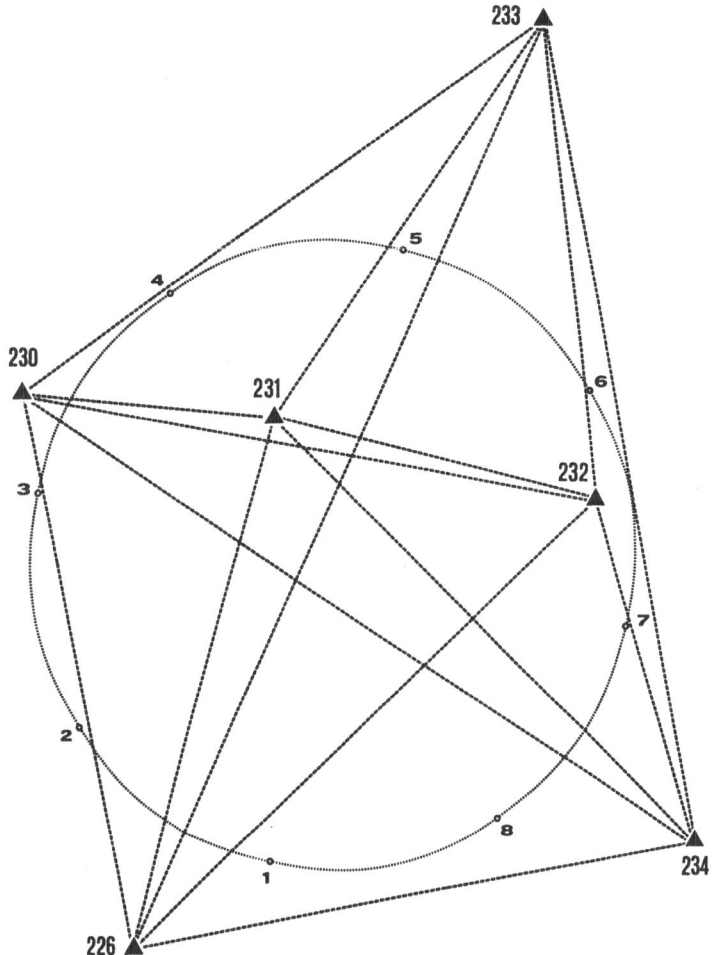

Fig. 7   LEP test geodetic network

At each point the pillar is equiped with a forced centring reference socket. The Terrameter has a locating cylinder with an expanding locking system. Each distance was measured 100 times. One measurement is displayed every ten seconds, which means a little more than a quarter of an hour per distance. Measurements were taken at different times of the day and over several days.

Table 1

LEP test geodetic network, June 1983;
Comparison of reciprocal observations with the Terrameter
(without addition of the instrument constant)

Pillars	-) (-	Observed distances (m)	Distance difference (mm)	Precision x 10^{-7}
226 – 230	-) (-	7482.33805 7482.33803	0.02	0.03
226 – 231	-) (-	7314.84645 7314.84691	0.46	0.6
226 – 232	-)	8808.13369		
226 – 233	-) (-	13663.40556 13663.40524	0.32	0.2
226 – 234	-) (-	7755.90733 7755.90860	1.27	1.6
230 – 231	-) (-	3543.57321 3543.57230	0.91	2.6
230 – 232	-)	8110.89990		
230 – 233	-) (-	8869.18119 8869.18148	0.29	0.3
230 – 234	-) (-	11001.47017 11001.46911	1.06	1.0
231 – 232	-)	4593.22481		
231 – 233	-) (-	6541.98778 6541.98669	1.09	1.7
231 – 234	-) (-	8080.72270 8080.72221	0.49	0.6
233 – 232	-)	6458.90541		
233 – 234	-) (-	11294.64935 11294.64892	0.43	0.4
234 – 232	-)	4861.91680		

Note : Pillar 232, water tower of Collex-Bossy, non-stationned

Table 1 shows the comparison of all reciprocal distance measurements. The r.m.s. value of these differences, expressed as a relative value is $0.9 \cdot 10^{-7}$. Therefore, <u>the relative precision of independent measurements carried out under different conditions is as stated by the manufacturer</u>.

The network has been computed by a tridimensional least squares adjustment programme constraining at each point the measured altitude above the XY plane of the CERN cartesian system. This transformation makes use of a local reference ellipsoid with the parameters recommended by IUGG (Canberra 1979) and includes the computation of geoidal heights.

The residuals between measured and adjusted distances are less than 2.5 mm and the r.m.s. values of these residuals is 1.3 mm. The axes of the confidence ellipses obtained from the covariance matrix are less than 1.5 mm (Table 2).

Table 2
Results of the LEP geodetic measurements

Fixed point and orientation point

Pillars	X (m)	Y (m)	Z (m)	DX(mm)	DY(mm)	EQX(mm)	EQY(mm)
226	1288.2656	723.7339	2500.8843				
233	-1150.7310	14165.7641	2729.9655	-13.0	71.5		1.5

Fixed points in Z

Pillars	X (m)	Y (m)	Z (m)	DX(mm)	DY(mm)	EQX(mm)	EQY(mm)
230	-4176.6449	5828.7078	2745.4066	-29.0	13.3	1.7	1.8
231	-1130.9828	7626.9147	2527.4834	-15.4	48.7	1.8	1.5
232	3132.0103	9336.7623	2491.8598	32.0	40.3	1.8	1.6
234	6815.9375	6164.0805	2456.1840	37.5	0.5	1.6	2.0

Distances

Pillars	Observed distances	Residues(mm)	Compensated distances
231 – 226	7314.87420	-0.60	7314.87360
231 – 230	3543.60020	-0.82	3543.59938
231 – 232	4593.25230	-1.01	4593.25129
231 – 233	6542.01470	-1.22	6542.01348
231 – 234	8080.75000	-0.67	8080.74933
230 – 232	8110.92740	1.83	8110.92923
226 – 230	7482.36550	0.24	7482.36574
226 – 232	8808.16120	1.72	8808.16292
226 – 234	7755.93550	-0.57	7755.93493
234 – 230	11001.49710	2.51	11001.49961
234 – 232	4861.94430	-0.01	4861.94429
233 – 226	13663.43290	-1.44	13663.43146
233 – 230	8869.20880	2.05	8869.21085
233 – 232	6458.93290	0.90	6458.93380
233 – 234	11294.67660	-0.69	11294.67591

The actual surface geodetic network for LEP (Fig. 8) was measured in October 1983. It is not possible to publish here the comparison of reciprocal distance measurements between all the pillars of the network, but the r.m.s. difference is 1.3 mm and the axes of the confidence ellipses are less than 1.5 mm. These results are the same as those obtained in the measurement of the test network.

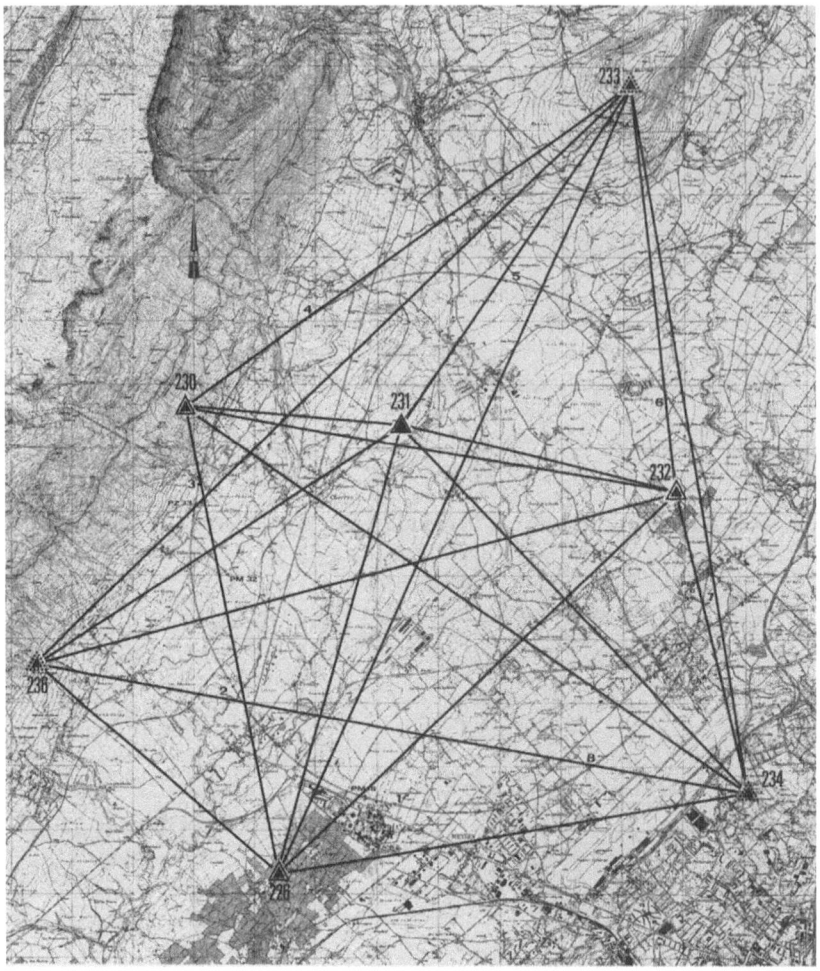

Fig. 8   LEP geodetic network (1983)

This network was adapted to that of the SPS, which shares five geodetic points including 226. The displacement vectors range from 0.25 mm to 6.10 mm (8 m pillar), showing the quality of the two networks and the stability of the points, as the SPS surface network has not been measured since 1976.

In summer 1985, a new measurement of the LEP geodetic network (Fig. 9) has been made, in order to check the stability of points and to determine new pillars at the vicinity of access pits to the tunnel.

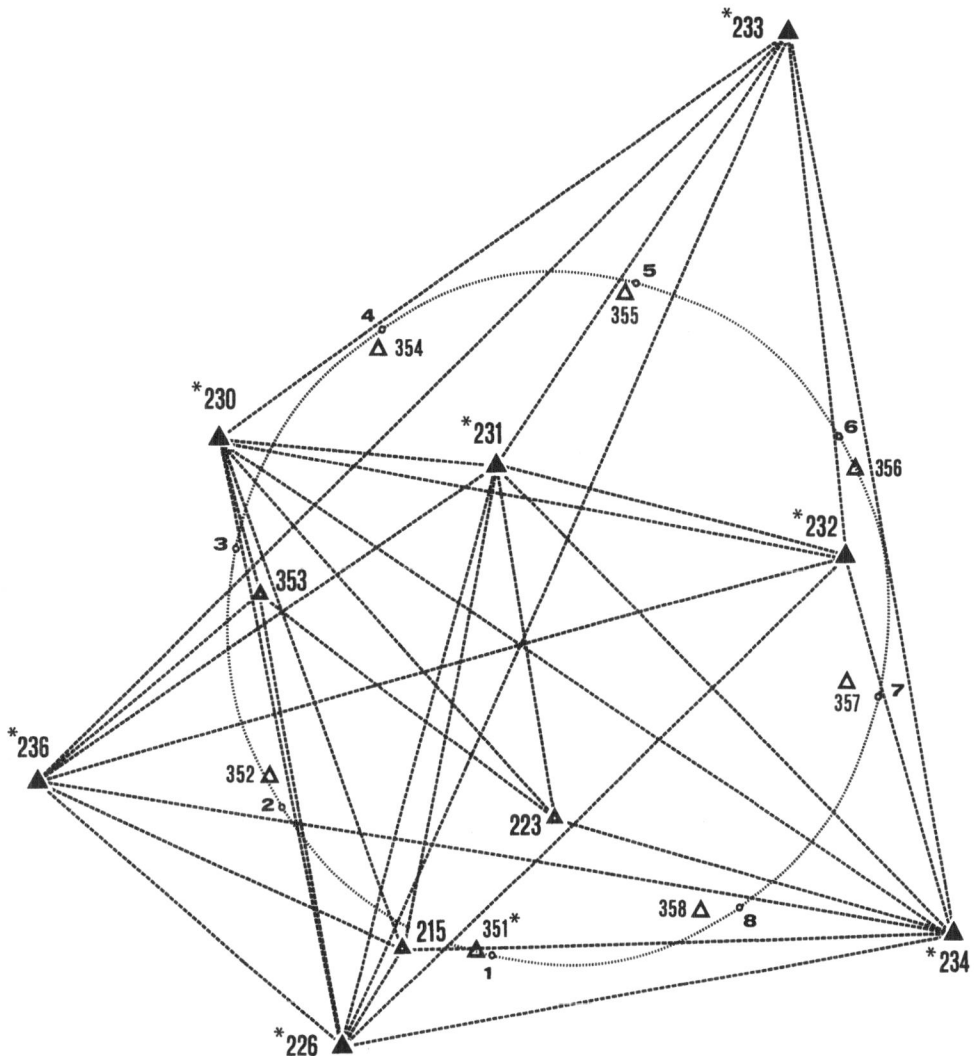

▲ : Points of the principal network

△ : Points situated near the pits

* : Points measured by G.P.S.

Fig. 9   LEP geodetic network (1985)

The main results of this trilateration campaign are shown in Tables 3 and 4.   The comparison with 1983 network is given in Table 5.

Table 3

LEP test geodetic network, 1985

Free figure adjustment

Fixed point :   226
Orientation point :   233
Number of observations : 31
Number of unknowns :   18
All points constrained at their measured altitude H

[1]	[2]	[3]	[4]	[5]	[6]
Point	X (m)	Y (m)	Z (m)	SX (mm)	SY (mm)
215	1200.67301	2077.57816	2484.5225	1.23	.32
223	1821.90857	4555.77730	2476.2915	1.10	1.53
226	1288.26220	723.73090	2500.8724	–	–
230	−4176.63933	5828.70653	2745.4962	.88	.92
231	−1130.98254	7626.91154	2527.4999	1.01	.73
232	3132.00429	9336.76067	2491.8226	.80	1.19
233	−1150.72829	14165.76389	2729.9498	.15	.84
234	6815.93367	6164.07295	2456.1702	1.02	1.50
236	−3886.99035	1148.46141	2602.9765	.98	1.14
353	−2796.83180	4648.59131	2527.0570	1.17	1.23

Columns [2] and [3] : adjusted coordinates

Columns [5] and [6] : standard-deviation (1σ) on adjusted coordinates

<u>Table 4</u>

LEP test geodetic network, 1985

Adjusted Distances of LEP network

[1] Point 1	[2] Point 2	[3] Observed (m)	[4] Sigma (mm)	[5] Adjusted (m)	[6] Residual (mm)
223	230	6138.0287	1.43	6138.0287	- .01
223	231	4260.7574	.84	4260.7574	.02
223	353	4619.9517	.93	4619.9517	- .05
226	215	1356.8156	.38	1356.8156	- .02
226	230	7482.3613	1.43	7482.3636	-2.34
226	231	7314.8718	1.25	7314.8723	- .51
226	234	7755.9323	2.94	7755.9314	.86
226	236	5193.6560	1.37	5193.6558	.25
226	353	5665.0880	1.05	5665.0867	1.20
230	215	6561.7170	1.69	6561.7173	- .28
230	232	8110.9225	.58	8110.9226	- .13
230	233	8869.2101	1.46	8869.2111	- .93
230	353	1828.7323	1.22	1828.7308	1.50
231	230	3543.5988	.79	3543.5983	.48
231	215	6019.4543	1.79	6019.4545	- .18
231	226	7314.8740	2.73	7314.8723	1.68
231	232	4593.2468	.65	4593.2465	.32
231	233	6542.0165	1.75	6542.0154	1.10
231	234	8080.7475	2.62	8080.7464	1.11
231	236	7040.7109	1.85	7040.7096	1.30
233	226	13663.4334	1.43	13663.4330	.36
233	232	6458.9298	.98	6458.9299	- .07
233	234	11294.6758	1.96	11294.6764	- .65
233	236	13302.3839	1.11	13302.3839	.07
234	215	6944.8709	4.13	6944.8711	- .25
234	223	5246.6471	1.30	5246.6472	- .08
234	232	4861.9497	.43	4861.9496	.02
236	215	5173.2320	1.40	5173.2318	.16
236	353	3666.7583	1.61	3666.7586	- .28
236	234	11820.7631	3.84	11820.7654	-2.25
236	232	10785.4941	.72	10785.4942	- .16

Mean value of residuals : 0.07 mm

Standard-deviation of residuals : 0.9 mm

Column [4] : standard-deviation (1 σ) of measurements

Column [6] : difference between observed and adjusted values

Table 5

Comparison between 1985 and 1983 networks (Helmert Transform)

Scale Factor	Rotation	Translation
Ech = 1.000000	Alpha = − .00001 (gon)	TX(m) = − .000 TY(m) = .001

Transformed coordinates of active (stable) points

[1] Point	[2] X (m)	[3] Y (m)	[4] dX (mm)	[5] dY (mm)
215	1200.6729	2077.5789	−1.05	.35
223	1821.9083	4555.7780	.91	.74
226	1288.2622	723.7316	.25	−1.80
231	−1130.9831	7626.9120	.45	− .56
233	−1150.7293	14165.7644	− .73	.99
234	6815.9333	6164.0741	−1.42	− .70
236	−3886.9903	1148.4617	1.06	1.19
353	−2796.8321	4648.5917	.53	− .22

Transformed coordinates of passive (unstable) points*

[1] Point	[2] X (m)	[3] Y (m)	[4] dX (mm)	[5] dY (mm)
230	−4176.6397	5828.7068	2.40	− .30
232	3132.0036	9336.7615	−6.00	1.10

Columns [2] and [3] : transformed 1985 coordinates

Columns [4] and [5] : dX and dY with respect to 1983 coordinates

---

* The unstable character of some points results from both external considerations and iterative runs of the Helmert transform program. The point 232 is at the top of a 36 m high water-tower. The point 230 is located on a steep side of the Jura mountain and its displacement vector, oriented along the line of greatest slope may correspond to a slippage of the ground.

5. <u>CONCLUSION</u>

The results shown in Table 5 are self-explanatory and show the quality of the LEP trilateration network. These results will be compared with those obtained from the satellites of the Global Positionning System but, to be able to make this comparison it is necessary to take into account the Newtonian attraction of the nearby and distant mountains around the site so allowing the calculation of the true heights above the reference ellipsoid.

*   *   *

<u>BIBLIOGRAPHY</u>

P.L. Bender & J.C. Owens, "Correction of Optical Distance Measurements for the Fluctuative Atmospheric Index of Refraction", Journal of Geophysical Research, Vol. 70, No 10, (p. 2461-2462), The American Geophysical Union, Washington D.C., 1965.

J. Cruset, "Cours d'Optique Appliquée et de Photographie", Vol. 6, (p. 501-508), I.G.N., Paris, 1948.

K.D. Froome & L. Essen, "The Velocity of Light and Radio Waves", Academic Press, London, 1969.

J. Gervaise, "Applied Geodesy for CERN Accelerators", Chartered Land Surveyor / Chartered Minerals Surveyor, Vol. 4, No 4, RICS Journal Limited (p. 10-36), London, 1983.

J. Gervaise, "Instruments électroniques de Mesure de Distances à deux Longueurs d'Ondes", Proceedings XVII Congrès de la FIG, No 503.2, Sofia, 1983.

J. Gervaise, "Premiers Résultats de Mesures Géodésiques avec le Terramètre - Appareil électronique de Mesure de Distances à deux Longueurs d'Ondes", XVIII Assemblée générale de l'Union Géodésique et Géophysique Internationale, Hamburg, 15-27.9.1983.

J. Gervaise, "First Results of the Geodetic Measurements carried out with the Terrameter, Two-wavelength Electronic Distance Measurement Instrument", Geodätisches Seminar über Electrooptisch Präzisionstreckenmessung, München, 23.9.1983.

J. Gervaise, "Results of the Geodetic Measurements carried out at CERN with the Terrameter, a two-wavelength Electro-magnetic Distance Measurement Instrument", FIG, Commission 6, Technical Papers p. 23-32, Washington D.C., March 10-11, 1984.

J. Gervaise, "Résultats de Mesures géodésiques avec le Terramètre, Appareil électronique de Mesure de Distances à deux longueurs d'Ondes", Vermessung, Photogrammetrie, Kulturtechnik, Mensuration, Photogrammétrie, Génie Rural, Vol 6.84, 189-194, Zürich, 1984.

J. Gervaise, "Applied Geodesy for CERN Accelerators", Seminar on "High Precision Geodetic Measurement", Facolta di Ingeneria di Bologna, October 16-17, 1985.

M.T. Prilepin, "Light Modulating Method for Determining the Average Index of Refraction of Air along a line", Translation, Institute of Geodesy, Aeronomy and Cartography, URSS, No 114, (p. 127-130), 1957.

E.N. Hernandez & G.R. Huggett, "Two Color Terrameter - Its Application and Accuracy", Technical papers, The American Congress on Surveying and Mapping, Washington D.C., 1981.

DEVIATION OF THE VERTICAL

Werner Gurtner

Astronomical Institute, University of Berne, Switzerland

Beat Bürki

Institute for Geodesy and Photogrammetry, ETH Zurich, Switzerland

ABSTRACT

The presentation gives a short overview of the most important
formulae, definitions used in the computation of vertical deflections
and geoidal undulations, a summary of the known computation methods.
As an example the results of the respective computations for the CERN
LEP control network are presented.

1. INTRODUCTION

The ultimate goal of the control survey for a particle accelerator is to ensure that
the particle beam will follow a path as close to the theoretical figure as possible.

One critical point is the realisation of a true geometrical plane of the size necessary
for the Large Electron Positron Collider (LEP, diameter ca. 8.5 km) using classical geodetic
methods as precise levelling: Since levelling is related to the equipotential surfaces of
the earth's gravity field, any undulations and divergences of these surfaces will show up in
a distortion of the plane deduced from the levelled heights, if they are not known or not
taken into account.

In the presented paper we first summarize the most important definitions and formulae
of the physical geodesy used or referred to later on. Then we will give an overview of the
methods of observation and computation of the vertical deflections and the geoidal heights.
As an example we will show the main results of the respective computations performed for the
control network of the LEP.

2. THE GRAVITY FIELD OF THE EARTH

The notations used here are mostly taken from Ref. 1.

The potential W of the **gravity** is the sum of the potential V of the **gravitational force**
and the potential Φ of the **centrifugal force**:

Potential of the gravitational force:

$$V = k \cdot \iiint \frac{\rho}{1} \cdot dv . \tag{1}$$

Potential of the centrifugal force:

$$\Phi = \frac{1}{2} \omega^2 \left(x^2 + y^2 \right) . \tag{2}$$

Potential of the gravity:

$$W = W(x,y,z) = V + \Phi \tag{3}$$

with: k : gravitational constant
       ρ : density of the volume element dv
       1 : distance between dv and the point P(x,y,z)
       ω : angular velocity of the earth

$\sqrt{x^2+y^2}$: distance from the axis of rotation.

The gradient $\vec{g}$ of the gravity field is called **gravity**, its direction the **direction of the plumb line** ("Vertical").

$$dW = \text{grad } W \cdot \vec{dx} = \vec{g} \cdot \vec{dx} \qquad \text{with: } \vec{dx} = (dx,dy,dz). \tag{4}$$

The equipotential surfaces W(x,y,z) = const are called **level surfaces**, so:

$$dW = \vec{g} \cdot \vec{dx} = 0 \qquad (\vec{dx} \in \text{level surface})$$

i.e. $\vec{g}$ is perpendicular to the level surface.

The **geoid** is a unique level surface ~ coinciding with the surface of the oceans.

The direction of the plumb line in a Point P is usually expressed by two angles, the **latitude** ϕ and the **longitude** λ. (Since the plumb line is slightly curved, ϕ and λ change with the height of P.)

The height of P above the geoid is called **orthometric height h**.

The difference between the values of the gravity potential at P and at its footprint in the geoid is called **geopotential number**. It can be computed as:

$$C = W_0 - W = \int_0^H g \cdot dh \qquad (W_0 \text{ at the geoid}) \qquad (5)$$

or

$$h = -\int_{W_0}^W \frac{dW}{g} = \int_0^C \frac{dC}{g} \qquad (6)$$

ϕ, λ, h or ϕ, λ, C are called **natural coordinates**.

A first approximation of the earth by a mathematical surface is a sphere. A much better approximation is an **ellipsoid of revolution**. We can look for a "best" fitting ellipsoid either locally (e.g. "Bessel-Ellipsoid" in Switzerland) or globally (e.g. NWL8-E, WGS-72).

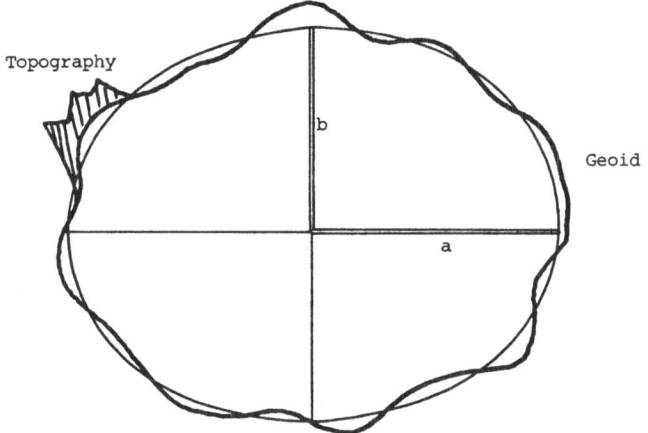

Fig.1 Approximation of the Geoid

If the ellipsoid is an equipotential surface of a **normal gravity field** it is called **level ellipsoid**:

$$U = U(x, y, z) \quad : \quad \text{Potential of normal gravity}$$
$$\gamma \quad : \quad \text{Normal gravity}.$$

U is completely determined by:

- the ellipsoidal dimensions a, b
- the total mass M
- the angular velocity ω

Ellipsoidal Coordinates:

The reference ellipsoid be given by its semi-major axis a and the excentricity e, with:
$e^2 = (a^2 - b^2)/a^2$

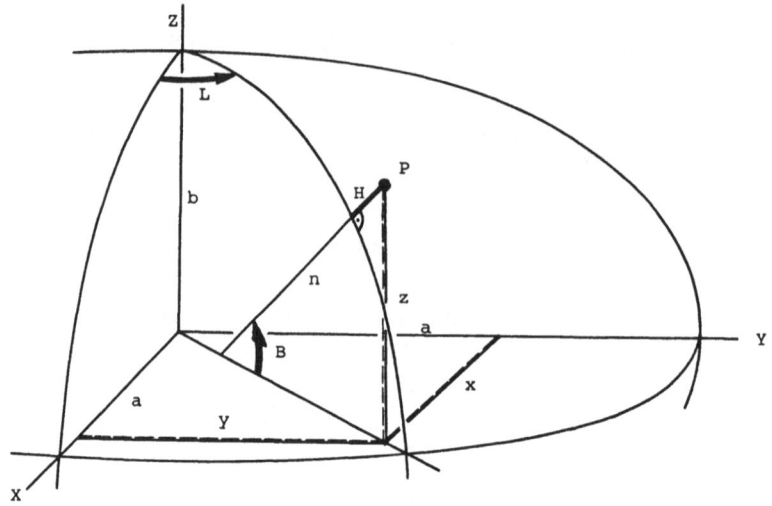

Fig. 2 Ellipsoidal coordinates

$$P(x,y,z) = P(B,L,H)$$

with: (B,L) : Direction of the normal n

H   : Ellipsoidal height

$$x = (N + H) \cdot \cos B \cdot \cos L$$
$$y = (N + H) \cdot \cos B \cdot \sin L \qquad\qquad (7)$$
$$z = (N \cdot (1-e^2) + H) \cdot \sin B$$

$$N = a \cdot \left(1 - e^2 \cdot \sin^2 B\right)^{-\frac{1}{2}}$$

Undulations, deflections, anomalies

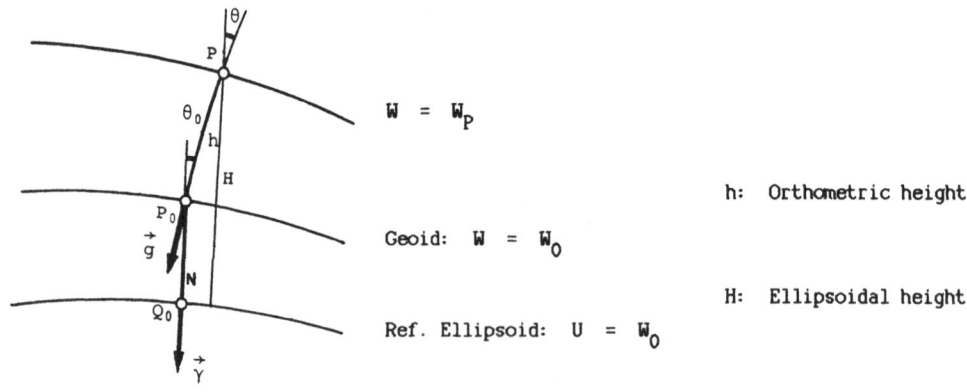

$$W = W_P$$

Geoid: $W = W_0$

Ref. Ellipsoid: $U = W_0$

h: Orthometric height

H: Ellipsoidal height

Fig. 3

Some definitions:

$$N \qquad : \quad \text{geoidal height (undulation)}$$

$$\vec{\Delta g} = \vec{g}_{P_o} - \vec{\gamma}_{Q_o} \quad : \quad \text{gravity anomaly vector}$$

$$\Delta g = g_{P_o} - \gamma_{Q_o} \quad : \quad \text{gravity anomaly}$$

$$\theta_o, \theta \qquad : \quad \text{deflection of the vertical}$$

$$\theta - \theta_o \qquad : \quad \text{curvature of the plumb line}$$

$$T = W - U \qquad : \quad \text{disturbing potential}$$

$$\left. \begin{array}{l} \xi = \phi - B \\ \eta = (\lambda - L) \cdot \cos B \end{array} \right\} : \quad \begin{array}{l} \text{components of the deflections} \\ \text{in NS-, EW-direction} \end{array}$$

An important relation between the disturbing potential T and the geoidal height is the so called **Bruns Theorem**:

$$N = \frac{T}{\gamma} \tag{8}$$

Divergence of the Equipotential Surfaces

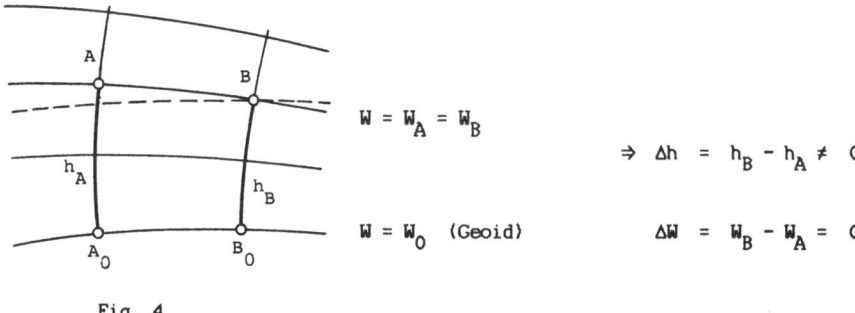

$$W = W_A = W_B$$

$$\Rightarrow \Delta h = h_B - h_A \neq 0$$

$$W = W_0 \text{ (Geoid)} \qquad \Delta W = W_B - W_A = 0$$

Fig. 4

Important consequences of the non-parallelity of the equipotential surfaces are the dependancy of levelled height differencies on the chosen path of the levelling; the mis-closure of a levelled loop:

$$\oint dh \neq 0$$

and the curvature of the plumb lines:

$$(\phi_B, \lambda_B) \neq (\phi_{B_o}, \lambda_{B_o})$$

$$(\xi_B, \eta_B) \neq (\xi_{B_o}, \eta_{B_o})$$

3. SOME EXAMPLES

3.1 Detailed structure of the geoid

The first example shows a small part of the geoid in the Bernese Oberland. Undulations on the decimeter level over a few kilometers can be found.

(Equidistance: 5 cm, altitudes in m above sea level)

Fig. 5 Geoidal undulations

3.2 Deflections in a profile

The figure shows a north-south profile of the deflections (ξ-component) at the earth's surface and in the geoid. Variations of 10 seconds in the surface values can be found within a few kilometers only.

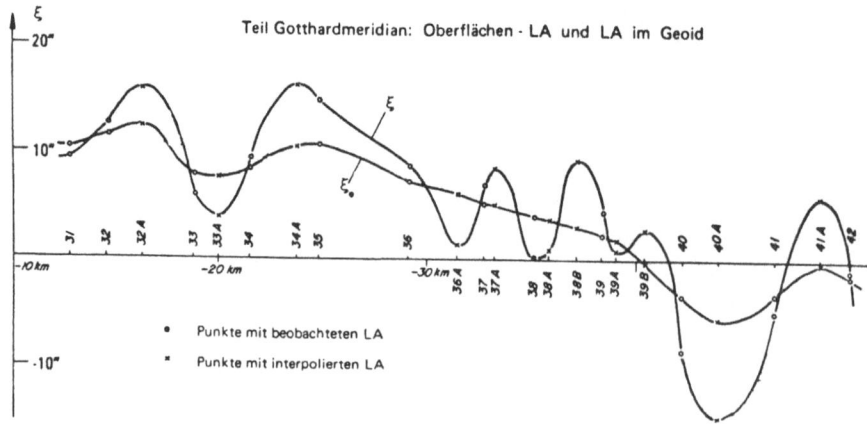

Fig. 6: North-South profile "Gotthard" (Deflections ξ)

3.3 Gravity anomalies

The largest gravity anomalies in Switzerland are caused by the Ivrea mass anomaly.

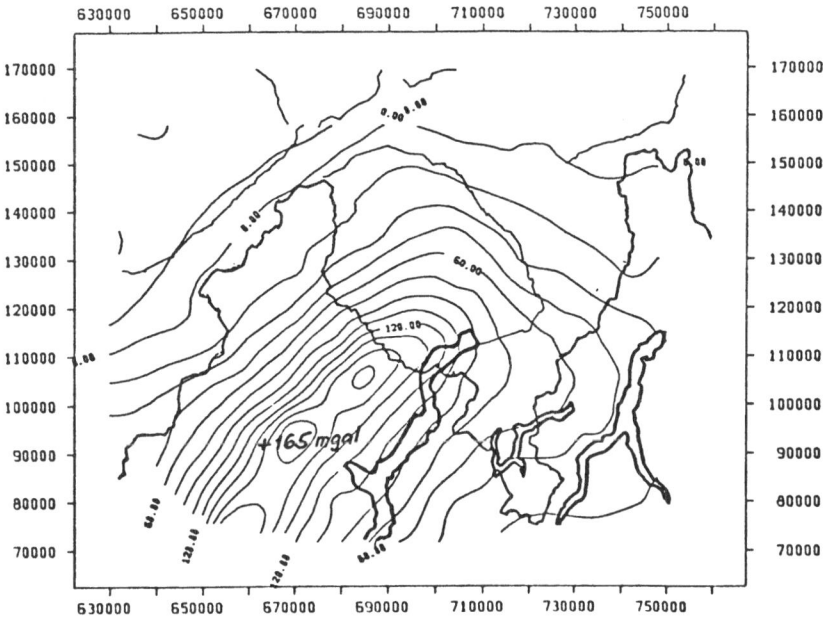

Fig. 7: Gravity anomaly caused by the Ivrea Body

Why do we need to know the geoid and the deviations of the vertical? Most of all because of the difference between the "physical" and the "geometrical world". A few examples:

- Levelling is related to equipotential surfaces
- Zenith and horizontal angles measured with a theodolite are related to the local directions of the plumb line
- Astrogeodetic point positioning gives the direction of the physical plumb line
- Network adjustments are done in the geometrical space: The natural coordinate systems have to be bound together!
- Only distances are not influenced by the disturbing potential
- Every site has its own local (natural) coordinate system
- GPS-results are given in a pure three dimensional coordinate system. GPS "sees neither geoidal undulations nor deviations of the vertical".
- Particles in an accelerator will not follow the equipotential surface but they have to orbit in a true plane.

4. DETERMINATION OF (ξ,η), Δg, N

4.1 By Observation

4.1.1 Overview

Quantity	Instrument	Reference	Result
ϕ , λ	Theodolite Zenith camera	B,L	ξ_A , η_A
g	Gravimeter	γ	Δg
ΔH Δh	Doppler, GPS Levels	ξ_M^R , η_M^R	ΔN

4.1.2 Example

As an example we give a short description of the transportable zenith camera used by the Institute for Geodesy and Photogrammetry of the ETH Zurich.

Principles of operation: By taking a photograph of the fixstars by means of a special camera in a vertical setup the astronomical parameters ϕ and λ of the observation point can be determined. Two essential quantities must be known to fulfill the requirements of the method:

- The epoch (local sidereal time) corresponding to the exposures (start and end of each single exposure as fed back from the shutter)
- The readings of the electronic levels mounted in perpendicular position on the camera tube to determine the spacial orientation of the camera and its optical axis with respect to the local plumb line.

The camera as it can be seen in Fig. 8 and 9 is equipped with a set of electronic levels, a high precision internal clock, an electronic shutter and a dedicated microprocessor system for the monitoring of the camera and the data access and storage.

After development, the films (size 6.5 x 9 cm) are measured on a high precision comparator to obtain the plate coordinates of the star images. Applying an appropriate transformation model between the theoretical and the measured plate coordinates of the known stars one can determine a set of astronomical parameters ϕ and λ referring to the camera position. The components of the vertical deflection are easily obtained by the equations $x=\phi-B$, $\eta=(\lambda-L)\cos B$.

Fig. 8

Transportable zenith camera of the Institute of Geodesy and Photogrammetry, ETH Zürich. Beside the optical tube a pair of electronic levels of "Talyvel 2" type is mounted in perpendicular orientation.

Fig. 9

Complete camera system in "ready-to-use" position. The electronics on the right hand side contains the power supply, the clock with time-signal receiver, the level electronics, the hardware to monitor the shutter and a microprocessor system for the data handling and storage.

4.2 By computation

4.2.1 Overview

Input	Method	Result
Δg	Vening-Meinesz $$\begin{Bmatrix} \xi \\ \eta \end{Bmatrix} = \frac{1}{4\pi G} \iint_\sigma \Delta g \cdot \frac{dS}{d\psi} \cdot \begin{Bmatrix} \cos\alpha \\ \sin\alpha \end{Bmatrix} \cdot d\sigma$$	ξ_I, η_I
Δg	Stokes $$N = \frac{1}{4\pi G} \iint_\sigma \Delta g \cdot S(\psi) \cdot d\sigma$$	N_I
ξ_A, η_A	Astrogeodetic Levelling $$\Delta N = - \int_A^B \theta \cdot ds$$	ΔN_I
z	Reciprocal Zenith Distances	θ
$\xi_A, \eta_A, \Delta g$	Astrogravimetric Levelling: Combination of Vening-Meinesz and Astrogeod. Levelling	
ξ_A, η_A $(\Delta g, N)$ Mass Model	Indirect Methods using Collocation	$\xi_I, \eta_I, \Delta N_I$ (Δg_I)

4.2.2 Example: Indirect Computation Using Collocation

This method has been described in more detail e.g. in Ref. 2 or Ref. 3: Provided a mass model of (a part of) the earth's crust be given, it is possible to compute the differences between observed derivatives of the disturbing potential (deviations of the vertical, gravity anomalies etc.) and the corresponding values defined by the mass model. These differences show a much smoother behaviour than the original observations. Therefore they are more suited for interpolation. Using collocation (appropriate covariance functions between all the quantities are necessary) we can not only interpolate between the given types of observation, but compute e.g. equipotential surfaces δN^I of the reduced disturbing potential. Adding the geoidal heights given by the mass model (N_M^I) leads us to the final geoidal heights N^I.

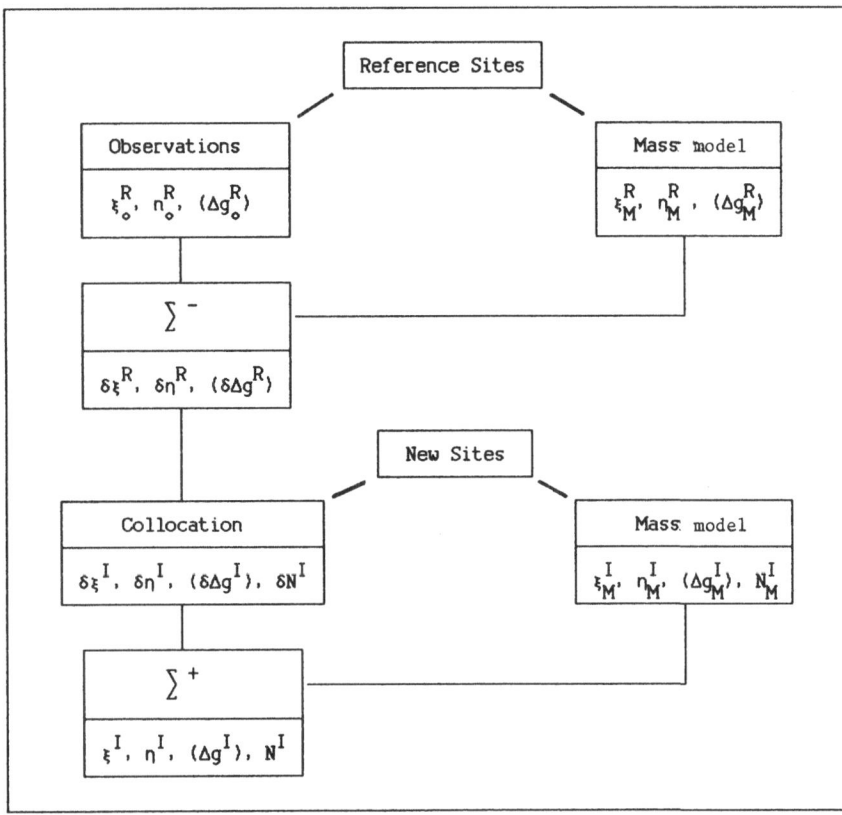

The astrogeodetic geoid in Switzerland has been computed using (Ref. 2):

- a set of more than 200 components of observed deviations of the vertical
- a mass model of the earth's crust in and around Switzerland, containing information about the topography, the depth of the Mohorovicic discontinuity, and the most prominent mass disturbancy, the "Ivrea Body".

5. APPLICATIONS ON THE CERN-LEP CONTROL NETWORK

In order to increase the relative accuracy of the interpolated deviations and geoidal heights, the Geodetic Institute of the ETH Zürich observed additional deviations of the vertical in the LEP area using the above described transportable zenith camera. The field measurements within the LEP-area took place in August 1983. All observations were carried out in only two nights. This particular work within the LEP project was the first field experiment with the camera, which was built at the Institut für Erdmessung, Technical University Hannover. Meanwhile the system has been used in the southern part of Switzerland and Northern Italy where, so far, a set of about 100 vertical deflection points have been measured (c.f.Ref. 6).

These additional vertical deflections in the LEP area have been introduced as additional reference values into the computation program. The results of the computations can be found in detail in Ref. 4 and 5. They can be summarized as follows:

- The vertical deflections range from 0 to 15 arc seconds at surface level and from 0 to 9 arc seconds at zero level.
- The resulting separation between the local reference ellipsoid and the geoid reaches 200 mm at 10.5 km from the origin.
- The altimetric corrections along the LEP machine - to obtain a true plane in space - vary from -40 mm to +100 mm.

Figures 10, 11, and 12 show the locations of the astrogeodetic stations, the total deflection intensities, and the equipotential heights in the LEP area. (The local reference ellipsoid coincides with the geoid at P_0.)

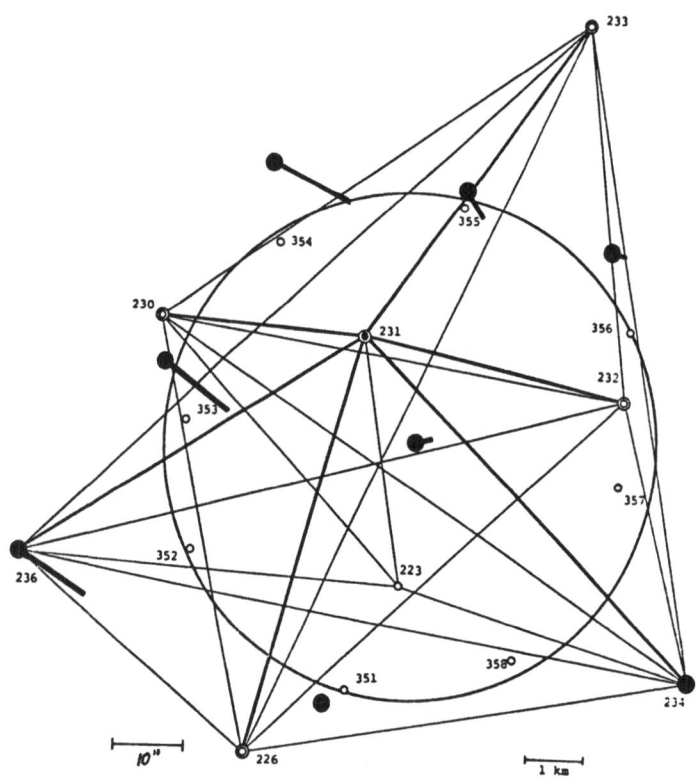

Fig. 10: Astrogeodetic stations

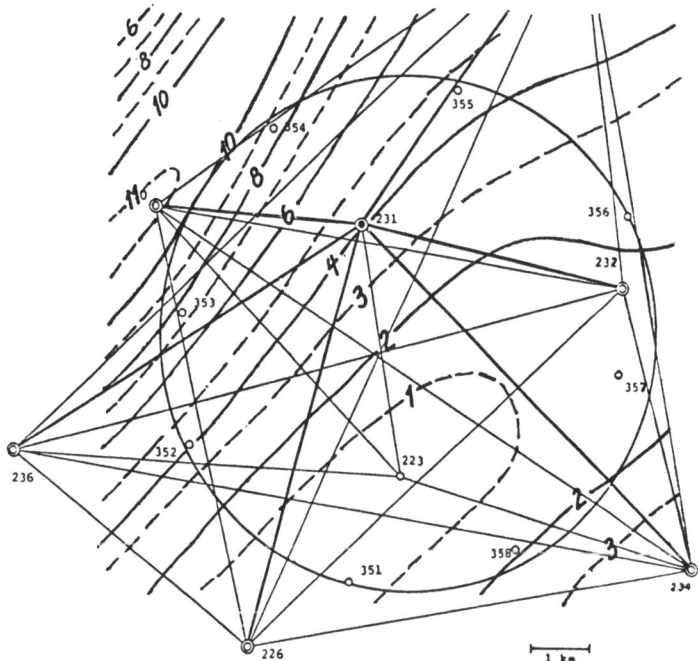

Fig. 11: Total deflection intensity (arc seconds)

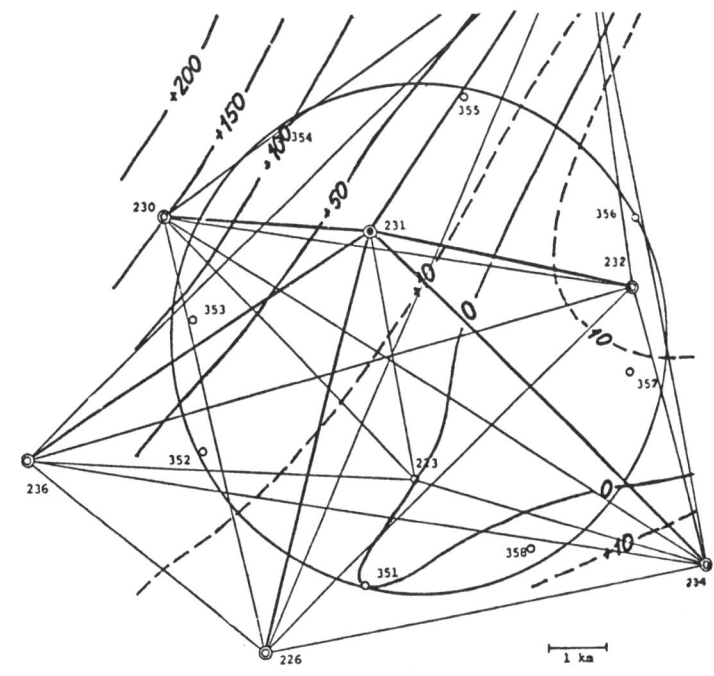

Fig. 12: Equipotential heights (mm)

REFERENCES

1) W.A.Heiskanen, H. Moritz, Physical Geodesy, W.H. Freeman and Co., San Francisco and London, 1967.

2) W. Gurtner, Das Geoid in der Schweiz, Astronomisch-geodätische Arbeiten in der Schweiz, Band 32, 1978.

3) W. Gurtner, A. Elmiger, Computation of Geoidal Heights and Vertical Deflections in Switzerland, XVIII General Assembly of the IUGG, Hamburg, 1983.

4) B.A. Bell, A Simulation of the Gravity Field around LEP, Private communication, 1985.

5) J. Gervaise, M. Mayoud, G. Beutler, W. Gurtner, Test of GPS on the CERN LEP Control Network. Proceedings of the Joint Meeting of FIG Study Groups 5B and 5C, Inertial, Doppler and GPS Measurements for National and Engineering Surveys. München, 1985.

6) B. Bürki, H.G. Kahle, W. Torge, Plans for the Application of the Hannover-type Zenith Camera System in the Central Alpine Region. Proceedings of the 2nd International Symposium on the Geoid in Europe and Mediterranean Area, Rome, 1982.

COMPARISON BETWEEN TERRAMETER AND GPS RESULTS -- AND HOW TO GET THERE

G. Beutler

Astronomical Institute, University of Berne, Switzerland.

ABSTRACT

After a short historical introduction accuracy limiting influences on GPS-derived engineering type networks are discussed in section 1, sections 2 and 3 are devoted to the 1984 Macrometer- and the 1985 Sercel- GPS Campaigns conducted on the CERN - LEP control network.

1. MEASURING ENGINEERING TYPE NETWORKS

In the early days of GPS (1982-1984) measurement campaigns usually were organized with two receivers operating simultaneously. The processing software was baseline- and session-oriented, which means that one three-dimensional vector (from the first to the second receiver) was estimated for each session. The scientific objectives may be summarized as follows:

- Repeatability demonstrations of baseline measurements
- Study of figure misclosures
- Accuracy demonstrations over very short baselines (invar-wire measurements)

Accuracy demonstrations of the order of 3 to 5 ppm were considered a success. The Ottawa-Macrometer campaign[1] may serve as a typical example of a "pioneer-age" campaign.

If a GPS campaign is organized today, it is almost certain that more than two receivers are operated simultaneously. Also, the processing software went through a remarkable development. The First International Symposium on Precise Positioning using GPS in Rockville 1985 and the Fourth International Geodetic Symposium on Satellite Positioning in Austin 1986 clearly showed the trend from baseline- and session- to network- and campaign- oriented software systems[2-4].

The scientific objectives may be summarized as follows:
- Repeatability demonstrations on entire networks (no figure misclosures)
- Subdivision of a large network: If a network has to be observed in several sessions, how many points of the network should be in common from session to session?
- Accuracy demonstrations on terrestrial high precision networks.
- Studies of accuracy-limiting effects such as troposphere, ionosphere, orbits and receiver clocks.

The ultimate accuracy-goal for GPS-established engineering type networks certainly is 0.1 ppm -- but it is doubtful whether this will be achievable at all. What certainly is possible on a routine basis with good processing software are accuracies of the order of 1 ppm in networks of the order of 10 km x 10 km.

Let us now have a look at the accuracy-limiting influences on GPS derived small scale networks.

1.1 Orbits

The influence of orbit errors on baseline and network estimates is one of the topics discussed in the contribution "GPS Orbit Determination" also included in these proceedings. It is therefore sufficient to summarize here our experiences with different orbit and processing types in small scale applications.

- Two orbit types are -- more or less -- readily available: Broadcast ephemerides transmitted by the satellites and precise ephemerides produced by the U.S. Department of Defence in an a posteriori analysis. It has been our experience during the past few years that broadcast ephemerides usually are of a surprisingly good quality. Generally one may expect orbit accuracies of the order of 20 meters or better, the exceptions being satellites with oscillator problems (such as space vehicle 4) where errors of the order of 100 meters may be present in broadcast ephemerides. The few specimens of precise ephemerides we had at our disposal do not allow a general judgement of their orbit quality, but there are indications that the accuracies are of the order of 10 meters for all satellites.

- Our parameter program has the capability of estimating orbits[5]. According to our experience one may not expect significantly better network coordinates if this program option is used if, at least, orbits of the broadcast quality are available. It must be pointed out that this statement does not hold for networks larger than ~ 10km x 10km.
- For the 1984 CERN Macrometer campaign we had both orbit types at our disposal. No significant difference in the quality of the GPS solution could be detected when replacing one orbit type by the other. This again demonstrates the high standard of the broadcast orbits.
- This comfortable situation could change considerably if, in future, broadcast ephemerides were artificially deteriorated. Then it would be mandatory to estimate at least some of the orbital elements in order to produce high quality results.

1.2 Ionosphere

As probably first pointed out by Campbell et al[6] the main systematic effect of neglecting a "well behaving" ionosphere (no dramatic changes in space and time) when processing GPS carrier phase data (single frequency) is a scale-factor in the GPS-derived baselines or networks. Baselines measured with single frequency GPS receivers tend to be short by a few tenths to -- may be -- three ppm depending on the electron content in the ionosphere. It should be mentioned that an agitated ionosphere (e.g. travelling ionospheric disturbances) may have an unpredictable influence on the coordinates derived from single frequency GPS observations. We had a closer look at this scale-phenomenon when we processed the 1984 CERN Macrometer campaign. This part of the 1984 CERN GPS analysis may be summarized as follows:

- The study was motivated by an unexplained scale factor of 1.8 ppm between the GPS and the terrestrial solutions when the ionosphere was completely neglected.
- It seemed appropriate (night time observations in moderate latitude) to model the ionosphere for the Cern campaign with the single layer model, assuming that all electrons are concentrated in a thin layer at a height H above ground. Moreover it was assumed that the electron density E was a constant (in time and within the layer).
- Four program runs were made using electron densities of
  $E = 0*10^{16}$, $10*10^{16}$, $20*10^{16}$, $30*10^{16}$ electrons per m^2,
  and it could be demonstrated that each change of $\Delta E=10*10^{16}$ introduced a change of 0.5 ppm in the scale of the GPS solution.
- It should be stated clearly that we had no hope of estimating the electron content from the Macrometer observations (only marginal differences in the estimated rms errors of double differences for different electron contents occured). It seemed most unlikely however, that an electron content of more than $E=10*10^{16}$ should have been present.

More conclusive statements concerning the ionosphere could be made when analyzing the material of the 1984 Alaska GPS Campaign[7]. The situation was quite different in this case:
  (a) the size of the net was ~ 1000km x 2000km,
  (b) the observations were made during the afternoon,
  (c) both carrier frequencies L1 and L2 were measured by the TI-4100 receivers.

It turned out that the L2 derived solution had a scale of 1.9 ppm and the L1 1.4 ppm with respect to the solution using the so called ionosphere-free linear combination of L1 and L2. (For the sake of completeness I should mention that the scale of this combined solution was identical with the scale of VLBI). From our point of view these results show that an ionosphere induced scale factor has to be expected if baselines or networks are estimated from single frequency phase data and if the ionosphere is neglected.

1.3 Troposphere

When processing GPS observations, usually the models developed for the TRANET - Doppler system are used; the best known ones may be the Hopfield- and the Saastamoinen models. In order to give an impression of the order of magnitude of the tropospheric refraction correction we reproduce here the Saastamoinen formula[8]:

$$\Delta r = 2.28 * (P + (1255/(273+T) + 0.05) * e)$$

where Δr is the tropospheric refraction correction in zenith direction in mm and P,T,e are atmospheric pressure (mbar), temperature (°C) and water vapour pressure (mbar).
The water vapour pressure in turn may be expressed as a function of relative humidity H (expressed in %) and temperature:

$$e = H/100 * \exp(-37.2465+.21366 * (T+273)-.000256988 * (T+273)^2).$$

Using the two formulae it is quite easy to compute the bias $\Delta(\Delta r)$ introduced into the GPS observable (in the zenith direction) by an error in temperature of $1^{\circ}C$, in pressure of 1 mbar and in humidity of 1 %, at 0° C, 1000 mbar and 50% humidity:

$$\Delta(\Delta r) = \quad 3 \text{ mm} \quad , \quad 2 \text{ mm} \quad , \quad 0.6 \text{ mm} \text{ for errors of } 1^{\circ} C \quad , \quad 1 \text{ mbar} \quad , \quad 1$$

These biases are even more pronounced if the surface temperature is rising. If we have for example $T=30^{\circ}C$, an error of $1^{\circ}C$ would introduce a bias of 14 mm while an error of 1% in humidity would introduce 14 mm into the observable in the zenith direction! It may be shown that these meteorological biases in the observable induce biases in the estimated height coordinates, where the above figures will be amplified by a factor of about 3. (For a more detailed treatment see Ref. 9).

Effects of this order of magnitude are of no importance in the "TRANET-Doppler world", but they are of vital importance in GPS where we are hunting the millimeter! Keeping in mind the accuracy limits of standard meteo equipment and the severe problems involved with this kind of measurement (e.g. surface effects), it is not really surprising that in GPS process-ing of small scale networks it is usually preferable not to use the weather data recorded at the receiver sites directly, but to use them to compute some kind of local "standard atmosphere". This has been our experience and that of others too[10]. During the last few months we have tested atmosphere models of the kind used for classical EDM – measurements (e.g. model by Essen and Froome). Although first applications are looking promising[11], we are not yet in a position to draw any conclusions.

For the two CERN campaigns to be discussed in the subsequent pages we used the surface weather data to generate a local atmosphere, where temperature, pressure and humidity gradients were assumed to be known as a function of the heights of the sites. Consequently the meteorological measurements uniquely served to define temperature, pressure and humidity at a reference height H_0.

2. THE 1984 CERN MACROMETER CAMPAIGN

A detailed report concerning this campaign was presented at the 1985 FIG meeting in Munich[12]. Therefore I only summarize the main characteristics and results.

Points P226, P230, P231, P232, P233, P234 and P236 of the CERN-LEP control network (see Fig. 1) were observed with three Macrometer V-1000 Suveyor instruments from 11 December to 13 December 1984. The daily observation periods started at ~ 0^h and ended at ~ 3^h30^m (UT).

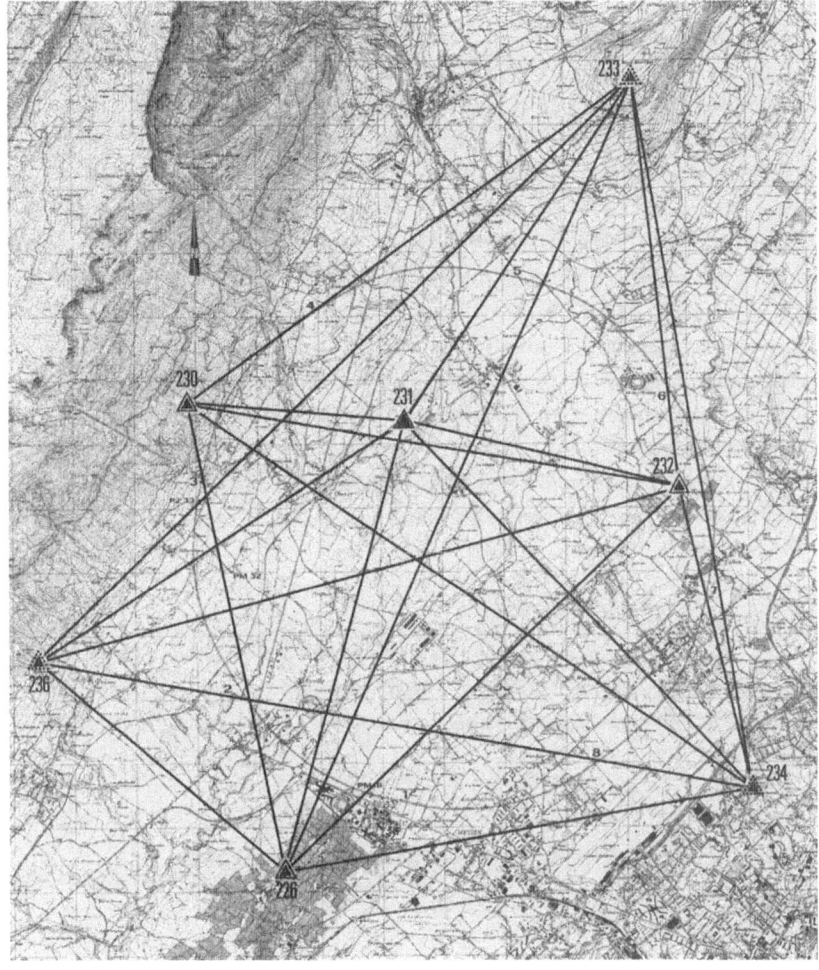

Fig. 1 CERN - LEP Control Network

Each observation period was divided into two observation sessions: The first lasted for two hours, the second for 80 minutes. Therefore two independent baselines could be measured per session, four per day and twelve during the entire campaign. Roger Jones from GEO/HYDRO Inc. organized the campaign in order to measure twice those six baselines starting from the central point P231. (This implies that P231 was permanently occupied by one of the receivers).

Since Navstar 4 was moved by the U.S. DoD during the week preceeding the campaign, the Macrometer instruments could not find this satellite. The consequence was that usually only three satellites could be observed simultaneously, a fourth satellite being visible only during 10-15 minutes at the end of the first and during the first 10-15 minutes of the second, session. Considering five to six simultaneously observed satellites to be an optimum, it is clear that the Macrometer Campaign was severely handicapped by the "absence" of Navstar 4.

The observations were processed with the Bernese GPS Software Package[a]. The main characteristics of the processing step may be summarized as follows:

- Double differences were used as basic observables.
- Precise ephemerides could be used to produce the final solution. No attempt was made to improve these orbits.
- The central point P231 was held fixed. Its WGS-72 coordinates were transformed from the Swiss Datum "CH - 1903" using TRANSIT-Doppler derived parameters.
- Initially 54 parameters (6*3 coordinates, 36 ambiguities) had to be estimated.
- Of these 36 ambiguities 33 could be resolved (related to integer numbers) by our automatic ambiguity resolution algorithm.
- After ambiguity resolution the adjustment was repeated with only 21 unknowns (6*3 coordinates, 3 ambiguities) to give the final solution.
- Four different program runs (with different electron contents) gave four different solutions. In this report we only present the solution for $E = 10 * 10^{16}$ electrons per m^2, which seems to be a hypothesis that makes sense.

In Table 1 we reproduce what we believe is "the best" GPS solution. The 1 σ errors for the coordinates (actually the coordinate differences with respect to point P231) are not given in this Table. They are typically 1 mm for latitude, 2 mm for longitude and 3 to 4 mm for height.

Table 1

Results of 1984 CERN Macrometer Campaign

Datum: CH 1903

Station	Latitude			Longitude			Height
P226	46	13	36.24140	6	01	54.44865	498.821
P230	46	17	32.26122	6	00	36.07847	747.743
P231 *)	46	17	25.41472	6	03	21.02502	428.774
P232	46	16	54.05284	6	06	50.77772	494.033
P233	46	20	21.28416	6	06	11.26072	740.512
P234	46	14	22.10559	6	07	50.34466	457.019
P236	46	15	21.30264	5	58	45.18231	603.758

*) Fixed point

This brings us to the last and most interesting part of the analysis, the comparison between the GPS and the Terrameter results. In order to make a meaningful comparison, the terrestrial coordinates -- given in a local Cartesian system -- had to be transformed into the WGS - 72 coordinate system. The parameters of this transformation are:

(a) astronomical latitude and longitude of the z - axis.

(b) astronomical azimuth of the x - axis and

(c) WGS - coordinates (Doppler derived) of the origin of the local system.

For a more explicit definition of these parameters see Ref. 12.

The transformation between the GPS- and the Terrameter- derived WGS - 72 coordinates was then performed through a seven - parameter Helmert transformation (three translation parameters, three rotation parameters, one scale factor). The latter four parameters (transformation GPS --> Terrameter) were estimated as:

$$\omega_{ns} = 0.25" \pm 0.08" \quad \text{(rotation about ns - axis)}$$
$$\omega_{ew} = 0.52" \pm 0.08" \quad \text{(rotation about ew - axis)}$$
$$\omega_{z} = 0.72" \pm 0.05" \quad \text{(rotation about z - axis)}$$
$$s = 1.3 \text{ mm/km} \pm 0.3 \text{ mm/km} \quad \text{(scale factor)}.$$

The three rotation angles are all smaller than 1", which means that the GPS based orientation and the orientation deduced from the terrestrial observations match quite nicely. The scale factor of 1.3 ppm still lacks an explanation.

Table 2 gives the residuals in the coordinates that remain after the Helmert transformation between the terrestrial and the GPS - solutions. The figures in Table 2 demonstrate the high consistency and integrity of both the Terrameter and the GPS solution.

<u>Table 2</u>

Residuals of Helmert Transformation (in mm)

Station	Latitude	Longitude	Height
P226	3	-3	6
P230	1	2	4
P231	3	3	-6
P232	-3	-8	-1
P233	-3	2	3
P234	1	2	-1
P236	-2	2	-4

3. THE 1985 CERN SERCEL CAMPAIGN

Eight points of the CERN - LEP control network -- actually the seven points already surveyed by Macrometer in 1984 plus point "PM 15" (see Fig. 1) -- were equiped with three Sercel TR5S receivers[13] from 14-18 October, 1985. The campaign was organized in a way very similar to that of the CERN Macrometer: The daily observation period (from 3^h30^m to 7^h UT) was divided into two sessions, each lasting for approximately 90 minutes. P231, the central point, was always occupied and always by the same receiver unit. As opposed to the 1984 campaign, the Sercel measurements lasted for five nights. Since the receivers were not moved between the two daily sessions, three points could be surveyed per night (namely P231, P230, P236 on the 14th, P231, P226, P234 on the 15th, P231, P232, P233 on the 16th, P231, P236, PM 15 on the 17th and finally P231, P226, PM 15 on the 18th of October).

Although the three Sercel receivers had no problems tracking all GPS satellites available at the time of the campaign, the configuration was not much better than during the 1984 campaign, due to low maximum elevations of two of the satellites.

Moreover it became obvious -- unfortunately only during the processing phase four months after the campaign -- that one of the receivers had had serious oscillator problems. As a matter of fact these problems turned out to be so serious that its phase observations could not be used to produce high quality coordinates. Being familiar with E. Murphy's law I was not surprised to see that the faulty receiver was the one always used at the central point P231! Therefore it is clear that the Sercel campaign could not produce a network comparable to the Macrometer net.

What remained is summarized in the following Table, where we only compare baseline lengths (established with Terrameter, Macrometer and Sercel) from that part of the net which could be covered by the two properly working Sercel receiver units.

The conclusions are quickly drawn:

(a)   If the Sercel receivers are working properly they promise to produce high quality results (the formal errors being of the same order of magnitude as those for Macrometer).

(b)   In order to demonstrate the Sercel receivers' capability to produce GPS derived networks of a consistency and integrity comparable to the Macrometer V-1000 receivers it is certainly necessary for Sercel to repeat the CERN campaign (or a campaign in a net with a comparable ground truth.

<u>Table 3</u>

Comparison of Baseline lengths (meters)

(T = Terrameter, M = Macrometer, S = Sercel)

Baseline	S	S - T	S - M	M - T
P226-P234	7755.932	-.002	.017	-.019
P226-P236	5193.633	-.022	-.012	-.010
P232-P234	6458.931	-.004	.003	-.007
P234-P236	11820.738	-.032	.000	-.032
P226-PM15	2204.744	-.002		
P234-PM15	5757.046	-.005		
P236-PM15	6248.884	-.021		

REFERENCES

1) G. Beutler, D.A. Davidson, R. Langley, R. Santerre, P. Vanicek, D.E. Wells, Some theoretical and practical aspects of geodetic positioning with the Global Positioning System using carrier phase difference observations, Dept. of Surveying Engineering Technical Report No. 109, University of New Brunswick, Canada, (1984).

2) W. Gurtner, G. Beutler, I. Bauersima, T. Schildknecht, Proc. First Int. Symp. on Precise Positioning with GPS, Vol. 1, Rockville, 1985 (National Geodetic Survey, Rockville, 1985), pp. 363-372.

3) C.C. Goad, Ibid., pp. 347-356.

4) Y. Bock, R.I. Abbot, C.C. Counselman, R.W. King, A.R. Paradis, Processing of GPS Data in the Network Mode, Fourth International Geodetic Symposium on Satellite Positioning, Austin, Texas (to be printed 1986).

5) G. Beutler, W. Gurtner, I. Bauersima, R. Langley, Proc. First Int. Symp. on Precise Positioning with GPS, Vol. 1, Rockville, 1985 (National Geodetic Survey, Rockville, 1985), pp. 99-111.

6) J. Campbell, H. Cloppenburg, F.-J. Lohmar, Int. Symp. on Space Techniques for Geodynamics, Vol. 1, Sopron, 1984 (Hungarian Academy of Sciences, Sopron, 1984), pp. 196-206.

7) W. Gurtner, GPS papers presented by the Astronomical Institute of the University of Berne in the Year 1985, Mitteilungen der Satellitenbeobachtungsstation Zimmerwald No. 18, 53-70, University of Berne (1985).

8) I. Bauersima, Navstar/Global Positioning System (GPS) II, Mitteilungen der Satelliten-beobachtungsstation Zimmerwald, No. 10, University of Berne (1983).

9) G. Beutler, W. Gurtner, Influence of Atmospheric Refraction on the Evaluation of GPS Carrier Phase Observations, Berichte der Satellitenbeobachtungsstation Zimmerwald, No. 16, University of Berne (1986).

10) Y. Bock, private communication.

11) M. Rothacher, G. Beutler, A. Geiger, W. Gurtner, H.G. Kahle, D. Schneider, The 1985 Swiss GPS Campaign, Fourth Int. Geodetic Symp. on Satellite Positioning, Austin, Texas, (to be printed 1986).

12) J. Gervaise, G. Beutler, W. Gurtner, M. Mayoud, Proc. of FIG Groups 5b and 5c on Iner-tial, Doppler and GPS Measurements for National and Engineering Surveys, Munich, 1985, (Universität der Bundeswehr, München, 1985), pp. 337-358.

13) C. Boucher, G. Nard, Proc. First Int. Symp. on Precise Positioning with GPS, Rockville, 1985 (National Geodetic Survey, Rockville, 1985), pp. 135-146.

GEODETIC NETWORKS FOR CRUSTAL MOVEMENTS STUDIES

P. Baldi and M. Unguendoli
University of Bologna, Italy.

ABSTRACT

Starting from a short review of the geophysical phenomena representing the major causes of crustal movements, we analyze some problems related to the geodetic networks used for the detection of the movement them selves. Some examples of real networks set up in Italy are also shown and discussed.

1. INTRODUCTION

The application of Geodesy to the observation of crustal deformation can be divided into two main categories: the continental scale monitoring of the relative movement of tectonic plates, and the regional or local scale observation of deformation connected with seismicity, volcanic activity and subsidence. On an even smaller scale one might study the phenomena of landslides and land settlement etc.

Since the relative movement of continental blocks is estimated to be within the order of a few centimetres a year, the relative positions of the points need to be defined with an accuracy of $10^{-8} \div 10^{-9}$. This is now made possible by the use of Satellite Laser Ranging and Extragalactic Source Radio Interferometry (Very Long Base Interferometry).

On the other hand, the use of artificial satellites, particularly those using the GPS (Global Positioning System), will in the next few years furnish us with an exceptionally versatile instrument for small and medium scale geodetic applications; the use of a constellation of 18 satellites permitting one to determine the relative positions of points on the Earth's surface with an accuracy in the order of 1 cm $^{1 \div 3}$.

However, classical geodetic techniques, i.e. the repeated measurements of distances, angular height differences, direction and intensity of the gravity field, remains, at least for the time being, the most effective way of observing and checking deformations associated with seismic activity and/or other local or regional scale geophysical phenomena. Here crustal movement is generally subdivided into horizontal and vertical deformations. The techniques commonly employed consist of triangulation and trilateration of reasonably complicated networks for the former, whilst vertical deformations are studied by means of geometric or trigonometric levelling integrated with gravimetric, with clinometric, extensimetric and mareographic measurements. In some cases, such as mountainous areas, plano-altimetric networks can also be used, integrating the various types of measurement, and including astronomic observations in order to obtain three-dimensional networks.

There is no doubt about the importance of studying surface deformations that may be connected with geological and geophysical phenomena taking place beneath the Earth's crust; one need only think, for example, about the possibility of monitoring in detail perhaps the most fascinating, though still not well-defined, phase of the seismic cycle taking place before an earthquake and characterised by the presence of the phenomena known as precursors. This phase, observed prior to several earthquakes, can be connected with changes in the physical properties of rock subjected to tectonic forces, according to models amply dicussed by various authors. Without doubt, anomalous deformations of the terrain seem to be the most easily recorded signals of such activity among the various other effects observable, (namely the velocity variations of seismic waves, changes in rock resistivity, etc.).

Monitoring of the deformations in volcanic areas using geodetic techniques in association with seismic and geochemical measurements etc., is particularly important and useful when it facilitates the making of "deterministic" previsions.

Let us finally remember the possibility of monitoring and thus modelling subsidence phenomena, which, if not forecast or controlled, may result in damage to or destruction of surface facilities, the irrigation system and ground water regime, etc.

2. SOME EXAMPLES OF GROUND DEFORMATIONS

2.1 Seismic areas

The pattern of crustal deformation in a seismic area can provide essential information on the state of stress, and thus furnish further details concerning the danger level of the area. Generally speaking, a stationary deformation may be correlated with the seismic risk of the zone, the level of which depends upon the speed of deformation, and the time elapsed since the preceding earthquake; in those cases where the entire deformation characteristics over the whole area are known, the crustal movements measured can be used to estimate the period in which such events might recur [4]. Clearly an affirmation of this nature would be based upon the hypothesis that the deformation cycle of large seismic events is not complicated by complex phenomena, and is repetitive. In the schematisation of Fig. 1 this is shown in the presence of a post-seismic transient and permanent deformations modify the simple scheme proposed by Reid [5]. At the present moment the data available for various seismic areas in the world indicate a post-seismic transient in the overall coseismic deformation, notwithstanding the fact that a part of the period concerned shows constant deformations. Many cases seem to show that a large number of seismic deformations take place in a short time (several years), and are then followed by a slow and gradual transition to pre-seismic deformation levels.

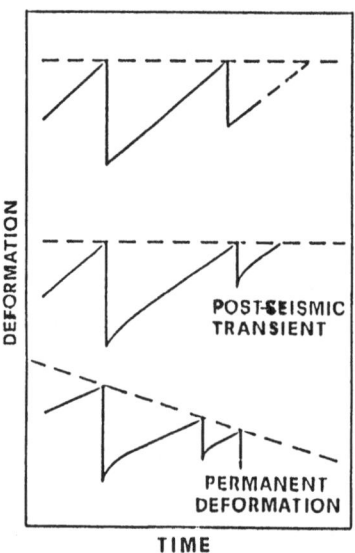

Fig. 1 Diagrams of seismic cycle

Many authors have developed models to explain time-dependant crustal deformation observed in regions of ongoing tectonic activity. They found evidence for the existance of visco-elastic relaxtion in the asthenosphere. In particular, visco-elastic stress relaxation was found to be the cause of the long-term inter-seismic/aseismic deformation occuring between major earthquakes in several regions of Japan. In this case the deformation takes the form of subsidence in the source region, with a compensating uplift on the flanks of the subsiding region. The width of the subsidence profile seems to be roughly proportional to the elastic plate thickness, and its time development was found to depend on the characteristic relaxation time ($\tau = 2\eta/\mu$), where η is the asthenospheric viscosity and μ is the asthenospheric shear modulus).

Information concerning structural deformation on a geological scale allows one to point out permanent deformations. It is, however, necessary to bear in mind the fact that some of the deformation over the centuries may not be strictly connected with the storing of elastic energy, and that anomalous crustal movements, defined as precursor phenomena, may be adding their presence to the deformation cycle; a presence which in many instances shows different characteristics, overall behaviour and duration from that of the characteristic seismotectonic changes in the area.

As an example of the possible complexity of deformation see Fig. 2 in which are laid out the altimetric variations in the region of Honshu (Japan). Three different fields of deformation following three different processes can be identified: the dominant process produces a constant subsidence related to the westerly subsidence of the Pacific Plate along the Japanese trench. Local effects are added to this movement, namely uplift of the Quaternary structure, and sinking in the locality of the fault associated with the 1896 earthquake (M=7.3). In Fig. 2 the heights over time variations are repeated in relation to two different zones A and B. Whilst zone B, which is far from the fault, shows a linear process, in zone A the deformations agree with those forecast by the theory of stress relaxation in a visco-elastic asthenosphere following a faulting [6].

Important features of the earthquake mechanism include the size and shape of the rupture surface, its orientation, the faulting motion on this surface, and the time history of the process. Seismic data, in general, provide information about seismic source mechanisms; geodetic surveys could produce important advances in our knowledge of earthquake source models.

Theoretical expressions for surface and subsurface deformations accompanying faulting, obtained on the basis of the elastic theory of dislocation [7,8] have been given by Press [9], Mansinha and Smylie [10] and others. For example, the horizontal displacement components are derived for a finite rectangular strike-slip fault in a uniform elastic half-space with the geometry shown in Fig. 3. In particular the horizontal displacement component U along a median line perpendicular to the fault strike is shown in Fig. 4.

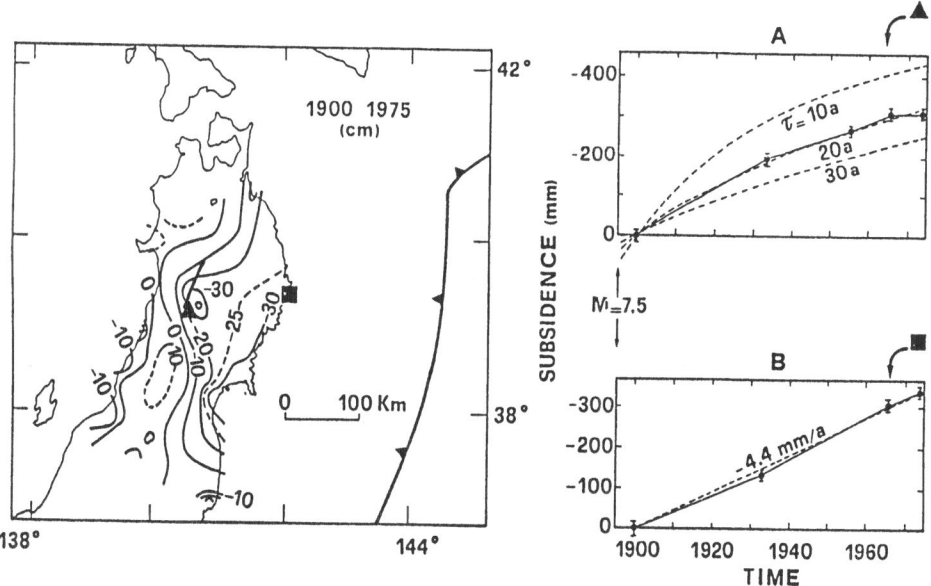

Fig. 2   Level changes, 1900-1975, in Northern Honshu, Japan, and region of surface faulting in the 1896 M=7.3 Riku-u earthquake (Thatker, 1981)[4]

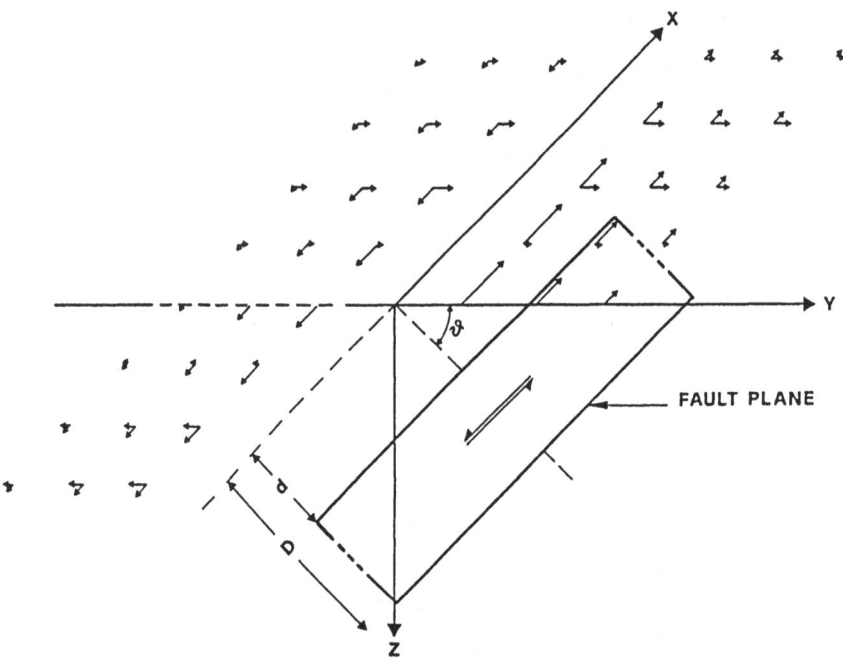

Fig. 3   Horizontal components of displacement, for a finite rectangular strike – slip fault

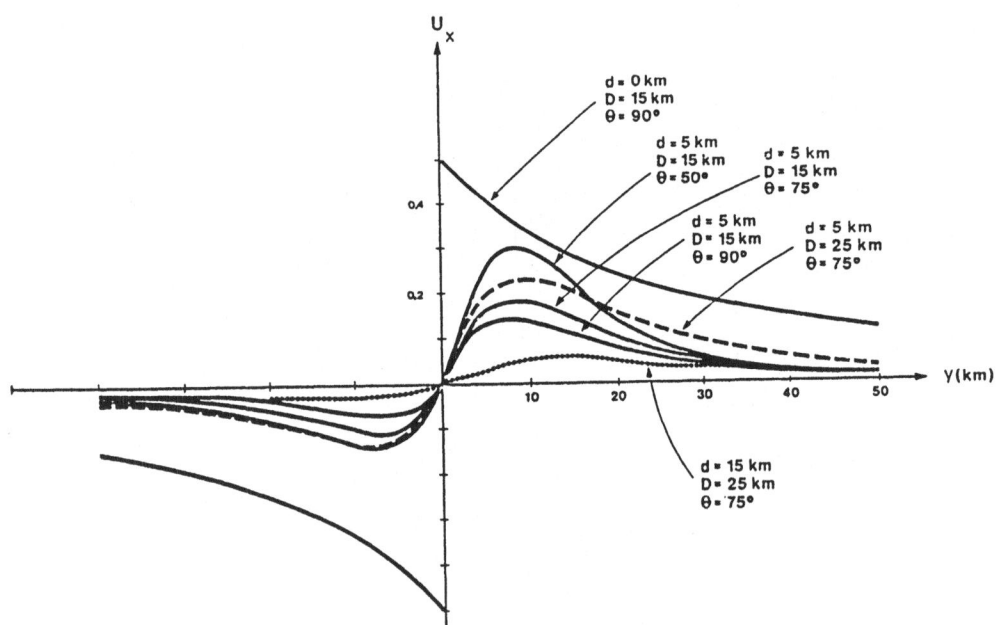

Fig. 4   U component of displacement for different parameters of a strike-slip fault . The displacement is given as a fraction of the slip of the fault.

In general, the assymetry of the deformations at the two sides of the fault is strictly correlated to the slip and inclination of the fault plane, whereas the displacement fields near to and distant from the fault are indicative repectively of the upper and lower limits of the fault.

Obviously, if we consider a more realistic model, in which the uniform elastic half space is substitued with a variable shear modulus, we obtain different results. In fact assuming that rigidity (z) is a continous and smooth function of depth:

$$\mu(z) = C (1 - Te^{-z/z_o})$$

where the value of T is derived from seismic data (T=0.9) as well as from laboratory measurements (T=0.5) [11,12]; Mahrer and Nur [13] obtain theoretical displacements almost identical for different values of T in the case of surface faults; in contrast, in the case of a buried fault, a notable dependance is present (Fig. 5).

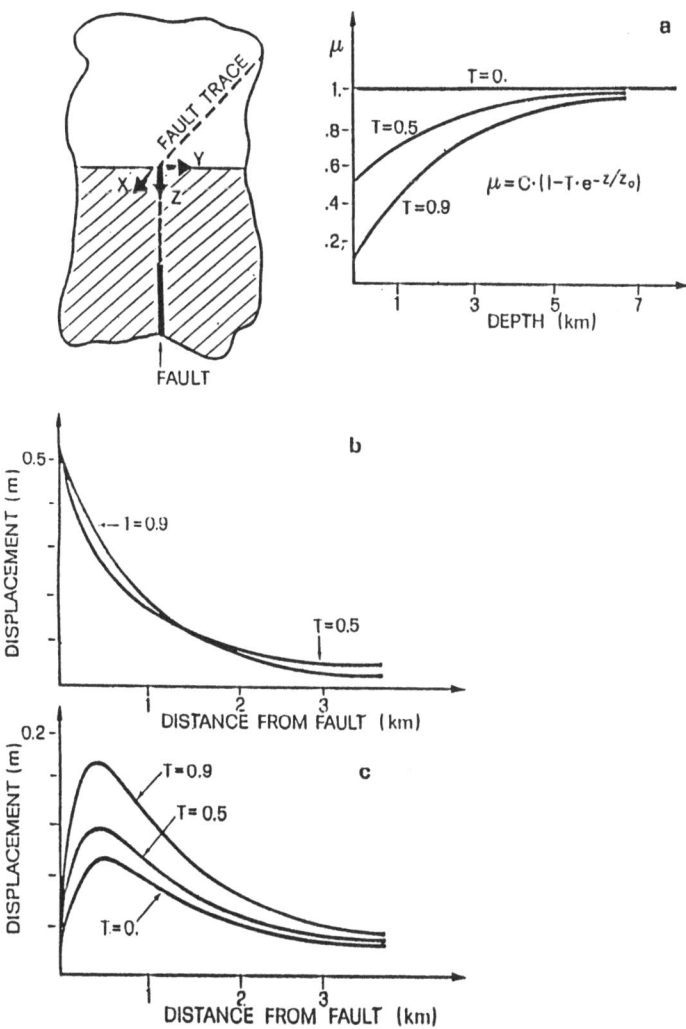

Fig. 5   Normalized shear modulus as a function of depth (a);surface displacement as a function of distance from the fault:(b) surface fault case;(c) buried case. (Mahrer and Nur, 1979) [13]

It is evident that the quantitative information obtained by geodetic measurement should be interpreted in the light of one's total knowledge regarding the phenomenon being studied, and of the various factors that might strikingly influence the surface effects associated with a state of stress in the crust, or more simply a slip along a fault plane. One thinks, for example, of the possible correlations between morphology and disturbances in surface deformations using a simple elastic model. Starting in from a model with topography of small slope, McTigue and Stein [14] analyse those deformations associated with a variation in the uniform tectonic force (Fig. 6), establishing that topography gives size to local departures from regional strain of the order of the characteristic slope ($\frac{H}{L}$). Considering instead structural discontinuities caused by surface structures (alluvial deposits), one sees that these can introduce notable disturbances (dependant upon rigidity contrasts) into the general field of surface deformations, even if, in general, they are limited to areas of similar dimensions to those of the disturbed tract Fig. 6.

To these sources of disturbance one should add the local, changing effects produced by the weather, and related to rainfall, changes in atmospheric pressure and temperature, etc., which can generate a background noise detectable using modern geodetic instrumentation.

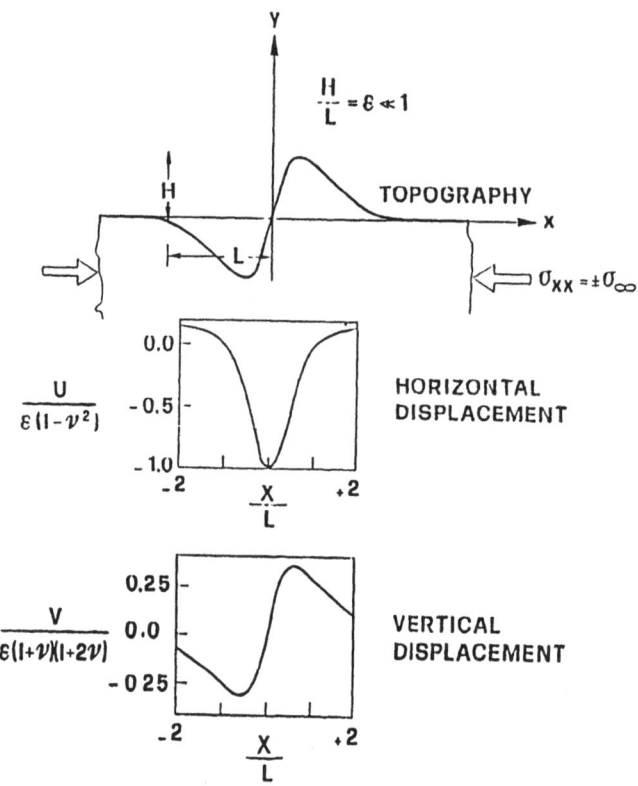

Fig. 6 Example of dimensionless perturbation displacement field, due to a far-field tectonic stress (compression); the surface relief is assumed to be characterized by a horizontal length scale, L, and a vertical length scale H. (from McTingue and Stein, 1984) [14]

2.2 Volcanic areas

A classic case where the integration of a multitude of different techniques is indispensable, is undoubtedly the control of deformations related to volcanic activity, due to their speed of change, and the real possibility of making previsions. Subsurface movements of magma generally cause rapid movements of the surface, with uplifting and subsidence that alternate and vary the magma level and pressures involved.

We have laid out the case of Campi Flegrei (in the region of Naples, Italy) as an example where periods of rapid uplift, accompanied by intense seismic activity, alternate with periods of calm. The main features of the expansion can be summarised as follows: the expanding area of quasi-circular shape, and radius of about 7 km, shows maximum uplift near the town of Pozzuoli (Fig. 7). The uplift practically vanishes at the edge of the caldera. The speed of uplift (as high as 4–5 mm/day) and its change over time constrained the operators following the phenomenon both to set up mareographic and clinometric networks in the area, which permit continual monitoring of the phenomenon, and to carry out the periodic checks by geometric levelling using a high number of working teams.

Horizontal measurements provided additional information on the strain field. The displacement vector referred to a point located in the area of maximum uplift, appearing to be roughly perpendicular to the contour lines of ground uplift, with an average horizontal strain value of about 10^{-4}.

To this, microgravimetric measurements were added because of the magnitude and speed of the deformations; these being extremely useful for monitoring, as well as for supplying an important quantitative element in the definition of the source model.

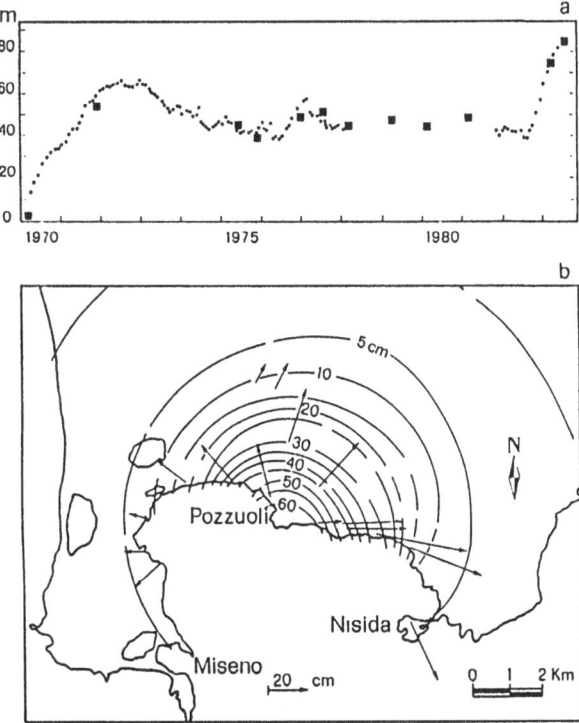

Fig . 7 (a) Vertical ground deformation at Pozzuoli (Italy) during 1970–83 period ;
(b) area distribution of vertical ground deformation, and horizontal displacement for the period 1980–83 (from Berrino et al., 1985)[15]

2.3 Subsidence

It is a well known fact that extensive withdrawals of fluid from subsurface resources is the major cause of land subsidence. This is due to a subsurface compaction which represents the response of a compressible, porous, medium to changes in the fluid flow field operating within it. The artificial withdrawal of fluid modifies the   natural balanced condition by causing a decrease in pore pressure and a consequent increase in intergranular stress. Under the influence of this increase in the effective stress, the formations compact and the land surface subsides. Subsidence may in general be connected with vertical and horizontal surface displacements, cracks and fissures in soft ground, changes in the hydrological regime, contamination of fresh water aquifer, changes in local tectonic regimes due to altered stress and strain conditions with a consequential risk of increased seismicity, etc.

The qualitative and quantitative determination of the parameters governing the subsidence process involve measurements and analyses of geological, geotechnical and geophysical quantities, as well as surface and near surface measurements of vertical and horizontal movements, such as levelling and trilateration surveys, tilt-meter and extensometer surveys.

If a complete set of the above-mentioned information is available, the complex natural system will be a mean of those mathematical and numerical prediction models formulated from the equation of subsurface flow combined with the elastic equilibrium equations.

Major subsidence in oil and gas fields, or due to ground water pumping, has been extensively covered in the literature [16]. For example, a surface lowering of about 8m from 1937 to 1962 has been observed in the harbour area of Los Angeles and Long Beach. A more recent case of particular interest is the subsidence of the Po Plain (Italy) and Venice [17], where the lowering, though modest, is nevertheless a matter of concern because it compounds the effect of the exceptionally high tides in the city.

3. GEODETIC MEASUREMENTS

Having started with this short review of the geophysical phenomena representing the major causes of crustal movement, we will now move on to consider the geodetic method used for the detection of such movement, basing our lecture more on practical than on theoretical considerations. We will thus describe classical networks, notwithstanding our awareness that soon much will change with the extensive use of spatial techniques (GPS), since at the present moment we do not have sufficient experience with such techniques to be able to supply exhaustive answers to all the problems related to their application. As we have mentioned, we will also subdivide our exposition into three different subjects: planimetric, altimetric and plano-altimetric networks.

3.1 Planimetric networks

Horizontal crustal deformation may be measured by means of a survey involving angles and distances; electromagnetic distance-measuring instruments are able to   determine distances with an accuracy of 1 ppm or less, according to the extent to which one is able to estimate the average propagation speed of light in the atmosphere, using either meteorological measurements [18,19], or by simultaneously measuring optical path length at two or three different wavelengths [20,21]. With a view to developing adequate geodetic networks, various criteria, such as accuracy, reliability and expense, need to be balanced in order to optimise the network. Such optimisation is usually classified in four different orders, starting with a parametric adjustment based on the observation equation:

$$Ax - d = v$$

which, by means of a least-squares solution and following the free adjustment scheme, gives:

$$X = (A^T PA)^{-1} A^T Pd$$

or following the free adjustement scheme

$$X = (A^T PA)^+ A^T Pd$$

and with the co-variance matrix:

$$K_{xx} = \sigma^2 (A^T PA)^+ = \sigma^2 Q_{xx}$$

where A is the configuration matrix, P the weight matrix, σ^2 is the variance of the unit weight and ( )$^+$ denotes the Moore-Penrose inverse.

In the case of planimetric networks, for example, disregarding the "Zero-order design" relating to the optimal reference system, the optimization process is carried out by means of 'First, Second and Third-order design", which relate respectively to the configuration of the network and observation plan, to the search for an optimal distribution of observation weighting, and to the improvement of the network by including additional points and/or observations [22].

One of the criteria used to evaluate the reliability of a network, is based on the analysis of the diagonal elements r of the matrix R, where:

$$R = A Q_{xx} A^T P - I \quad .$$

These elements may vary from 0 to 1; good reliability is obtained for high values of z, as well as with a homogeneous distribution of such values throughout the entire network.

As regards "First-order design", the theoretical approach is severely limited by topography and the available information relating to the geophysical phenomenon in question.

As an example of theoretical trilateration networks we show you those in Fig. 8 which correspond to differing needs: the first is used when it is impossible to localise the direction of expected movements, and the others to concentrate research on the deformation along a traverse of the strike fault.

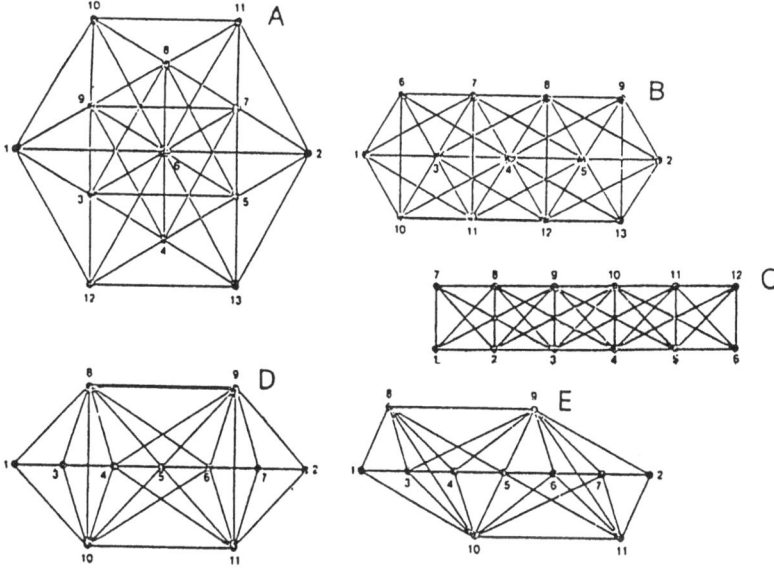

Fig. 8 Different schemes of trilateration networks.

On the basis of point distribution, measurements should be planned seeking the greatest possible reliability and precision, without forgetting the question of cost. The most practical approach is the interactive simulation.

Once the network has been drawn up and the theoretical aspect of the measurements established, one can take steps to improve the accuracy of single measurements, with a view to obtaining a better overall network design. This may be done by seeking a method for weighting the measurements themselves which respond more satisfactorily to the needs for which the measurements are taken (second-order design).

In effect, the analytic approach is founded on the solution of the following equation, as it relates to P:

$$(A^T PA) = Q_{xx}$$

in which the criterion matrix Q_{xx} must be defined in advance.

In general, a geodetic network should tend towards homogeneity and isotropy. In this case, Q_{xx} is normally constructed using the Tailor-Karman structure solution [23], the simplest case of which may be reduced to the equation:

$$Q_{xx} = I$$

where I is the unit matrix, involving the drawback that the co-variance between the coordinates is assumed to be zero.

A different approach, which appears to yield results more in line with real needs, consists in the use of the singular value decomposition of the matrix Q of the network:

$$Q_{xx} = V \Lambda V^T$$

where the diagonal matrix Λ contains the eigenvalues of Q_{xx}, and V is the matrix of the corresponding orthonormalised eigenvectors. A possible derivation of the criterion matrix might consist in contracting the eigenvalue spectrum, without changing the eigenvectors, with a consequent diminution in the semi-axes of the error ellipses. A different approach, of particular interest in the study of deformations, might be to obtain a variation in the orientations of the error ellipses semi-axes by rotating the eigenvectors matrix V [24].

In any case, these methods are severely limited by the shape of the network and also, especially in distance measurements, by instrument limitations. For these reasons practical results are generally negligible, but second-order design nevertheless supplies us with much useful information for the planning and execution of networks.

We have extensively used second-order design and reliability criteria in an undogmatic way to reduce the measurements in a rational manner involving only a slight loss in accuracy, with a consequent reduction in costs and execution time. In trilateration networks, for example, starting with the maximum number of measurements possible, we eliminate, on the basis of second-order design, the least weighted sides one after another, even if they are characterised by high reliability. This method, which may be easily programmed for automatic processing, gives good results and can be stopped when the points error becomes greater than a pre-fixed value and/or the minimum reliability is less than an acceptable value (Fig. 9).

Second-order design can also be useful for solving some of the problems related to the third order, such as adding measurements of different kinds.

As an example of this, see trilateration network E of Fig. 8 in which the error ellipses are very flat, and may therefore be improved by the addition of angular measurements, in this example the angles around points 4,5 and 6 improve the isotrophy of the net. They were found by applying an empirical trial and error procedure, taking into account only those angles with maximum weighting.

When a network has different types of observation, (angles, distances, Laplace azimuth...), correct weighting of the observations presents a problem. This is a very difficult matter and at present we think that the best method is an empirical approach based on experience, also if we look with interest to the method proposed by Prof. Kubik [25], for an a posteriori estimate of the various measurements' weighting. This method has been tested in my institute with good results [26].

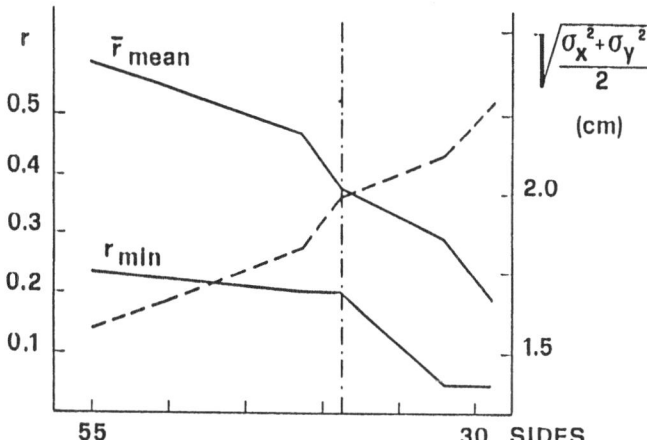

Fig. 9   Effect of the reduction of measured sides
on reliability and precision (net B, Fig. 8)

We shall describe at length the measurements themselves because they are so strictly related to the instruments used. For instance, atmosphere related problems can be overcome using a two-wavelength instrument such as the terrameter, while they represent the most limiting factor if a one wavelength model is used over long distances. Although much research has been devoted to this problem [27], we think that any atmospheric model used to correct measurements is likely to be far from perfect.

We have obtained good results using the method proposed by Robertson, based on the relation between two lengths measured from the same point in two directions with a similar topographical profile in the same period of time [28,29]. For short or medium sides one tends towards limitation of the instrument's base error by using a graduated slide for the reflectors [30,31].

As regards the adjustment and analysis of data we have tended towards procedural rationalisation using a series of well organised programmes:

1) Handling and organisation of the data.

2) Adjustment by indirect observation. For the free net we use the bordering technique or the expression $N^+ = N (NN)^{-1}$ or the singular value decomposition $N = USV^T$, $N^+ = V^T SU$. For the inversion of the regular matrices we use the modified Cholesky method.

3) The detection of the outlayers can be performed using data snooping and considering the weighting matrix to be a diagonal using the variable

$$W = \frac{v_i}{\sigma v_i}$$

if $|W_i| > F^{\frac{1}{2}}_{1-\alpha\,;1,\,\infty}$ (normally 3.29) the observation is eliminated and the adjustment is repeated. It is possible to have an interactive version of this procedure. One can also use an alternative approach based on the minimisation of the sum of the absolute value of residuals [32], which in some situations seems to be more efficient [33].

4) The analysis of the network is carried out by considering the absolute and relative error ellipses, distance error, and that of the bearings between pairs of points. The ellipses can be plotted automatically.

5) Other programs for similarity transformation, for testing the significance of supposed movements, for studying the local strain field etc. are needed.

From amongst the various networks installed in Italy to study crustal movement, let us show you as an example the one we set up in the Cassino Area (south of Rome) in 1984.

Starting form the geophysical need to connect three mountainous blocks separated by two supposed fault zones, the work sequence has been as follows:

1) to search, using suitable cartography, for those intervisible points which might be used to form a very compact network (within a circle),

2) to mount a field campaign in order to verify the supposed possible connections by means of powerful lamps; to verify the possible routes and transportation needs (special cars, mules, helicopters..); to obtain permission from the owners of the land; to study the geological and lithological features of those places selected in order to choose the foundations of the bench-marks (reinforced concrete pillars); and so on.

3) After that we planned various possible networks on the basis of optimisation criteria, taking into account all the information collected during on-the-spot investigations. In this manner we chose the network in Fig. 10 consisting of three points on every three blocks.

4) Another field campaign was necessary to construct the pillars, built upon rock foundations, and with a self-centring device on the top.
On a geologically stable block we installed two pillars in order to have a base line for the calibration of the instruments before, during and after every measurement campaign.

5) During the first measurements campaign we experienced difficulty in measuring some of the long sides, and thus decided to add another three points in order to eliminate such long connections. The final network is that shown in Fig. 11 in which the error ellipses are laid out. As you can see the network is characterised by very good homogeneity and isotrophy, which is most important since the direction of the expected movements is not known.

3.2 Altimetric networks

A series of high-precision levelling campaigns is the most fruitful method of investigating vertical crustal movements; using this technique it is possible to examine in detail and with great accuracy displacements extending over even the largest areas. Vertical control networks should be planned bearing in mind a considerable number of parameters, such as the area affected by deformation, the characteristics of the displacement and its supposed magnitude and velocity, the type and location of the bench marks and their density, the possible level routes and the degree of accuracy expected, also taking into account the high cost of such measurements. It is not possible here to give a detailed examination of all the problems associated with precise levelling, such as instrumentation, operation methods, refraction, magnetism, rod calibration, bench-mark stability and so on. We will only emphasise the fact that interest in this technique has increased to such an extent that entire conferences are dedicated to it, many interesting suggestions and ideas are given in their proceedings [34,35,36]. As far as accuracy is concerned, we have no problems in the detection of coseismic deformations, large scale subsidence or bradeisism. However, we

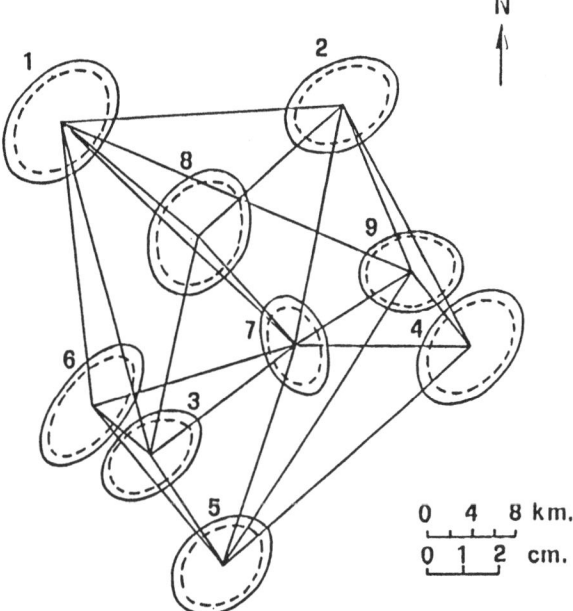

Fig. 10 Scheme of the trilateration network of Cassino and
comparison between the error ellipses obtained in the first
campain and the expected ones (dashed lines)

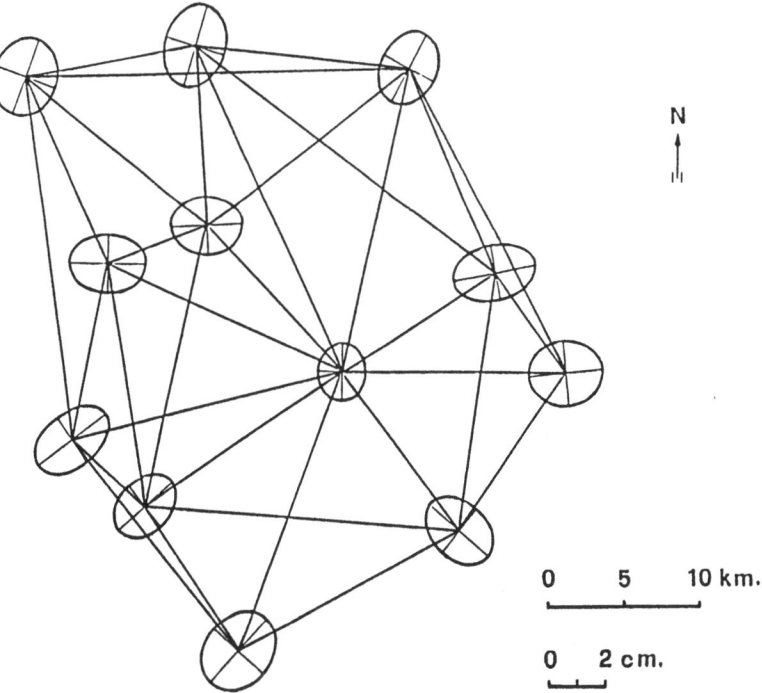

Fig. 11 Scheme of a modified network in Cassino

encounter some difficulties in other cases, such as pre-seismic or post-seismic deformations, or with a slow subsidence or uplift where the small degree of movement concerns a large area.

Amongst the various sources of error we shall consider the following two in particular: refraction, and movement occuring during the measurement phase.

Regarding refraction, the greatest limiting factor for lines in hilly areas, there are essentially two ways in which it may be taken into account: the first is based on the well-known Kukkumaki's formula

$$R = -10^{-6} \ A \ (\frac{L}{50}) \ \delta h \ \delta t$$

L = length of sight in metres
δh = difference in level in metres
δt = difference in temperature in degrees at two hights, normally 0,5 m and 2.5 m.

$$A = \frac{1190}{z_2^c - z_1^c} \ |\frac{1}{c+1} (z_1^{c+1} - z_2^{c+1}) - z_0 (z_1 - z_2)|$$

where z_0 is the height of the instruments, z_1 and z_2 are the sight heights, and C is a coefficient the value of which is normally assumed to be equal to $-1/3$. The second way is based upon the creation of atmospheric models which provide a means of adjusting the measurements using one temperature and general data regarding insolation, wind, etc.

Different techniques can be applied to check on any possible movements taking place during the measuring campaign, or to obtain a periodic indication of whether or not it is necessary to repeat the measurements over the entire network. One of these is based on the continous recording of tilt components by tiltmeter; another method we have used extensively is based on the use of double levels at night [37]. This technique makes it possible to ascertain with accuracy, and within a few hours, the difference in height between points at almost the same height, but located at a considerable distance (3-6 km) from each another.

By way of example we have set out the closing error of a control triangle used in various campaigns.

$\triangle$	1977	1978	1980
11 km	2,9	-0,9	1,4

It must be remembered that the length of the precise levelling lines necessary to join the three points is over 20 km, with routes characterised by average differences in height of 50 m/km.

Having an idea of the movements occuring during the measurement period, it is possible for one to take them into account during the data processing phase [38]. Neverthless, we think that the best method is to limit the entire network's measurement period to as short a time as possible. This can be achieved by contracting small areas of the network to a variety of companies working simultaneously.

The need to employ several firms is typical of levelling networks, and this fact should be taken into account when preparing the specifications for the contracting firms concerning instrumentation, rod calibration, observation procedures, refraction, tolerances for a single section, for a line, for a loop etc. [39].

In addition we must emphasise that when all the blocks, already analysed separately, are put together, other errors arise; it by no means being a simple matter to complete a serious data analysis or the adjustment of a whole network consisting of hundreds or thousands of measured sections.

As an example of this kind of work we present in Table I a flow diagram of the program prepared in my institute by Prof. Barbarella.

Table I
Flow diagram for network adjustment

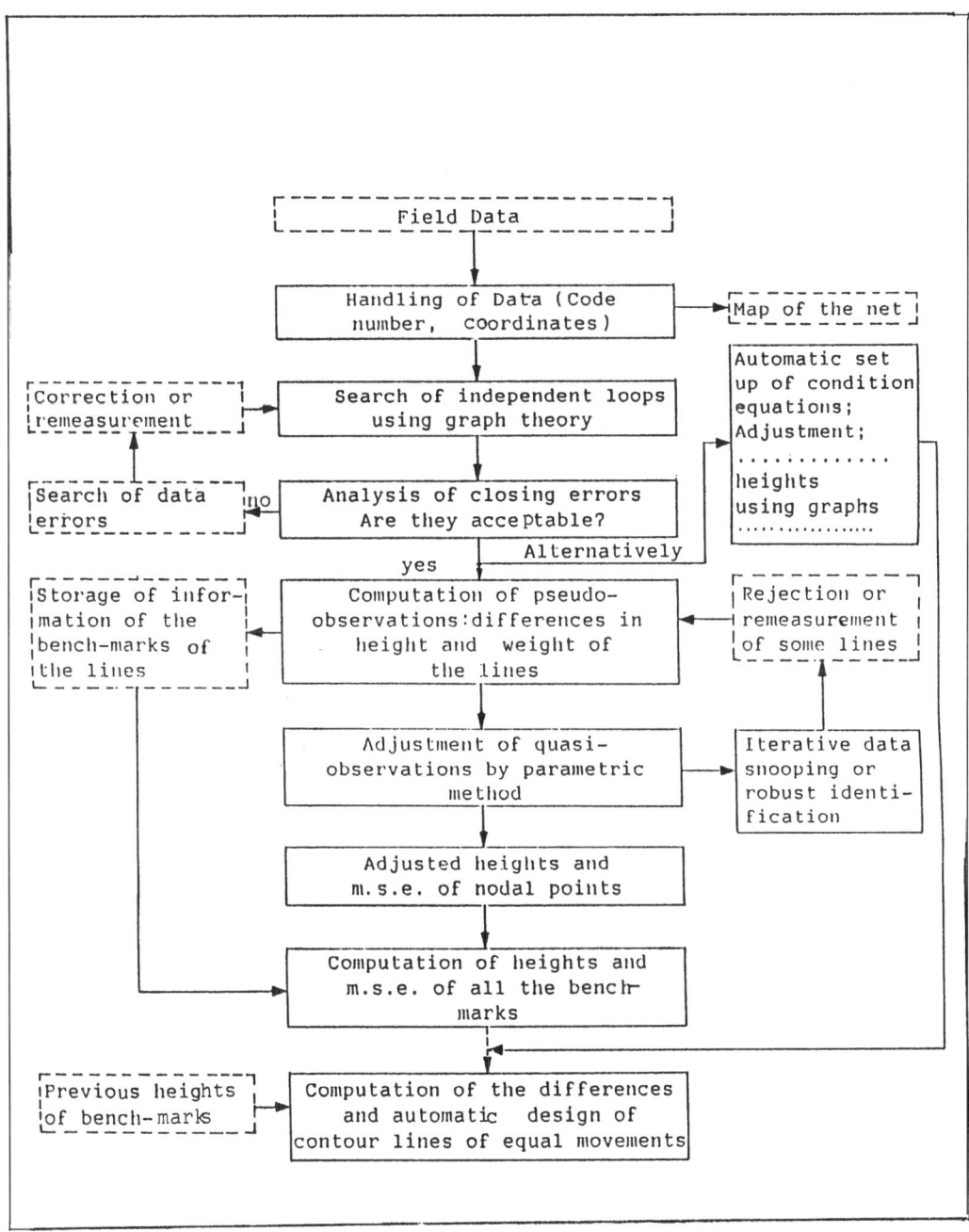

One can also apply other statistical tests for the study of error behaviour in order to obtain information about trends due to systematic errors. The tests are based on the random variable

$$x_{ij} = \frac{D_{ij}}{R_{ij}}$$

in which   is the discrepancy between the direct and reverse measurements of the relative height of two consecutive bench-marks, and R is the distance between them. The method is based on tests concerning each line, and cumulative tests regarding the entire network, preferably using non-parametric tests because of their generality. Many tests can be used, such as, for example: randomness tests, normality tests, the Wilcoxon test, the Person P. Test, the Kurskal Walis Test, the Bertlett Test [40].

At the end of this short dscription of altimetric networks we must briefly make one or two comments concerning gravity. Gravity remeasurements may to some extent be used as a substitute for a levelling survey. The accuracy obtainable is less satisfactory, but costs are lower and the method is quicker.

Simultaneous monitoring of elevation and gravity change provides us with further elements for the interpretation of the phenomenon in question [41]. For example, generally speaking the gravity variation corresponding to a vertical displacement caused by an accumulating stress field in a seismic area, can be expressed as a sum of the free air and Bauguer effects, plus the gravity contribution made by the subsequent density change in relation to the volumetric strain [42].

With the use of modern microgravimeters, (e.g.: la Coste-Romberg mod.D), standard errors smaller than 5.10^{-8} m/sec^2 may be obtained [43]. Improvements in accuracy depend mainly on the elimination of effects caused by earth tides, tidal loading, atmospheric mass movements, and so on.

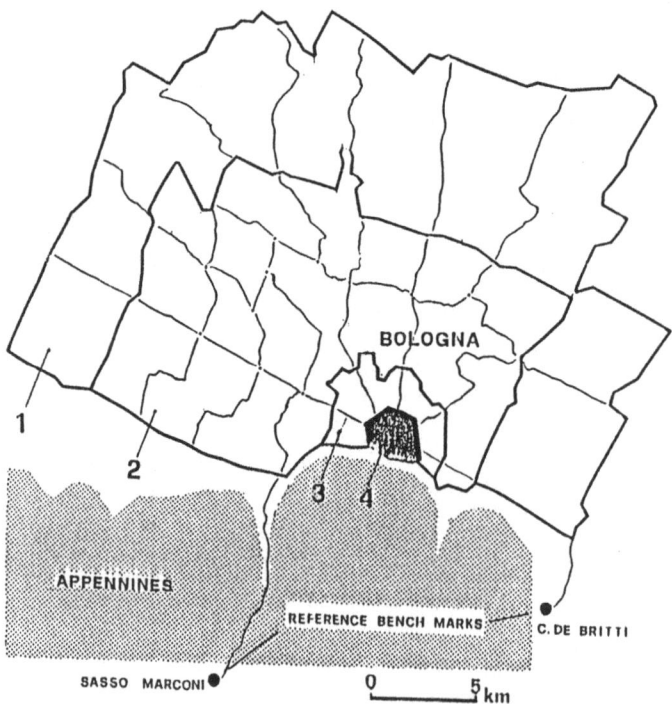

Fig.12 Scheme of the altimetric network in Bologna (Italy)

As an example of an altimetric net we shall show you the one set up in the Bologna area, (our home city), where over the last ten years we have had a differential subsidence velocity of at least about 11 cm/year, causing many problems regarding the stability of buildings, and water management. The network planned in my institute by Prof. Pieri and Dr. Russo was conceived in order to give different levels of information. In fact it consists of a large network, (460 km^2), in which the bench-marks are spaced from between 1 to 0.5 km along the lines, and the distance between the nodal points ranges from 7 to 3 km, the density increasing towards the town; (areas 1,2 and 3 of Fig. 12). Another network, (Fig. 13), covers the old town, with the bench-marks spaced at intervals of 0.25 km. along the lines, and with a distance between nodal points of about 0.5 km. Finally there are some local networks set up for the purpose of studying the movement of buildings or monuments, often in conjunction with other techniques, such as pendulums, tiltmeters and photogrammetry.

An example of these local networks and of the application of a variety of techniques is provided in the monitoring of two important monuments in Bologna: the two towers (Asinelli and Garisenda)[44] and St. James' Church [45] in Via Zamboni, where the differential subsidence is very strong (Fig. 14).

LOCAL CONTROL NETWORKS

● BENCH MARKS

0   500 m

Fig. 13 Scheme of the altimetric network in the town of Bologna

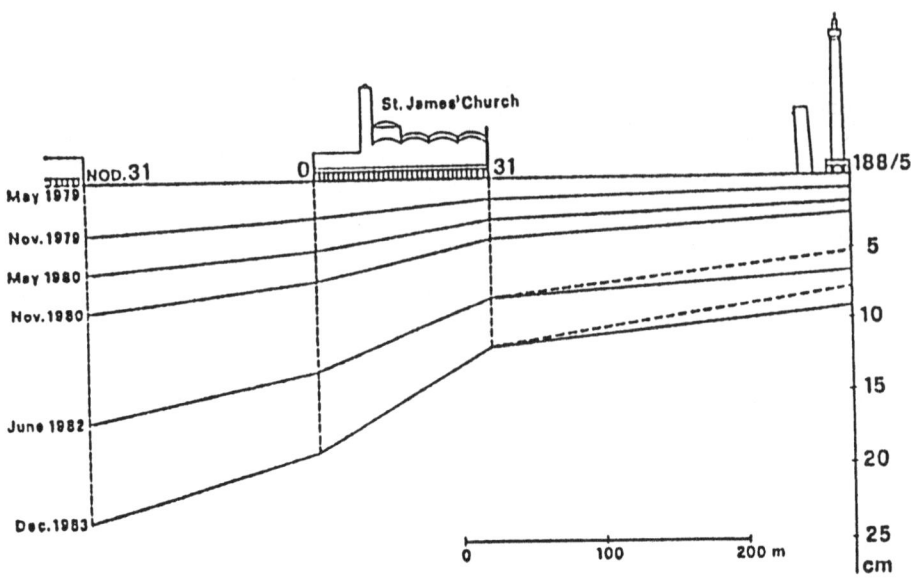

Fig. 14   Differential subsidence along a street of the old town (Bologna)

3.3 Plano-altimetric networks

     In order to give an idea of some of the problems related to plano-altimetric networks
we shall present an example of a network in which we have applied many different   techniques
in order to obtain the greatest possible information about movements in the seismic area of
Ancona, and to create a test network for the study of these various techniques. As you can
see in Fig. 15, it consists of a planimetric network of 10 stations, signalized by means of
reinforced concrete pillars using 30 optical connections which provide a very good
redundancy (1.76), furthermore using length measurements only performed by the AGA
geodimeter mod. 8. Seven of these stations are also connected by a levelling
network, consisting of over 80 km of levelling lines with more than 100 bench-marks, and
connected with the national first-order levelling network and with a tide gauge located at
Ancona. A gravimetric network was set up consisting of nine base stations situated on
levelling bench-marks. Another two stations were set up on a bell tower for the time
monitoring of the gravity gradient. The measurements are taken using a La Coste-Romberg mod.
D gravimeter.
     In addition to these base networks we have also set up two altimetric control triangles
using double levels (Zeiss Ni2).
     Astrogeodetic measurements have also been taken at five planimetric stations for the
direct measurement of vertical deflection.
     Consideration of all the measurements together in order to make a tridimensional
adjustment in a general or local framework presents many problems. However, this question
will be treated at length in a later lecture in these proceedings. Instead we should now
like to draw to your attention two problems that need to be solved when dealing with
trigonometric levelling. The problem is to improve trigonometric levelling so that this
technique may be more widely used. In some cases trigonometric levelling may be combined
with, or may even partially substitute, spirit levelling, with a considerable reduction in
costs, and with an increased rapidity of data acquisition. To accomplish this we have to
reduce the effects of refraction, and take into account the influence of vertical deflection
without astrogeodetic measurements.

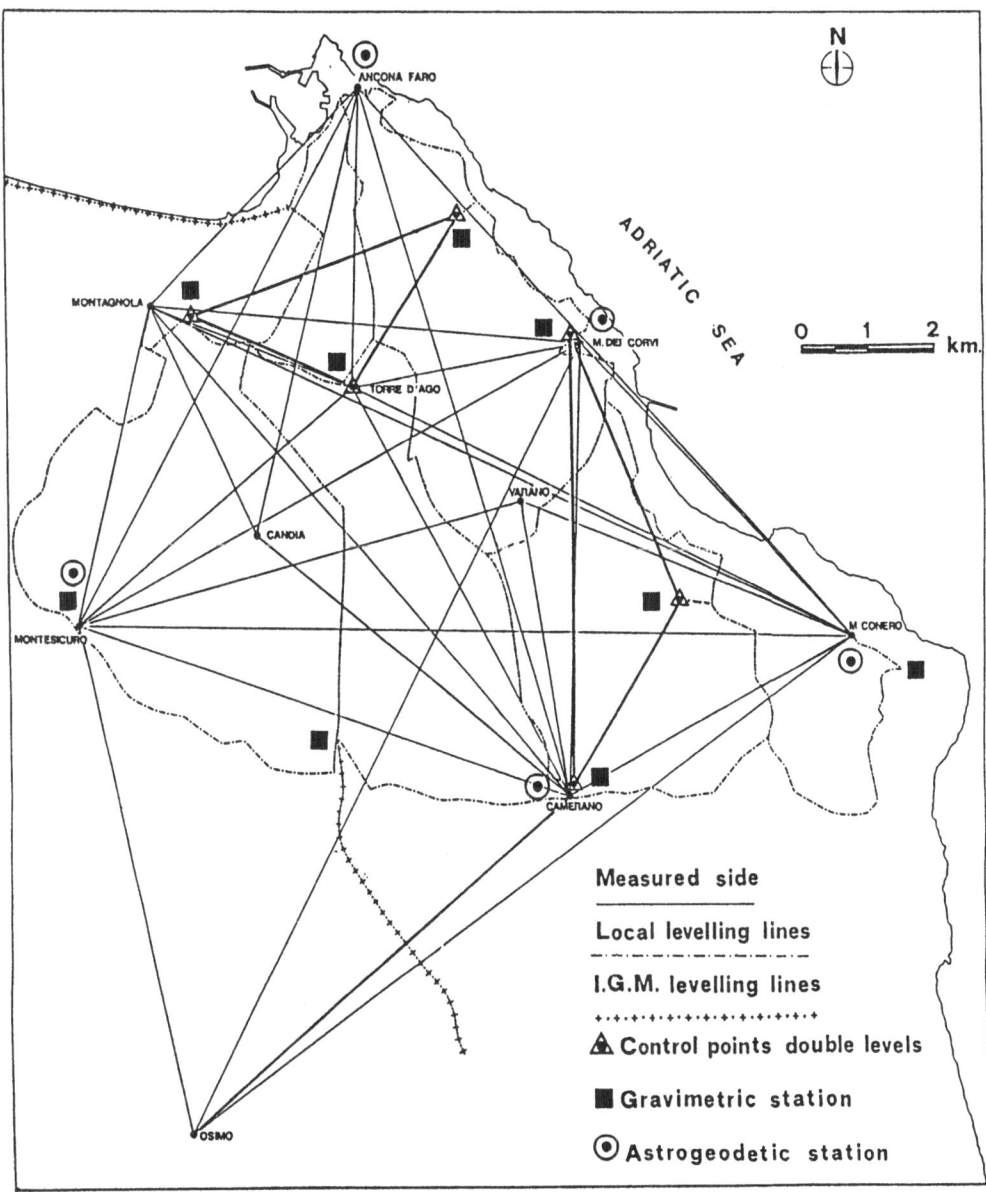

Fig. 15   Plano-altimetric network in the Ancona area

To carry out reciprocal and simultaneous zenithal angles we always make use of first-order theodolites, (Wild T3, DKM3), equipped with a small but powerful lamp for collimation at day or night. The lamp is mounted directly on the telescope and allows it to rotate around its horizontal axis. To link points that are at approximately the same height, (less than 1.5 m difference), we use, as we have said, double levels with special rods designed for collimation at night. Atmospheric data, (temperature, pressure, vapour pressure, wind speed), were collected during the measurement period by means of small meteorological stations.

As is well known, the accuracy of trigonometric levelling is severely limited by the effects of refraction. Essentially there are two possible ways of reducing these effects. The first is based on two wavelength angular measurements; and the second approach uses atmospheric models in order to find the variation of refractivity. Since the first method does not seem to be feasible for the near future [46,47], we have devoted our attention to the extensive use of the second.

The theory is well known. Starting with the formula

$$h_A - h_B = D (1 + \frac{h_A + h_B}{2R}) \tan \tfrac{1}{2} (Z_{AB} - Z_{BA} + \delta Z_{AB} - \delta Z_{BA})$$

following Prof. Moritz [48] we have

$$\delta Z_{AB} = \frac{10^{-6}}{S} \sin Z_{AB} \int_0^S \frac{dN}{dz} (S - x)\ dx$$

in which the refractivity gradient may be expressed:

$$\frac{dN}{dz} = -0,2696\ N_0\ \frac{P}{T^2} (0,0342 + \frac{dT}{dz}) + 11.25\ \frac{e}{T^2}\frac{dT}{dz} + \frac{11.25}{T}\frac{de}{dz}$$

where the temperature gradient is the most important factor.

As regards the evaluation of the temperature gradient, one can make use of the heat flux estimate, a parameter that can be deduced from the radiation budget estimate. This approach involves the consideration of a large number of additional meteorological and ground parameters, such as heat flux into and out of the ground, evaporation, wind spreed, cloudiness, sun height, terrain structure, vegetation and so on. We have applied this technique extensively following the suggestions of many authors well known for their interest in refraction problems, such as Angus-Leppan and Brunner [49-52], with good results.

Table II

Example of result of computation of refraction angles for recirpocal and simultaneous trigonometric leveling measurements (from Achilli, Baldi and Unguendoli, 1985) [54]

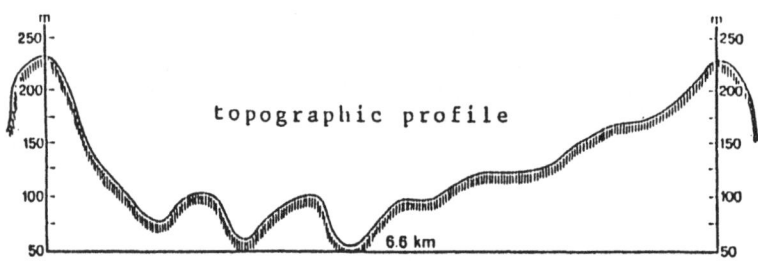

Equipment	Height differences (cm)	Time (h)	Ground Temperature (C)	a	b	Refraction correction (cm)	Corrected height difference (cm)	Mean of the observed values (cm)	Mean of the corrected values (cm)
Theodolite	822.9	13.0	30	−10.64	−8.61	− 1.19	819.7		
"	821.2	16.0	27	−10.64	−8.78	− 3.27	817.9		
"	812.0	20.0	25	−19.07	−22.03	4.65	816.7	816.0	819.0
"	810.9	2.0	20	−19.26	−25.84	10.33	821.2		
Double levels	813.2	21.0	13	−24.51	−28.43	6.16	819.4		

As an example of how the method may be applied, Table II sets out the results of five series of reciprocal and simultaneous measurements of zenithal angles relating to a link 6 km in length belonging to a triangle, taken in a variety of atmospheric conditions. The corrected values listed in Table II show that the deviations from the mean are much smaller than in the previous case. Furthermore, correcting the other two sides of the triangle in the same way, it should be pointed out that the mean value of the corrected determinations shows an appreciable improvement in the misclosure error.

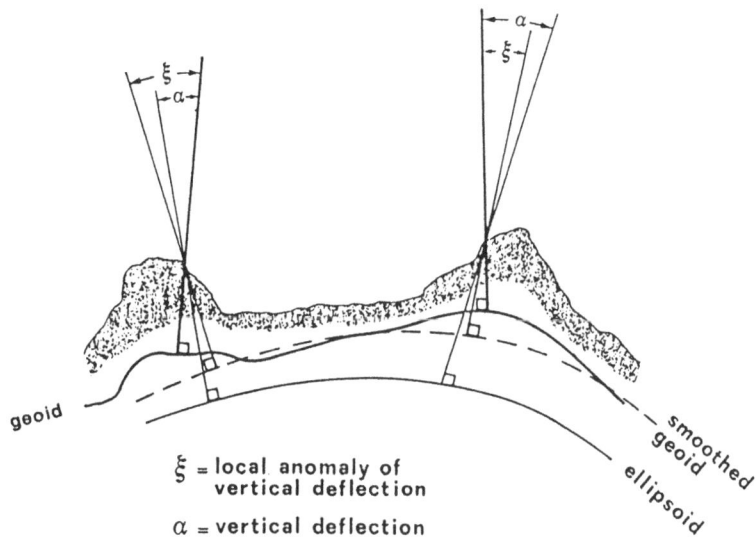

ξ = local anomaly of vertical deflection

α = vertical deflection

Fig. 16   Scheme of the local anomalies of vertical deflection (from Achilli, Baldi and Unguendoli, 1985)[54]

Strictly speaking, if we want to refer the trigonometric height to the same surface, (the ellipsoid), we have to know the vertical deflection, obtainable by astronomic measurement, which are a little troublesome. In control networks where the area involved is always small enough, the problem of the reference surface may be solved in a local way, taking into account only the local anomalies of vertical deflection, and obtaining a smoothed reference surface that we call a "smoothed geoid" (Fig. 16). These anomalies are mainly due to the distortion of the gravitational field caused by the distribution of the surrounding masses. For the computation of these anomalies we make the "mass model" with a program prepared by Achilli and Baldi [53], and consisting of an evaluation of the disturbing gravitational attraction exerted by the surrounding topography, and by density anomalies. The area is subdivided into a regular quadratic grid, with the side lengths constant or diminishing towards the point considered. The area is therefore represented by a set of parallelepipeds with a density determined from geological information, the gravitational components of which can be calculated by means of the formula:

$$T\zeta = \sum_{i=1}^{n} t_{xi}(P); \qquad T\eta = \sum_{i=1}^{n} t_{yi}(P)$$

in which:

$$t_x(\vec{P}) = \varrho G \int_{x_1}^{x_2} \int_{y_1}^{y_2} \int_{z_1}^{z_2} \frac{x}{|(x-a)^2+(y-b)^2+(z-c)^2|^{3/2}} \, dxdydz .$$

The components of the local anomalies in vertical deflection are thus:

$$\zeta = \arctan \frac{T}{g} \qquad ; \qquad \eta = \arctan \frac{T}{g} \; .$$

To verify the correction methods for refraction and local vertical deflection, we carried out, on a quadrilateral of the Ancona network, a series of trigonometric height measurements in different atmospheric conditions, paying particular attention during the measurement stage, and collecting atmospheric data. The sides range from 5 to 11.5 km., with a maximum difference in height of 400 m. The mean results of three series of measurements are listed in the table, and as you can see the closing error of the independent triangles improves markedly with the application of the corrections described above.

Table III

Differences in height obtained by means of trigonometric levelling and correction to take account of refraction and local anomalies in the deflection of the vertical (from Achilli, Baldi and Unguendoli, 1985)[54]

side	Δh(m) measured	correction (m)
1-2	386.89	-0.01
1-3	54.57	0.
1-4	115.79	0.01
2-3	-332.41	0.06
2-4	-271.18	0.09
3-4	61.19	0.03

MISCLOSURE ERRORS

triangle	measured (m)	corrected (m)
1-2-3	-0.09	-0.04
1-3-4	0.03	-0.01
1-2-4	-0.08	-0.01

As we have said, seven planimetric stations of the Ancona network belong to a precise levelling network, and their heights were obtained both by trigonometric and spirit levelling, so that one could make an interesting comparison between the two means of measurement. This kind of comparison is only possible if the two types of measurement refer to the same surface, which, given the small size of the area involved, can be identified with the smoothed geoid. We have used the mass model, which permits calculation of the local anomalies of the geoid (Fig. 17). By means of such geoidal undulation the heights obtained by spirit levelling were corrected, while the trigonometric measurements were corrected for refraction and for the local anomalies of vertical deflection. The results set out in the Table IV show a high level of congruence between the two height measurements, with a maximum difference of 2 cm.

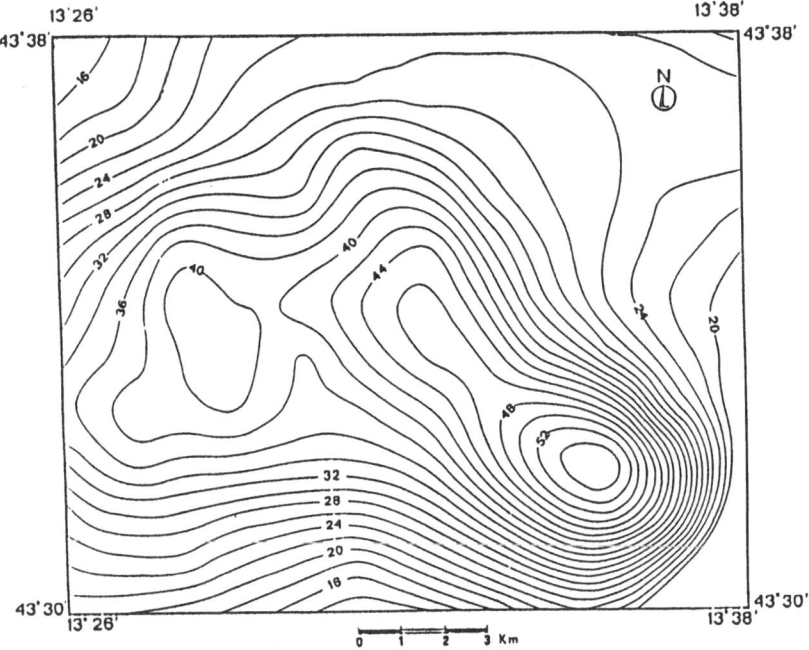

Fig. 17 Local anomalies of geoid in the Ancona area (mm)

<u>Table IV</u>
Comparison between height measurements obtained by means of trigonometric levelling and spirit levelling (from Achilli, Baldi and Unguendoli, 1985)[54]

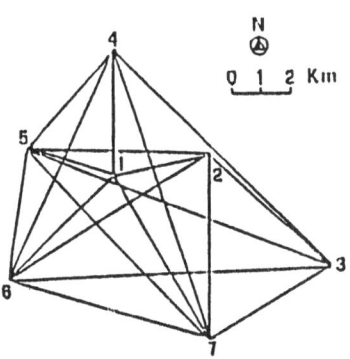

	TRIGONOMETRIC LEVELLING		SPIRIT LEVELLING
station	adjusted height (m)	corrected and adjusted height (m)	corrected (m)
1	195.83	195.83	195.83
2	238.23	238.19	238.18
3	582.77	582.74	582.75
4	121.57	121.55	121.55
5	251.14	251.12	251.13
6	311.61	311.65	311.67
7	250.40	250.41	250.39

4. CONCLUSIONS

As a conclusion to this short lecture on deformation measurement, we should like to consider the following:

Firstly, concerning methodology, the philosophy behind the mode in which we take our measurements is itself conceptually different from the classical determination of points location. In fact we measure large quantities to obtain small ones, and those problems related to precision, measurement conditions, instrumentation, and data analysis are more difficult and more delicate.

Secondly, considering the type of measurements to carry out, we think that the best way is to move towards an integrated application of temporally spaced measurements, (plano-altimetric networks, gravimetric networks, astrogeodetic measurements...) and continous recording, (tiltmeters, strain meters, hydrostatic levelling), in order to acquire a better knowledge of the phenomenon studied.

Finally, we think that in this field firm cooperation between different disciplines is absolutely essential. Geodesists, Geophysicists and Geologists have to cooperate strictly in the various steps of the work; i.e. in the planning and execution of measurements, and in the analysis and interpretation of data.

* * *

REFERENCES

1) Goad, C.C., Remondi, B.W. , Initial relative positioning results using the global positioning system, Bull. Geod. 58,(1984).

2) Bock, Y., Abbot R.I., Counselman,C.C., Gouzevitch, S.A., King, R.W., Paradis, A.R., Geodetic accuracy of the Macrometer Model V-100, Bull. Geod., 58,(1984).

3) Stansell, T.A., Chamberlain, S.M., Brunner, F.K, The first Wild-Magnavox GPS satellite surveying equipment: WM-101, WM Satellite Survey Company, (1985).

4) Thatcher, W., Crustal deformation studies and earthquake prediction research, Earthquake Prediction, American Geophysical Union, Washington,(1981).

5) Reid, H.F, Permanent displacements of the ground, The California Earthquake of April, 18, 1906, Report of the State Earthquake Investigation Commission, Washington, 2, (1910).

6) Thatcher, W., Matsuda, T., Kato, T., Rundle, J.B, Lithospheric loading by the 1896 Riku-u earthquake, northern Japan: implication for plate flexure and asthenospheric theology, J. Geophys.Res., 85, (1980).

7) Chimnery M.A. The deformation of the ground around surface faults, Bull. Seism.Soc. Am., 51, (1961).

8) Maruyama T, Statistical elastic dislocation in an infinite and semi-infinite medium, Bull. Earthquake Res.Inst., Tokyo Univ., 42, (1964).

9) Press F, Displacement, strains and tilts at teleseismic distances, J. Geophys.Res., 70, (1965).

10) Mansinha L. and Smylie D.E, The displacement fields of inclined faults, Bull. Seism. Soc. Am., 61, (1971).

11') Stewart S.W. and O'Neil M.E, Seismic travel and near surface crustal velocity structure bounding the San Andreas Fault zone near Parkfield, California, U.S., Geol.Surv.Prof.Pap., 800-C, (1972).

12) Brace W.F, Some new measurements of linear compressibility of rocks, J. Geophys. Res., 70, (1965).

13) Mahrer K.D. and Nur A, Strike slip faulting in a downward varying crust, J. Geophys.Res., 84, (1979).

14) McTingue D.F., Stein R.S, Topographic amplification of tectonic displacement: implication for Geodetic measurements of strain changes, J. Geophys.res., 89, (1984).

15) Berrino G., Corrado G., Luongo G. and Toro B., Crustal deformations and gravity changes accompanying the 1982 Pozzuoli uplift, Bulletin Volcanologique.

16) Polland J.F. and Davis G.H, Land subsidence due to withdrawal of fluids, Rev. Eng.Geol.2, (1969).

17) Caputo M., Pieri L. and Unguendoli M., Geometric investigation of the subsidence in the Po delta, Boll. Geofisica Teorica ed Applicata, 47, (1970).

18) Owens J.C., Optical refractive index of air: dependence on pressure, temperature and composition, Appl.Opt., 6, (1967).

19) Baldi P. and Unguendoli M., Some problems in monitoring crustal deformation by geodetic methods, Quaterniones Geodesiae, 3, (1982).

20) Slater L.E. and Hugget G.R., A multiwavelength distance-measuring instrument for geophysical experiments, J. Geophys.Res., 35, (1976).

21) Gervaise J., Resultats de mesures géodésique avec la terrameter, appareil électronique de mesure de distances a deux longueurs d'onde, Mensuration, Photogrammétrie, Génie Rural, 6, (1984).

22) Schmitt G., Optimisation of geodetic networks, Review of Geophysics and Space Physics, 4 (1982).

23) Grafarend E., Genavigkeitsmase geodatischer netze, Deutsche Geodatische Kommission, Reihe A., Heft 73, Munchen, (1972).

24) Crosilla F., A criterion matrix for the second order design of control networks, Proceedings of the Meeting on Survey Control Networks, Aalborg, ASPW, Munchen, (1982).

25) Kubik K., The estimation of weights of measured quantities within the method of least squares, Bulletin Gédésique, 95, (1970).

26) Barbarella M. and Pieri L., I pesi nella compensazione di reti topografiche, Boll. Geod. Sc.Aff.,3,(1983).

27) Baldi P. and Achilli V., Atmospheric models for geodetic measurements, Proceeding "High precision geodetic measurements", University of Bologna (M.Unguendoli Editor), (1984).

28) Robertson K., The use of line pairs in trilateration and traverse, Survey Rev., 165, (1972).

29) Robertson K., The use of atmospheric models with trilateration, Survey Rev., 186, (1977).

30) Allan A.L., The accurate measurements of a short line by geodimeter model G., Survey Rev., 145, (1967).

31) Mc Lean R.E., Accurate measurement by geodimeter: a practical application of the differential technique, Survey Rev., 153, (1969).

32) Fuchs H., Contributions to the adjustement by minimizing the sum of absolute residuals. Manuscripta Geodetica, 7, (1982).

33) Barbarella M. and Mussio L., A strategy for a robust identification of outliers in geodetic sciences, Statistics and decision, Supplement Issue 2. R. Oldenbourg Verlag; (1985).

34) Proceedings NAD Symposium 1980. Canadian Institut of Surveying, Ottawa.

35) Proceedings "NAVD Symposium 85". U.S. Department of Commerce. NOAA, Rockville.

36) Proceedings "Workshop on precise levelling", Dumnler Verlag, Bonn (H.Pelzer and W. Niemeir Editors), (1983).

37) De Sanctis R.G., Folloni G. and Gubellini A., Alcune esperienze di livellazione reciproca con livelli accoppiati, Bollettino Sifet, 2, (1973).

38) Pieri L. and Russo P., Considerazioni sulle livellazioni geometriche per lo studio dei movimenti verticali del suolo, Boll. Geod. Sc. Aff., 3, (1980).

39) Russo P., Some notes on specifications and recommendations for high precision levelling in Italy, Proceedings "High precision Geodetic measurements". University of Bologna (M.Unguendoli editor), (1984).

40) Chiarini A. and Pieri L., Statistical analysis of discrepancies in high precision levelling, Bulletin Géodésique, 99, (1971).

41) Baldi P. and Postpischl D., Gravity variations during pre-seismic deformations, Il nuovo cimento, 10, (1981).

42) Walsh J.B. and Rice J.R., Local change in gravity resulting from deformation, J. Geophys.Res., 84, (1979).

43) Baldi P. and Marson I., Gravity and geodetic networks for the study of crustal deformations in seismic area, Boll. Geod.Sc.Aff., 3 (1981).

44) Gubellini A., L'inclinazione della torre Asinelli: recenti rilievi e tendenza attuale, Technica Paper Istituto di Topografia, Geodesia e Geofisica Mineraria, Bologna, (1984).

45) Gubellini A., Lombardini G. and Russo P., Application of high precision levelling and photogrammetry to the detection of the movements of an architectonic complex produced by subsidence in the town of Bologna, Proceedings "Third International Symposium on land subsidence", Venice 1984 - Vol. 151 I.A.H.S. (in press),

46) Prilepin M.T.and Medovikov A.S., Effects of atmospheric turbulence on geodetic interference measurements: methods of its reduction, Geodetic Refraction. Springer-Verlag, Berlin, (1984).

47) Williams D.C.and Kahmen H. Two wavelength angular refraction measurements, Geodetic refraction, Springer-Verlag, Berlin, (1984).

48) Moritz M., Application of the conformal theory of refraction, Osterreichische Zeitschrift fur Vermessungswesen, Sonderband 25, (1967).

49) Angus Leppan P.V. and Webb E.K., Turbulent heat transfer and atmospheric refraction , XV General Assembly IUGG Moscow, (1971).

50) Angus Leppan P.V. and Brunner F.K., Atmospheric temperature models for short range EDM. The Canadian Surveyor, 34, (1980).

51) Brunner F.K., Vertical refraction angle derived from variance of the angle-of-arrival fluctuation, Proc. Intern. Astronomical Union, Symposium n.89, Uppsala , (1978).

52) Brunner F.K., Modelling of Atmospheric effects on terrestrial geodetic measurements, Geodetic Refraction, Springer-Verlag, Berlin, (1984).

53) Achilli V. and Baldi P., Computation of local anomalies of the vertical's deflection in Geodetic networks, Survey Rev., 205, (1982).

54) Achilli V., Baldi P.and Unguendoli M., Refraction and deflection of the vertical in trigonometric leveling. Proceedings NAVD Symposium 85, U.S. Department of Commerce, Rockville, (1985).

GYROSCOPE TECHNOLOGY, STATUS AND TRENDS

Wilhelm F. Caspary

Institut für Geodäsie, Universität der Bundeswehr München

ABSTRACT

After a sketch of history of gyroscope technology, the dynamic law
of a gyroscope and its specialization to the suspended gyrocompass
are outlined. The equations of motion of a north-seeking gyro are
reviewed forming the base for observation techniques. Following a
state-of-the-art report, the relations between the observables and
the final azimuth are fully discussed and actions and precautions
required for precise results are proposed. With modern instruments and observation methods it is presently possible to determine an azimuth with a standard deviation of 3" to 6" within less
than one hour. The principles of non-rotating optical gyroscopes
are presented and their potential as alternatives to conventional
instruments is discussed.

1.   INTRODUCTION

Gyroscopes as instruments for measuring azimuths have always had their place on the
fringe of geodesy. In the early days of this technology it really was a venture to carry
out observations, since the equipment was voluminous and heavy, the procedure time consuming and the accuracy of the results unsatisfactory. In addition to that, the average surveyor did not understand the way the gyroscope works. Not that the theory is difficult,
but it is beyond every day experience. By reason of this situation gyro technology acquired the reputation of being an exotic subject hardly of use for surveying applications.
This still exists although the scene has changed completely.

Modern mechanical gyros are handy and provide the azimuth with an accuracy of a few
seconds of arc in less than one hour. The dynamics of gyroscopes has been treated at
length in numerous papers, admittedly sometimes more confusing than enlightening, but the
interested surveyor can easily find suitable presentations of the theory. In spite of
these significant advances the acceptance has remained on a low level.

Presently optical gyroscopes have attracted widespread attention as an alternative to
mechanical gyros in navigation platforms. Their performance has already reached a level
which makes them rate as potential candidates for a completely new generation of north-
seeking instruments.

2. HISTORICAL BACKGROUND

The dynamic behaviour of a spinning wheel had been known for a long time, when the French physicist L. FOUCAULT made his famous experiment in 1852. He suspended a gimballed rotor from an earth-fixed support using a thread, such that the axis of the rotor was forced to remain in a horizontal plane. The objective was to prove that the earth rotates and that therefore, caused by the torque due to the precession in the earth's gravitational field, the spin axis would turn towards true north. Unfortunately the experiment failed because it was impossible at that time to maintain the required angular velocity (LAUF 1963[1])). Nevertheless, this was the birth of a north-seeking instrument for which FOUCAULT coined the name gyroscope and which has become increasingly important for navigation and geodesy.

FOUCAULT's experiment turned the attention of many famous scientists of the last century to the theory of the gyroscope. Authors like CAYLEY, EULER, KELVIN, KLEIN, SOMMERFELD and GRAY made valuable contributions to the dynamics of this instrument. But it took until 1908 before working gyrocompasses were constructed by ANSCHÜTZ-KÄMPFE in Germany and SPERRY in the USA. These single-gyro compasses were designed for the navigation of vessels, but it turned out that they were too sensitive to the ship's rolling. Further research and experiments led to two- and three-gyro compasses showing much better performance. The final breakthrough was achieved by ANSCHÜTZ in 1922 with a floated spherical gyroscope meeting all needs for use at sea.

Paralleling this development first studies and experiments were conducted aiming at the use of gyroscopes for bore-hole and tunnel measurements. These efforts were supported and advocated by the mining surveying authorities in Germany. The first experimental instrument was constructed by SCHULER in 1921 and tested in coal mines. Already in 1926 an improved gyroscope was built conjointly by ANSCHÜTZ and BREITHAUPT and in 1936/37 ANSCHÜTZ presented the third generation. All these instruments remained on the prototype level and were not adopted by the surveying community, because the accuracy was unsatisfactory, the weight was too heavy and the observational procedures were too complicated.

In 1948 a new start was made at the Mining University Clausthal by RELLENSMANN. The following period is noted by a steady progress and by noteworthy contributions of different companies, having led to the current arsenal of gyroscopic instruments for azimuth determination. There are basically two types of instruments: the gyroscope attachments and the gyrotheodolites. In Western Europe mainly three companies offer gyrocompasses for precision observations on the civil market which are suitable for geodetic applications:

Institute of Mining Surveying (WBK), Bochum, FRG

Hungarian Optical Company (MOM), Budapest

Wild, Heerbrugg, Switzerland

The instruments which are best documented and probably of most widespread usage are the attachment GAK 1 of Wild, the gyrotheodolites Gi-B1 and Gi-B11 of MOM and the MW 50 (MW 50 a) and Gyromat of WBK.

THE FREE GYROSCOPE

The free gyroscope consists of a fast spinning rotor which is mounted in a set of
mbals such that it is decoupled from all rotations of the support. Under the assump-
on that the gimbal bearings are free from friction and that the instrument is balanced
d strictly symmetrical so that the three axes coincide with the principle axes of iner-
a and intersect in the centrepoint of the rotor, the spin axis will indefinitely main-
in its orientation in inertial space. To an observer on the earth the spin axis appar-
tly moves on a cone with an axis parallel to the rotation axis of the earth. Figure 1
ows a gimbal mounted elementary (free) gyro.

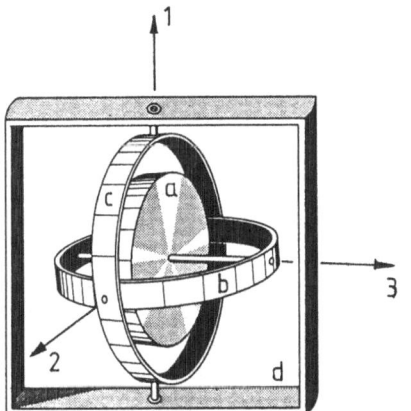

1	Outer Gimbal Axis
2	Inner Gimbal Axis
3	Spin Axis
a	Rotor
b	Inner Gimbal (Float)
c	Outer Gimbal
d	Frame (Base)

Figure 1   Schematic of a gimbal mounted free gyroscope
(FABECK 1980)

If a torque is applied in the direction of the 2-axis then the angular momentum of
the rotor changes its direction and the 3-axis begins a motion about the 1-axis which is
known as precession of the gyro. The free gyro tries to align the spin axis with the direction of the applied torque.

In the English literature this type of gyroscope is usually denoted as a two degrees of
freedom (TDF) gyro. It represents the basic concept of the gyros used to control the attitude of inertial platforms.

The suppression of one degree of freedom of the rotor relative to the frame yields
the single degree of freedom (SDF) gyro. A special version of this type is the north-
seeking gyroscope or gyrocompass. The gyroscope of Figure 1 could be converted into a

gyrocompass if the 1-axis is kept in the direction of the local vertical and the 3-axis in the horizontal plane, i.e. the rotational degree of freedom about the 2-axis is suppressed.

4.   THE SUSPENDED GYROCOMPASS

If a gyroscope is suspended from an earth fixed support such that the spin axis is constrained to remain in a horizontal plane (Figure 2), then a torque caused by the force of gravity acts upon the instrument. The force changes its direction relative to an inertial coordinate system with the angular velocity of the rotating earth. The rotor reacts with an angular motion (precession) towards the direction of the earth rate vector. The inertia of the system will prevent the spin axis from stopping when it reaches the meridian. Instead it will move further until the directional force exceeds the inertia and causes the system to swing back. Thus an oscillation

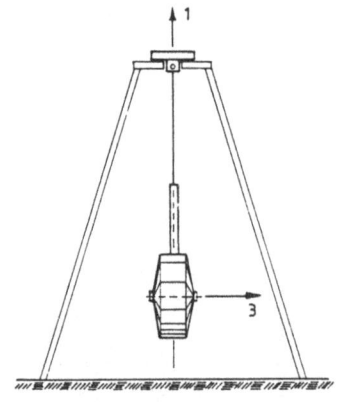

Figure 2   Suspended meridian-seeking gyroscope

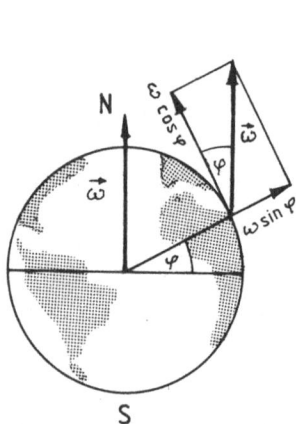

Figure 3   Decomposition of the earth rate vector in a local vertical and horizontal component as functions of the latitude

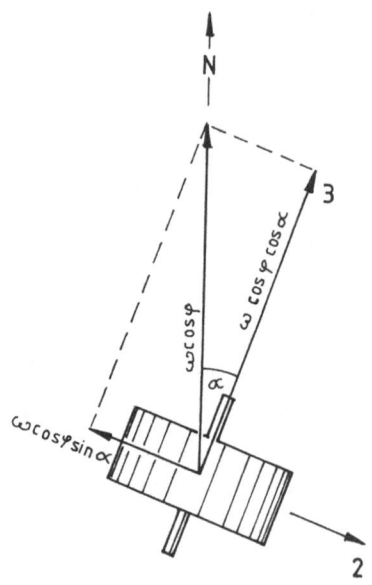

Figure 4   The horizontal component of Figure 3 resolved in directions of the 3- and 2-axis of the gyro

is built up, which would last indefinitely if the system was free of friction. Since the motion is constrained to the horizontal plane, only the component $\omega \cos \varphi$ of Figure 3 is effective, where ω is the earth rate and φ the latitude. Figure 4 shows the view from space at the horizontal plane. The horizontal component of the earth rate is decomposed in two components, the one pointing in the direction of the 3-axis (spin axis) is $\omega \cos\varphi \cos\alpha$ and the other one pointing opposite to the direction of the 2-axis is $\omega \cos\varphi \sin\alpha$ with α being the momentary angle between spin axis and local meridian. The exact equation of the precession of the gyroscope is rather involved. It expresses the resultant effect of the inertia of the rotor, the rotating gravity vector and the torsion of the suspension band. Since it is well documented in several text books e.g. DEIMEL (1950)[2]

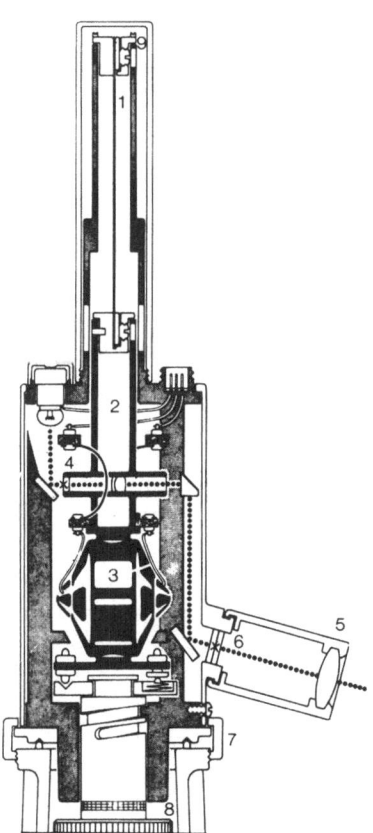

1 Suspension Tape (band)

2 Mast

3 Gyro Motor (rotor)

4 Illuminated Moving Mark

5 Collimator

6 Reading Scale

7 Forced Centering to the Theodolite

8 Clamping Ring

9 Upper Tape Clamp

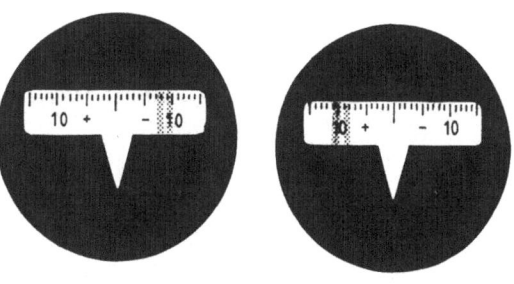

Observation window showing two positions of the moving mark

Figure 5   Cross-section of a gyro attachment (Wild GAK 1)

The moving mark (4) oscillates with the rotor (3) about the vertical axis. The centre line of the oscillation is the direction of the meridian except for some corrections explained later. The observation aims at the determination of this centre line, which is achieved by taking time and/or position information of the moving mark with respect to reading scale.

and FABECK (1980)[3] in this paper no derivation is given. The discussion of the equation in the next section follows basically the approach selected by VANICK (1972)[4].

For a better understanding at first, an actual gyroscope attachment is presented in Figure 5, showing some constructive details and introducing the relevant terms.

5. EQUATIONS OF MOTION

The motion of a spinning suspended gyroscope can be described by a pair of coupled differential equations. The 3-axis (spin axis) follows the curve of an elliptic spiral of the kind depicted in Figure 6. In reality the proportion $\beta : \alpha$ is much less and the damping much smaller, so that the position of equilibrium is practically never attained. If the β-component of the motion and other small effects are neglected and if the α-component is restricted to small angles by a good pre-orientation then the equation of motion has the simplified form

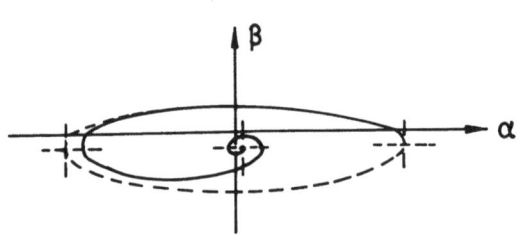

Figure 6   Plot of the motion of the spin axis, α horizontal, β vertical

$$(J + \frac{N^2}{G}) \ddot{\alpha} + k \dot{\alpha} + (F + B) \alpha - B \alpha_t = 0 \tag{1}$$

with

J (g cm^2)	moment of inertia toward 1-axis
N (g cm^{2}s^{-1})	gyroscopic moment torque of the rotor
G (g cm^{2}s^{-2})	moment of gravity
k (g cm^{2}s^{-1})	damping constant
F (g cm^{2}s^{-2})	directive moment torque of the earth's rotation
B (g cm^{2}s^{-2})	directive moment torque of the suspension
α_t	zero-torque direction of the suspension tape

Equation 1 can be decomposed yielding

$$J \ddot{\alpha} + k \dot{\alpha} + B (\alpha - \alpha_t) = 0 \tag{1a}$$

$$\frac{N^2}{G} \ddot{\alpha} + F \alpha = 0 \tag{1b}$$

where Eq. 1a describes the torsional pendular motion of the non-spinning gyro about the 1-axis and Eq. 1b expresses the motion of the torsion-free suspended spinning gyroscope.

The solution of Eq. 1 results in the expression for the horizontal pendular motion of the real spinning gyroscope

$$\alpha = \alpha_0 + C e^{-\lambda t} \cos(\omega_p t + c) \tag{2}$$

where C and c are integration constants depending on the initial conditions. The equilibrium position deviates from true north due to the twist of the suspension tape by

$$\alpha_0 = \frac{B}{B + F} \alpha_t . \tag{3}$$

The damping factor λ is given by

$$\lambda = k/2 (J + \frac{N^2}{G}) \tag{4}$$

and the frequency ω of the swing is

$$\omega_p^2 = \frac{B + F}{J + \frac{N^2}{G}} - \lambda^2 . \tag{5}$$

These equations describe the horizontal component of the oscillation of the suspended gyroscope with an accuracy better than required for the determination of azimuths, but they have to be modified for practical use, since α , α_0 and α_t refer to true north, while the observations are carried out with respect to zero of the reading scale of the gyro or of the theodolite.

6.   OBSERVATION METHODS

There are six methods of finding north using a suspended gyroscope (THOMAS 1982)[5]. The oscillation curve of the moving mark as observed through the collimator is depicted in Figure 7 as a function of time.

The equation of this damped simple harmonic motion is given by

$$\alpha = \Delta N' + A e^{-\lambda(t-t_0)} \sin((t-t_0) \frac{2\pi}{T}) \tag{6}$$

where $\Delta N'$ is the required angle between the zero line N' of the observation scale and the centre line M of the oscillation. A is the amplitude and T the period of the motion. Since the damping factor λ is usually neglectably small, three independent observations suffice to uniquely determine $\Delta N'$, A and T . The classical method is the observation of three consecutive reversal points, a_1, a_2, and a_3 for example. The well-known SCHULER Mean is then employed to calculate the required scale value of the centre line of oscillation

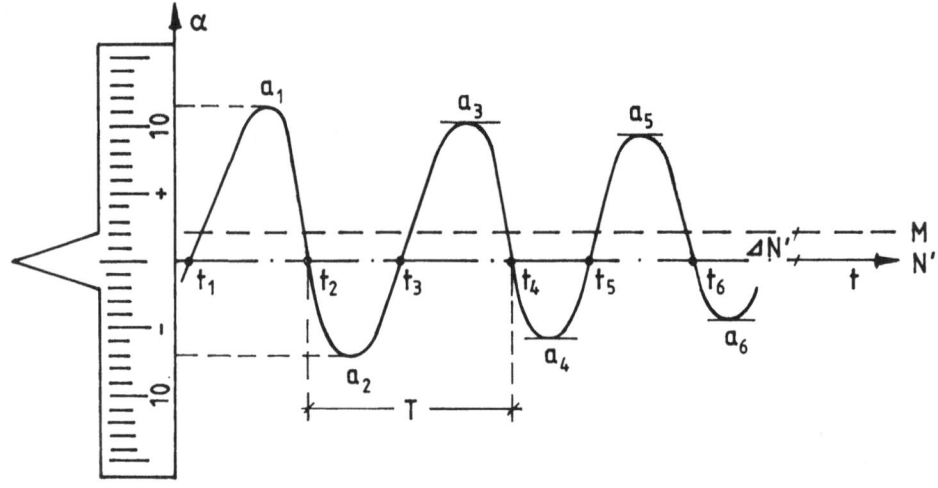

Figure 7   Oscillation of the moving mark plotted against time

$$\Delta N' = \frac{a_1 + 2a_2 + a_3}{4} \quad . \tag{7}$$

The observation of transit times was first proposed by SCHWENDENER (1964)[6]. If three consecutive transits through a selected graduation line $\bar{\alpha}$ are timed using a stop watch with lap timing facilities, then the difference $\Delta\bar{\alpha}$ between this line and the centre line of oscillation is computed from

$$t_{12} = t_2 - t_1 \quad , \qquad t_{23} + t_{12} = T$$
$$t_{23} = t_3 - t_2 \quad , \qquad t_{23} - t_{12} = \Delta t$$

$$\Delta\bar{\alpha} = A \frac{\Delta t}{T} \tag{8}$$

where again linear damping is assumed. The amplitude A is read at the turning point.

A combination of two transit times and one reversal point reading can be used to determine $\Delta N'$ as has been shown by THOMAS (1982)[5]. Let t_1 be the first transit time, next the turning point a_1 has to be read and finally the second transit time t_2 at the same selected graduation line $\bar{\alpha}$ is taken. If the period T is known then

$$t_{12} = t_2 - t_1 \quad , \qquad p = 2 \sin^2 \frac{\pi}{2T} t_{12}$$

$$\Delta\bar{\alpha} = (\bar{\alpha} + a_1[p-1]) / p \tag{9}$$

can be used to localize the centre line of oscillation on the reading scale, where linear damping has been assumed.

All three basic methods can be used in the clamped and in the tracking mode.

The tracking mode requires a special theodolite with a double tangent screw, so that the observer can rotate the theodolite continuously such that the gyro mark remains in co-incidence with the zero line of the reading scale. The reversing point readings and/or the transit times are taken at the horizontal circle of the theodolite. This observation method has the advantage that no torsion is applied to the suspension tape, so that the motion is governed by Eq. 1b. On the other hand the tracking requires much skill and patience and leads to systematic errors if not performed perfectly. The firm MOM has developed an automatic tracking system which is incorporated in the gyroscope series Gi-B2. In spite of the reportedly good results the manufacturing of these instruments has been given up.

It seems that generally the observations in the clamped mode are favoured. The disadvantage of having to consider the torsional oscillation of the system is outweighed by the absence of any interference of the observer with the oscillating system. To improve the accuracy of the results the basic methods as outlined above have been further developed. The reversal point method is usually employed taking more than the minimum of three turning point readings. The accuracy of the readings can be upgraded by using special optical micrometers and the redundant observations are evaluated in a strict least squares adjustment.

The same trend can be observed for the transit method. Precise recording watches are used which make it possible to take the times of transit at all graduation lines, so that as many as 60 observations become available for each swing. The times are taken to one hundredth of a second. A rigorous adjustment on the basis of Eq. 6 yields the required results (CASPARY/SCHWINTZER 1981)[7].

7. AUTOMATED OBSERVATION

Since the observation of transit times and turning points is wearisome, time-consuming and requiring experienced observers, efforts were soon made to automate the measuring procedure. First experiments were made in Canada with a MOM Gi-B1 and a WILD GAK 1 and similarly at CERN with the WILD attachment. The results were very encouraging as a considerable gain in accuracy and a reduction of observation time could be achieved. Today two automated gyroscopes of high precision are on the market.

MOM offers the Gi-B11 gyrotheodolite, which is equipped with two photodiodes. The transit of the moving mark through these diodes is automatically timed by a quarz-stabilized impulse counter. The results appear on a display and can be recorded or directly transferred to the computer (HP 41 C) if connected. The position of the centre line of the oscillation is computed from Eq. 10, requiring the observation of one complete period.

The observation has to be repeated at least once. Figure 8 explains the terms of Eq. 10.

$$\Delta N' = \alpha_0 \frac{\cos \tau_1 - \cos \tau_3}{\cos \tau_1 + \cos \tau_3} \quad , \quad \tau_i = \frac{t_i \pi}{T} \quad , \quad i = 1,3 \tag{10}$$

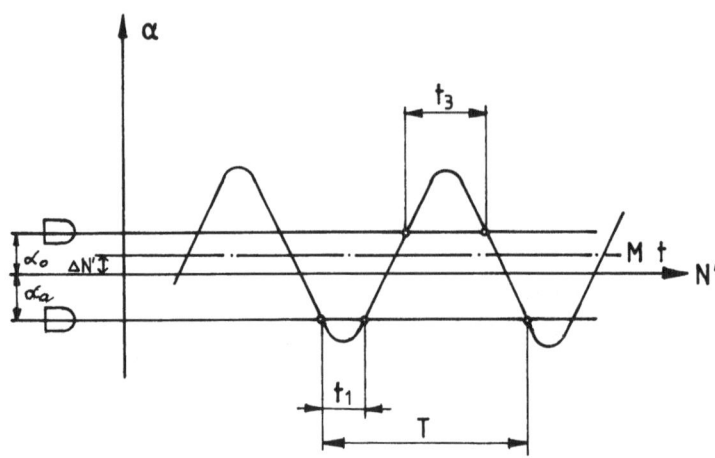

Figure 8   Automated time reading, MOM Gi-B11

WBK has automated the gyrotheodolite MW 77 and sells it under the name Gyromat. For the determination of the centre line of oscillation a novel integration method is used. The swing is continuously sensed by a photo-diode equipped with an opto-electronic pick-off. The signal of one period only is used. It is filtered and integrated, the result is a measure for the required angle $\Delta N'$ as can be seen from Figure 9. The observation process is automatically temperature corrected and controlled by certain functions.

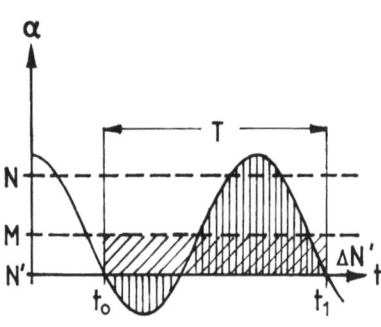

Figure 9   Integration method of signal sensing in the Gyromat (EICHHOLZ 1978)

On a lower level of accuracy, furnishing military requirements mainly, BODENSEEWERK GERAETETECHNIK (BGT) has developed an automatic gyro attachment. The directive moment of the spinning gyro is sensed and converted into a proportional current which generates a counter moment suppressing the precession. The same current drives a torquer which directs a reference mark towards true north to be picked off by the theodolite.

8. CALCULATION OF THE AZIMUTH

The astronomic azimuth A' relative to the true meridian is computed from Eq. 11. Figure 10 shows the meaning of the terms which are discussed in due detail later.

$$A' = \alpha_0 + \Delta N' + E + \beta_p + \beta_r \tag{11}$$

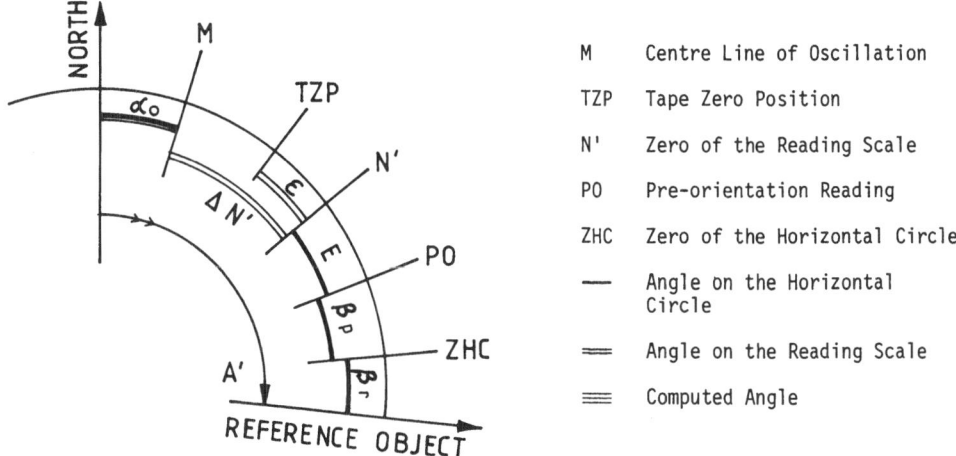

M	Centre Line of Oscillation
TZP	Tape Zero Position
N'	Zero of the Reading Scale
PO	Pre-orientation Reading
ZHC	Zero of the Horizontal Circle
—	Angle on the Horizontal Circle
=	Angle on the Reading Scale
≡	Computed Angle

Figure 10   Graph of the quantities required to compute the azimuth

The zero-torque direction of the tape according to Eq. 1 is given by

$$\alpha_t = \alpha_0 + \Delta N' - \varepsilon \tag{12}$$

substituting α_0 of Eq. 3 yields

$$\alpha_t - \frac{B}{B + F} \alpha_t = \Delta N' - \varepsilon$$

$$\alpha_t = \frac{B + F}{F} (\Delta N' - \varepsilon)$$

which finally results in

$$\alpha_0 = \frac{B}{F} (\Delta N' - \varepsilon) \quad . \tag{13}$$

Hence the computation requires the knowledge of ε and of the ratio $c = B/F$.

ε is the zero-torque direction of the tape on the reading scale. It can be determined by any of the methods outlined in sections 5 and 6 observing the oscillation of the non-spinning gyro. Since the tape zero position is not very stable it is usually measured before and after the observation of the spinning gyro. If the difference is small, then the mean is used for the correction of Eq. 13 otherwise the last value should be preferred. The ratio c is a constant which varies with latitude. The manufacturers provide a table of c as the result of a calibration for each individual instrument. As SCHWENDENER (1964)[6] has shown, the factor c is a direct function of the periods of the oscillations of the gyro in the tracking mode T_t and in the clamped mode T_c .

$$\frac{B}{F} = c = \frac{T_t^2 - T_c^2}{T_c^2} \tag{14}$$

Other methods of determination are discussed in VANICEK (1972)[4].

The angle E in Eq. 11 is an instrument constant which relates the zero line of the reading scale to the horizontal circle of the theodolite. This constant is usually determined on an astronomical reference line. The determination has to be repeated at regular intervals if accurate results are required. Modern instruments e.g. the MOM Gi-B11 and the Gyromat, possess an instrument fixed reference mirror enabling the measurement of changes of E together with each azimuth observation.

The last two terms of Eq. 11 are theodolites readings, β_r after sighting the reference object, and β_p after the pre-orientation of the gyroscope.

Since A' refers to the momentary direction of the earth's rotation axis, all standard reductions known from astronomic azimuth determination have to be applied to convert A' into a bearing defined in a selected coordinate system.

9. ACCURACY CONSIDERATIONS

On assessing the performance of gyroscopes it is particularly important to distinguish precision and accuracy and to consider the environmental circumstances at the observation site. Measurements in a laboratory or a tunnel yield more consistent results than those in the field. Most of the standard deviations being reported on in the literature are computed from repeat measurement and hence measures of precision. Figure 10 shows the number of elements, all affected by random and systematic errors, which are put together to make up the azimuth. It is obvious that the result cannot have an accuracy of one second of arc or better. But it is possible to create favourable conditions, to apply well designed procedures and to exercise care in order to get optimal results.

Two theodolite readings (β_p, β_r) belong to each azimuth, they should be taken twice at the beginning and twice at the end of each azimuth determination. Differences indicate instabilities of the set-up. Optimal results require observation pillars, tripods enable only second best results.

No azimuth can be more accurate than the instrument constant E . The methods of determination and of checking E have been outlined already in section 8. Experiments have shown that E changes with temperature, therefore a calibration in a temperature controlled laboratory should be carried out if field measurements are considered.

The determination of the centre line in the non-spin mode (ε) should be carried out before and after the observation of the equilibrium position of the spinning gyro ($\Delta N'$). If a transit method is employed, then three full swings for ε and two for $\Delta N'$ are adequate.

The constant c = B/F of Eq. 13 should be determined for each observation site. If a chronometric method is applied, then the period T_c of Eq. 14 is available without extra measurements, only T_t has to be measured, which can be done by timing four to five reversal points.

Some further rules to be exercised in order to get good results are:

Swinging of the gyro without spinning for half an hour eliminates torsions and deformations of the tape being imposed by clamping for transport.

After spin-up of the gyro a certain time is needed for the instrument to attain temperature equilibrium. Dependent on the air temperature this can require up to one hour for MOM and WILD instruments. The Gyromat needs less time since it only slightly warms up due to the low spinning rate.

The tape zero position (ε) should be regulated near to zero to minimize systematic effects on the oscillation. For the same reason the pre-orientation should be performed as well as possible to get a small value of $\Delta N'$.

Some authors recommend measuring with two different pre-orientations symmetrical to the zero line.

After these actions the actual observations can start. Regular checks and corrections of the vertical axis of the instrument contribute to the quality of the results.

A protection from certain environmental factors like magnetism, movement of air by ventilation or wind, and vibrations caused by heavy machinery or traffic is desireable. This can be achieved by carefully selecting the site and the time of measurements and by shielding the instrument.

If these rules are obeyed and the best observation procedures are employed then the observed azimuths have typically the following standard deviations. The WILD GAK 1 mounted on a single second theodolite using extended chronometric methods or special devices for reading the scale provides the azimuth with 5" to 6" standard deviation. The automated gyrotheodolites Gyromat and Gi-B11 perform slightly better, namely with an accuracy of 3" to 4".

The precision of results for measurements in a lab are usually better by a factor of two.

10. NON-CONVENTIONAL GYROSCOPES

Since the early 1960s optical gyroscopes have been under development. These instruments do not have moving parts, are of apparently somple design and seem to have the potential of better performance in respect to resolution and dynamic range, as compared with the conventional mechanical gyroscopes. But the first prognoses that these new instruments would soon completely replace the rotating machines had to be revised, since numerous problems were encountered in the development of prototypes. Today the laser gyro is on an operational level. It is used in strap-down platforms for aircraft navigation (Boeing 757, 767 and Airbus 310), where a wide measuring range is more important than a high precision.

The basic idea of optical gyroscopes is due to the French physicist SAGNAC, who built in 1913 the first ring interferometer to study the nature of light. If a ring interferometer is mounted on a turntable and rotated then a light beam travelling with the rotation will need more time to arrive back to the source than a beam travelling against the rotation. This is the so-called SAGNAC effect, which can easily be computed if minor corrections due to the general relativity theory are neglected. The difference ΔL of the optical path length of the two beams is

$$\Delta L = \frac{4F}{c} \omega \tag{15}$$

where F is the area enclosed by the path, c the velocity of light and ω the angular velocity of the ring with respect to inertial space. The path difference can be measured indirectly by observing the fringe pattern produced by the interfering beams, see Figure 11.

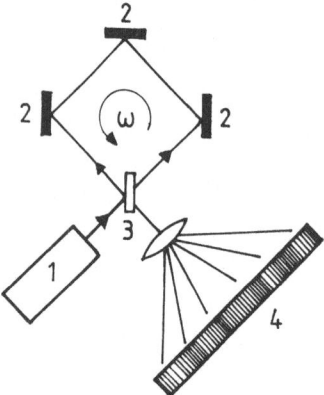

1   Light Source (LASER)

2   Mirror

3   Beam Splitter

4   Fringe Pattern

Figure 11   Schematic of a ring interferometer according to SAGNAC (RODLOFF 1982)[8].

Since c is constant, the path difference ΔL is directly proportional to the area F and the angular velocity ω of the set-up. The technical problem is to generate a measureable effect, which is achieved by two different approaches.

The fiber-optic gyroscope attains the required sensitivity by use of a large area F . Optical fiber with a length up to 1000 m is wrapped into a coil of very low volume. The first rotation rate sensors of this type are now being readied for application. Figure 12 shows the concept of a fiber-optic gyro as developed by SEL. It has a dynamic range of 400^0/s, a repeatability of 50 ppm and a drift of less than 3^0/h.

The laser gyro consists of a cavity as depicted in Figure 13. It converts the path length difference in a frequency signal, which can be observed more easily. The mirrors create a closed path calibrated to be an integral multiple of the wave length thus forming an optical resonator. Rotation of the ring laser induces a change of the eigenfrequencies of the two beams travelling in opposite directions. Their beat frequency is proportional to the rate of the resonator. It is measured by observing the fringe pattern produced by superimposing the beams. The earth's rate of 15^0/h generates a beat frequency of 1 Hz in a laser gyro of

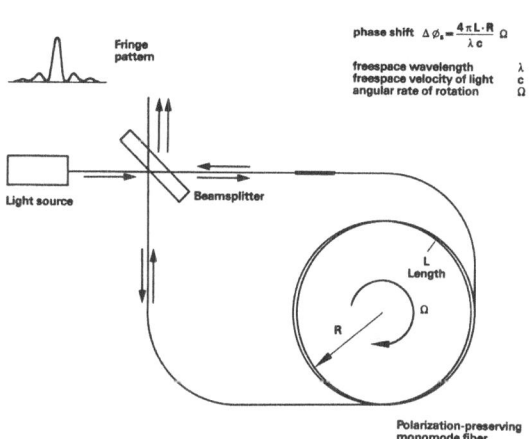

phase shift $\Delta\phi_s = \frac{4\pi L \cdot R}{\lambda c} \Omega$

freespace wavelength λ
freespace velocity of light c
angular rate of rotation Ω

Figure 12   Schematic of a fiber-optic gyroscope (SEL)

F = 100 cm^2 and λ = 0.633 μm , being easily measureable.

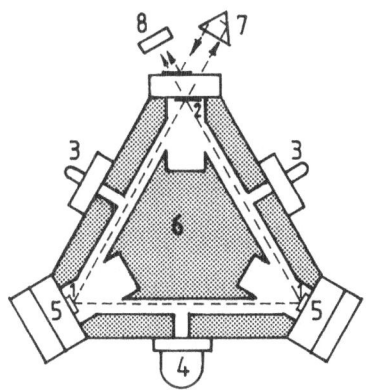

1   Mirror

2   Dielectric Mirror

3   Anode

4   Cathode

5   Length Control

6   Cervit Block

7   Corner Prism

8   Readout Detector

Figure 13   Schematic of a gyro block assembly (RODLOFF 1982)[8]

The main problem with the laser gyro is the so-called lock-in effect, which prevents the measurement of small angular velocities. The reason for this effect is the tendency of coupled oscillators to swing on a common frequency if the eigenfrequencies differ only a little. Thus input rates below a certain threshold produce a zero output. A remedy is to mechanically dither the laser block with a frequency above the lock-in rate.

As already mentioned, laser gyros are operational as attitude sensors for inertial platforms. They have an effective range of 10^{-2} $^\circ$/h to 1000°/s. Prognoses of today are that the sensitivity to smaller rates will improve and will become comparable to the best mechanical gyroscopes. Since their production is less expensive, they may possibly one day compete with conventional suspended gyroscopes. Suitably mounted, so that rotations about a horizontal and a vertical axis are possible, optical gyros are capable of sensing the true north direction and the geographic latitude.

11. CONCLUSION

Over the last two decades significant advances have occured in the technology of conventional suspended gyroscopes. The accuracy of determining north has improved considerably due to the advent of new instruments, the introduction of more effective chronometric observation methods and to automating the observation of the centre line of oscillation. The same actions have simplified the measurements and reduced the time required, so that it has become possible to measure an azimuth within 15 to 60 minutes with a standard deviation of 3" to 6".

In future, optical gyroscopes may replace the conventional instruments. Within 25 years laser gyros have reached a level of performance rendering them suitable for application in strap-down platforms. Fiber-optic gyros have been under development for 10 years, seemingly meeting now the performance requirements of inertial navigation. Both realizations of the SAGNAC effect hold the promise of giving higher performance at lower cost.

*   *   *

REFERENCES

1) LAUF, G.B.: The gyrotheodolite and its application in the mining industry of South Africa. Journal of the South African Institute of Mining and Metallurgy, pp. 349-386, 1963.

2) DEIMEL, R.F.: Mechanics of the Gyroscope. Dover, New York 1950.

3) von FABECK, W.: Kreiselgeräte. Vogel-Verlag, Würzburg 1980.

4) VANICEK, P.: Dynamical Aspects of the Suspended Gyrocompass. The Canadian Surveyor (26), pp. 77-83, 1972.

5) THOMAS, T.L.: The Six Methods of Finding North Using a Suspended Gyroscope. Survey Review (26), pp. 225-235, 1982.

6) SCHWENDENER, H.R.: Beobachtungsmethoden für Aufsatzkreisel. Schweizerische Zeitschrift für Vermessung, Kulturtechnik und Photogrammetrie (62), pp. 365-375, 1964.

7) CASPARY, W., P. SCHWINTZER: An Extension of Chronometric Gyroscope Observation Methods. The Canadian Surveyor (35), pp. 364-372, 1981.

8) RODLOFF, R.: Vom Pendel zum Laserkreisel. Ortung und Navitation, pp. 398-411, 1982.

BIBLIOGRAPHY

CASPARY, W.: Moderne Vermessungskreisel - Leistungsfähigkeit und Einsatzmöglichkeit. In Proc. of IX. Internationaler Kurs für Ingenieurvermessungen, Graz 1984.

CASPARY, W., H. HEISTER: Problems in Precise Azimuth Determinations with Gyrotheodolites. In Proc. of XVII. Kongress FIG, 609.2, Sofia 1983.

EICHHOLZ, K.: Moderne Vermessungskreisel - gerätetechnische Konzeption und Entwicklungsstand. In Proc. of IX. Internationaler Kurs für Ingenieurvermessung 1984

EICHHOLZ, K., R. SCHÄFLER: "Gyromat" an Automatic Gyro-Theodolite of High Precision, Influences of Interference Parameters on Northing Accuracy and Counter-Measures. In Proc. of 2nd ISS, pp. 613-625, Banff 1981

HALMOS, F.: High, Precision Measurement and Evaluation Method for Azimuth Determination with Gyrotheodolites. In Manuscripta Geodaetica (2), pp. 213-231, 1977

HALMOS, F.: Evaluation of Automatized Gyrotheodolite Measurements with Special Respect to MOM Gyrotheodolites. In Acta Geodeat., Geophys. et Montanist., Acad. Sci. Hungary (16), pp. 27-39, 1981

RÜEGER, J.M.: Inertial Sensors, Part I: Gyroscopes. Edited by K.P. Schwarz, Division of Surveying Engineering, The University of Calgary, Publ. 30002, Calgary 1982.

UNDERGROUND GEODESY

J.-C. Fischer, M. Hayotte, M. Mayoud, G. Trouche
CERN, Geneva, Switzerland

ABSTRACT

The tunnelling work, for a particle accelerator, must be carried out within very strict and tight tolerances. The CERN Applied Geodesy Group has the task to regularly control the geometry of the galleries and caverns constructed for the LEP project. For this purpose, different means and methods have been developed and implemented, related to the link between the surface and underground geodetic networks, the guiding control with accurate gyro traverses and the check of cross sections with photo-profiling or profile scanners. Instruments and methods are described by the authors and the results presented.

1. INTRODUCTION

The LEP tunnel constitutes an exceptional underground construction, due to its size (27 km circumference) as well as its position in a tilted plane. Its role is to accommodate a particle accelerator and this function involves some unusual geometric constraints.

The high precision required for the alignment of accelerator components [1,2], compared with the common accuracies of civil engineering, is such that the geometry of the accelerator must be considered as absolute, i.e. rigidly fixed by particle-beam optics. An accelerator tunnel being rather strictly dimensioned for economical reasons, its axis must follow the theoretical axis of the machine within tight tolerances.

These tolerances (Fig. 1) are derived from CERN experience in tunnelling. They fix the allowable limits of the errors which can affect the position and shape of the galleries [3] :

- errors of the surface geodetic network,

- errors in the determination of control points at the tunnel level,

- errors of the gyro-controlled traverse,

- errors in transfers from pillars to brackets for guidance by laser,

- deviations of the tunnel boring machines.

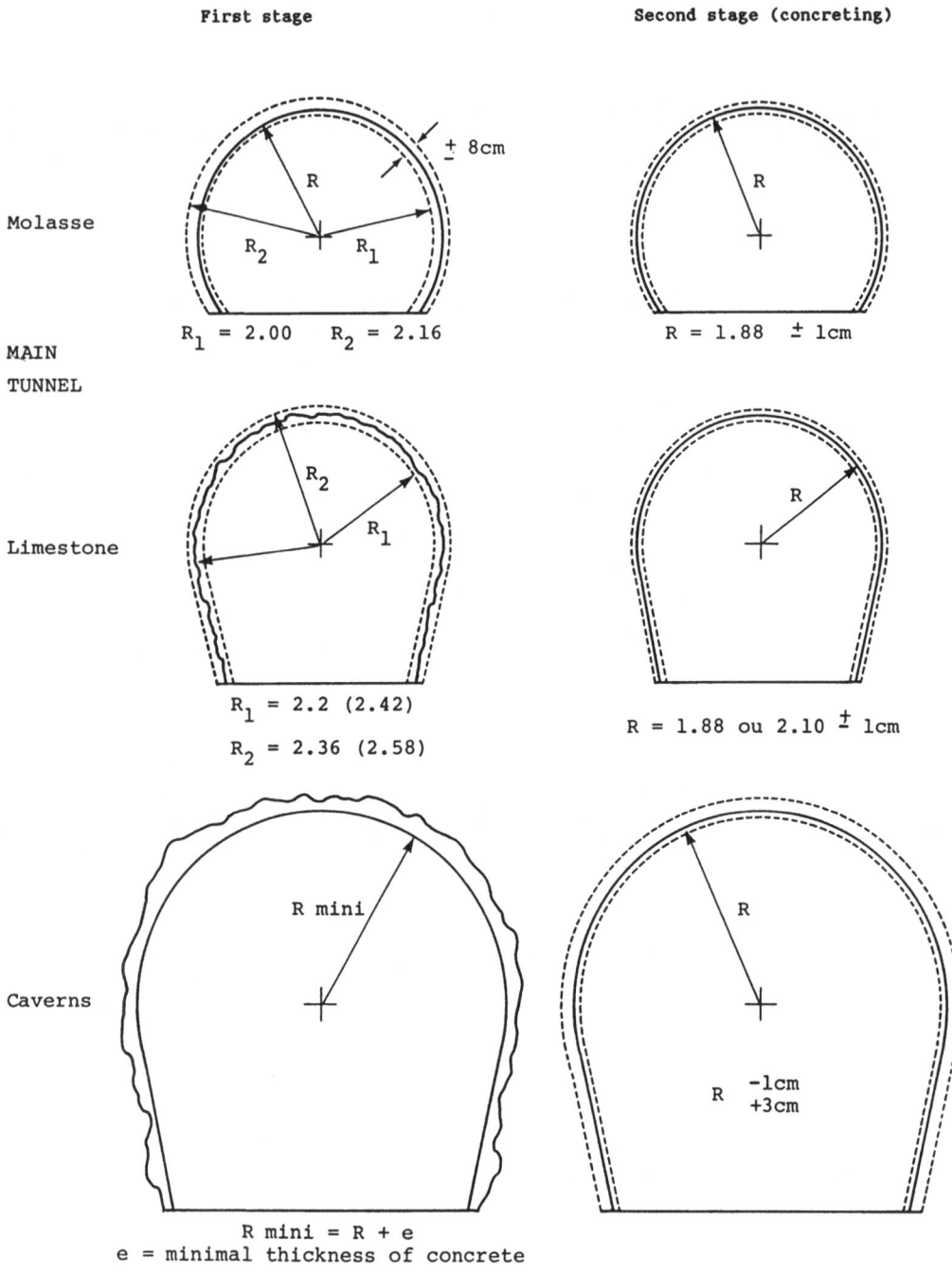

First stage

Second stage (concreting)

Molasse

$R_1 = 2.00 \quad R_2 = 2.16$

$R = 1.88 \quad \pm 1cm$

MAIN
TUNNEL

Limestone

$R_1 = 2.2 \ (2.42)$

$R_2 = 2.36 \ (2.58)$

$R = 1.88$ ou $2.10 \ \pm 1cm$

Caverns

R mini

R

R mini = R + e
e = minimal thickness of concrete

$R_1 \leqq R \leqq R_2$

Fig. 1   Tolerances of the tunnels and caverns

The tolerances also ensure a minimum thickness of concrete everywhere along the tunnels and caverns.

To avoid costly mistakes, conflicts or disputes, it is necessary to make controls at different stages of the construction, and to finally check the completed lining of the galleries and caverns. For such a long and difficult task, it was explicitely specified that the construction firms must have a competent team of surveyors. Respective roles of the firms and of the CERN Applied Geodesy Group have been defined and a close collaboration has been established. Thus, the main targets of underground geodesy are :

- to provide the basic geodetic framework defined in the specifications,

- to preserve the interests of CERN in the geometrical conformity of the construction,

- to limit the risks of errors by preventive operations and to facilitate the task of firms by giving convenient data.

2. LINK BETWEEN SURFACE AND UNDERGROUND GEODESY

2.1 Basic framework

For both the civil engineering works and the installation of the machine, CERN has implemented a surface geodetic network which covers the LEP site and provides facilities for the determination of control points near the eight pit areas. This double affectation of the network and the fact that it was measured with the Terrameter [4, 5] ensure that pillar locations are known to within a millimetric accuracy.

From that network, three top points are determined above each pit. They are used for a vertical transfer, down to tunnel level, in order to link the first reference pillars of the main gallery which constitute the basic underground framework that CERN was contractually engaged to provide.

2.2 Scheme of operations

The reference points above the pits are in the form of a conical hole in an aluminium plate, three of these plates being screwed to the curb-stone of each pit (Fig. 2). The conical hole has a centring function for either a standard CERN reference target or for half-spherical holders used for zenithal sights.

Generally, the nearest geodetic pillar of the surface network is always located between 100 and 500 m away from a pit. Linking measurements are made with a well calibrated AGA 140 total station and a WILD Na2 level. The three top points remain tied by accurate measurements in order to keep a good local homogeneity. The accuracy of this link is estimated to be $\sigma_{xy} \leq 2$ mm and $\sigma_H < 1$ mm.

184

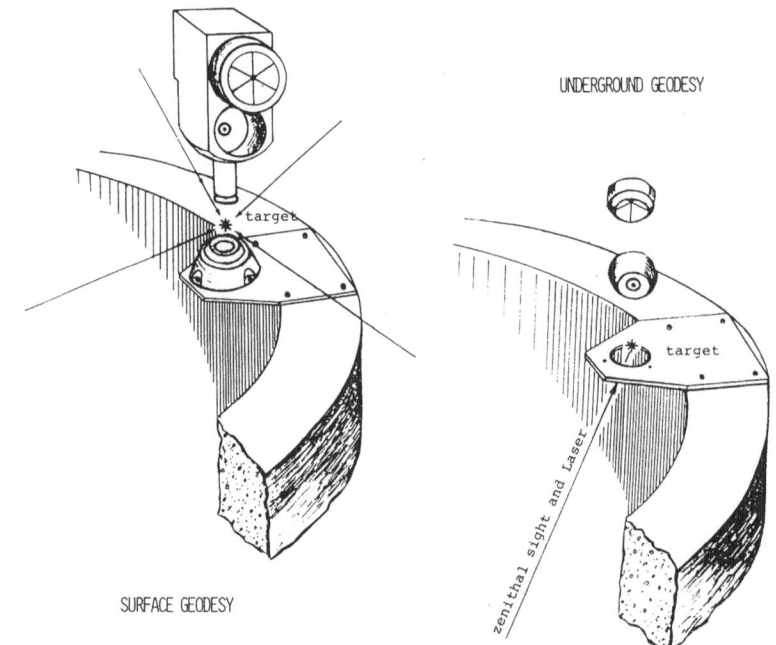

Fig. 2   Reference points above pits

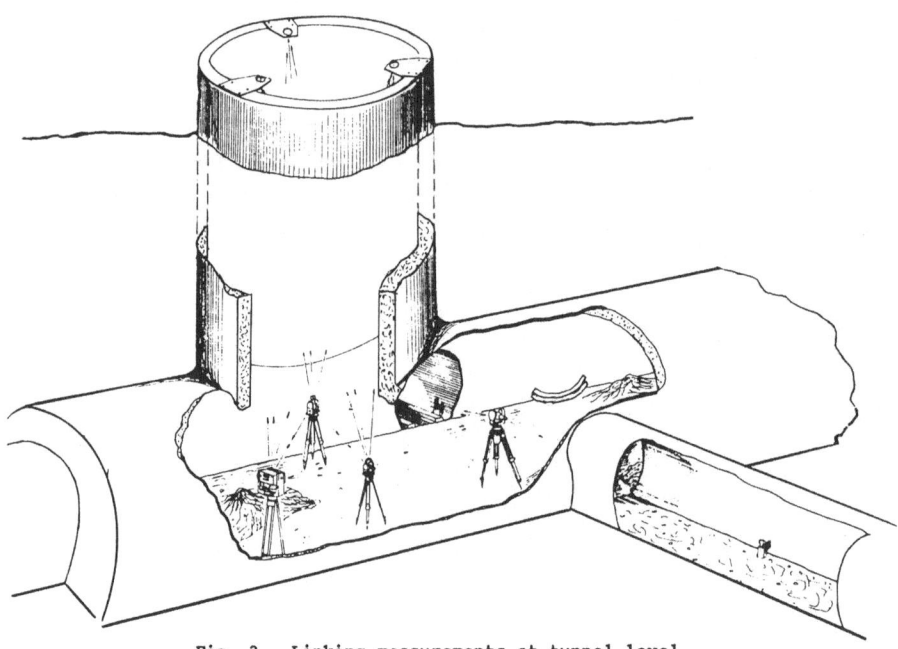

Fig. 3   Linking measurements at tunnel level

The second operation is the vertical transfer to tunnel level. It consists of the simultaneous determination of three points on tripods with respect to top points, completed with linking measurements to the first tunnel pillars (Fig. 3).

This operation involves setting up theodolites and their (calibrated) EDM, one MOM or WILD gyro, one WILD Na2 level and the special target or reflector holders. All horizontal and vertical angles are measured, as well as distances, orientations and height differences. Observations are processed as a spatial block by means of the CERN 3-D adjustment program [6]. Deflections of the vertical are taken into account to correct gyro masurements and to express the correcting vector due to the difference between the physical plumbline and the normal to the reference ellipsoid. For a pit as deep as 140 m, this correction can reach 6 mm. The accuracy of this second operation is estimated to be $\sigma_{xy} \leq 1$ mm and $\sigma_H < 2$ mm.

3. <u>GUIDING CONTROL WITH GYRO TRAVERSE</u>

3.1 <u>General principles</u>

Having guided the main and transfer tunnels of the SPS (more than 10 km of total length), for which the first semi-automatic gyroscopes were developed and used, the CERN Applied Geodesy Group has gained good experience in tunnelling control. Even with free traverses 3.3 km long for the LEP, instead of 1.1 km for the SPS, it was evident that the use of gyroscopes was the only reliable method to ensure the required guiding accuracy.

Control pillars are regularly spaced along the LEP tunnel every 39.50 m corresponding to the lattice of the machine (Fig. 4). Over the 3.3 km of an octant — from one geodetic point to the next — and with an orientation error $\sigma_\alpha = 40$ cc (0.0040 gon) on each gyro measurement on each pillar, the expected radial misclosure will not exceed 23 mm r.m.s. if distances between the pillars are carefully measured with invar wires. When adding the transfer and acquisition errors for laser piloting and the normal oscillations of the tunnel boring machines, it can be estimated that the resulting radial errors will be, at most, in the vicinity of 8 cm r.m.s. for the excavation stage. These figures are derived from CERN experience in using WILD GAK 1 gyros (manual or semi-automatic) and they have been used to define the tolerances in the specifications issued to contractors.

However, CERN development of fully automatic computerised gyros as well as new commercially available instruments promised better accuracies. This fact was not a sufficient argument to change the already tight tolerances, but it had a good practical consequence. The radial errors of a gyro traverse follow a quadratic law with respect to the number of stations, hence a better accuracy has the advantage of allowing an increase in the distance between measuring points (Fig. 5), thus giving a noticeable gain in the efficiency of such long and delicate survey operations.

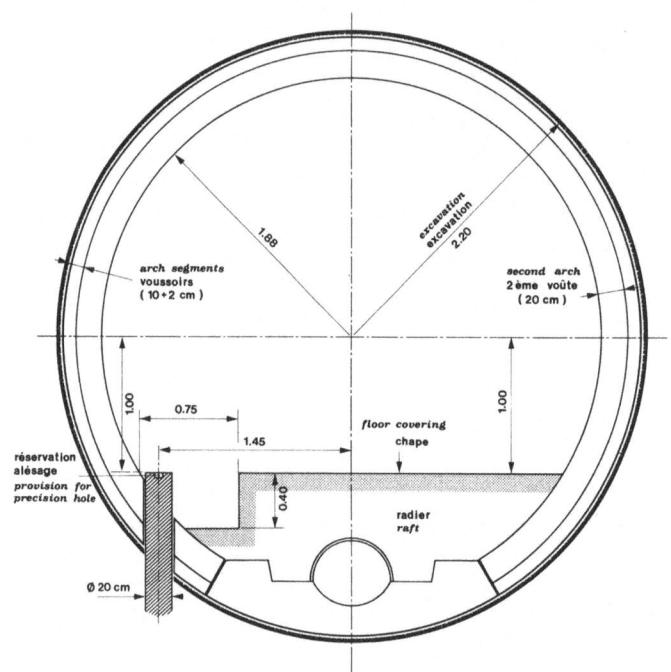

Fig. 4   Tunnel cross section and control pillar

POLYGONALE GYRO (Fermeture 22.8 mm)

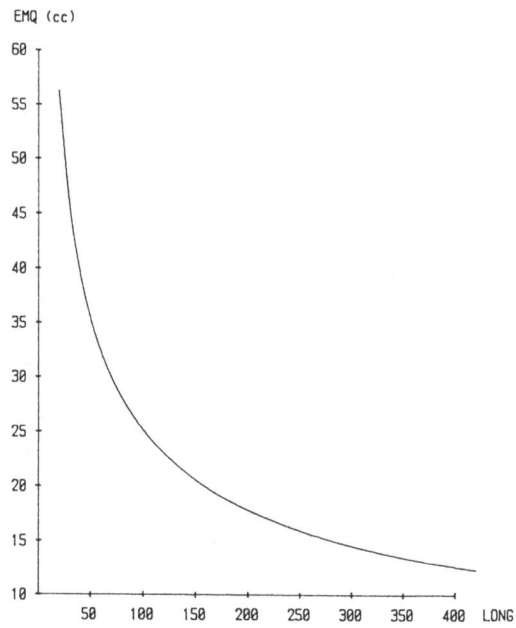

Fig. 5   Optimisation graph

In practice, considering the severity of the tolerances, the special experience of CERN in the domain and the interest of the contractors, a compromise has been reached between the two parties. The CERN Applied Geodesy Group carries out and computes the gyro traverses in a free and open collaboration with the firms, under the latter's responsability. While CERN is engaged to closely follow the progress of the tunnel boring machines, the transfer of the geometry from control pillars to wall brackets, for laser guiding, remains the task of the firms' surveyors.

3.2 Automation of a WILD GAK 1 gyro

The automation of the WILD GAK 1 gyro revolves around four adaptations (Fig. 6, 7):

- the replacement of the graduated scale by a CCD line of 1024 photodiodes, spaced at 13 μm, with some subsequent modifications of the optical system,

- the replacement of the manual adjustment knob by a motor-encoder set in order to achieve computer-controlled release of the spinning motor,

- the development and integration of a miniature microcomputer - also used in the new Distinvar [7] - based on a NSC 800 processor and constructed on a 177 mm x 55 mm multi-layer printed circuit board. This computer can act on all electro-mechanical functions of the instrument and leaves an 8 kb space for the BASIC programs which control the measuring procedures and process the data,

- the replacement of the GKK3 converter by a small, flat 400 Hz power supply housed inside the cover of the instrument.

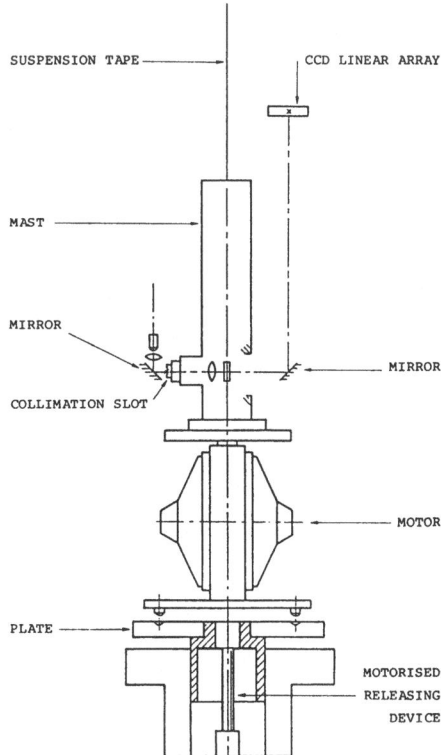

Fig. 6   Schema of the modified GAK 1 gyro

Fig. 7   The modified GAK 1 gyro

The aim of these modifications was to produce a compact and fully automatic gyro. A single 4-pin jack links the instrument to any terminal display or portable computer : two pins for the power supply and two for RS232 communication. Technological details will be given in a future report on this instrument [8].

The computation of North with the in-board computer is based on the least-squares fitting of a damped sinusoid onto a sample data set of the motion, provided by the CCD line. This slow but efficient method has been finally retained after unsuccessful attempts at numerical integration.

The integration of all CCD data [9] would certainly have been the best algorithm; fast and exhaustive. Nevertheless, when looking for North with the motor on, mechanical micro-oscillations tend to generate a lot of parasitic data. Buffering and filtering problems related to CPU and RAM limits made real-time processing impossible.

The calibration of the instrument must be carefully and rigourously carried out. For the highest accuracy, this operation is not so simple because the torque of the suspension tape cannot be neglected [10]. Starting from the basic equation of gyros, the ratio between the torque Mr of the tape and that of the spinning motor can be expressed as :

$$k = Mr \: / \: B \: \theta \: \Omega \: \cos \varphi \qquad (1)$$

where B is the inertial momentum of the spinning motor with respect to its axis, θ the rotation speed of the motor and Ω that of the earth, and φ the latitude of the observation station.

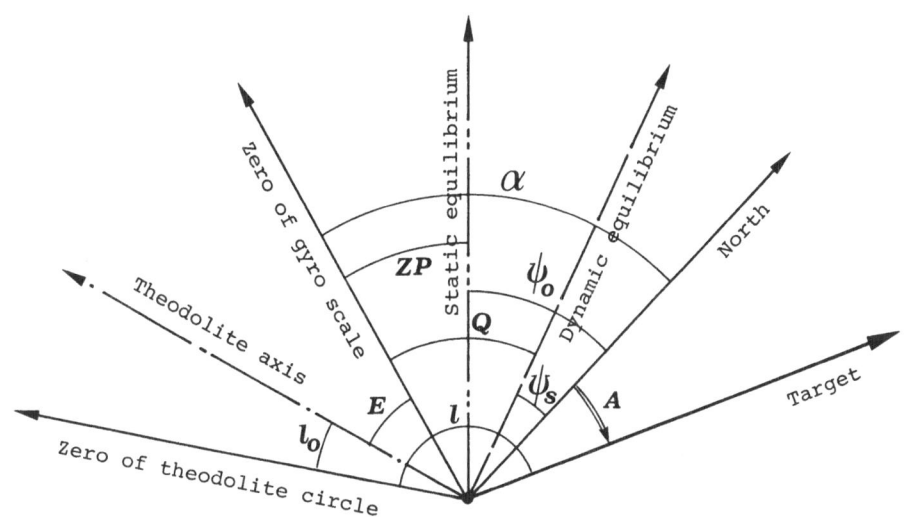

Fig. 8   Parameters of a gyro-theodolite measurement

The above figure shows the angles and directions which need to be considered for the expression of the azimuth A of a target point :

$$A = \ell - (\ell_o + E + \alpha). \qquad (2)$$

Starting from Eq. 1, the value of α can be expressed in linear readings of the CCD line :

$$\alpha = (1 + k)Q - k.ZP. \qquad (3)$$

Introducing a conversion factor μ, this value is turned into angular units :

$$\alpha = \mu \ (1 + k)Q - \mu.k.ZP. \qquad (4)$$

Its solution requires three steps :

1) successive measurements around North give a collection of linear values $Q = f(\ell_o)$ with slope $- 1/\mu \ (1 + k)$;

2) the k ratio must be determined by measuring the periods T and Te of the motion, respectively with and without torque in the suspension tape. The last operation is made by following-up the reference index with the slow motion screw of the theodolite. It can be shown that $k = (Te^2/T^2) - 1$;

3) the angle E, external constant of the gyro, is obtained by outdoor calibrations with respect to the local bearings in the CERN system. A special calibration hut has been made over a pillar of the geodetic network for this purpose.

A strong dependency of the GAK 1 performance on the internal temperature of the instrument has been observed, and this phenomena has been modelled (Fig. 9). The improvement on the results is shown in Figs 10 and 11. The standard deviation of an identical set of calibration measurements is reduced from 55 cc to 11 cc.

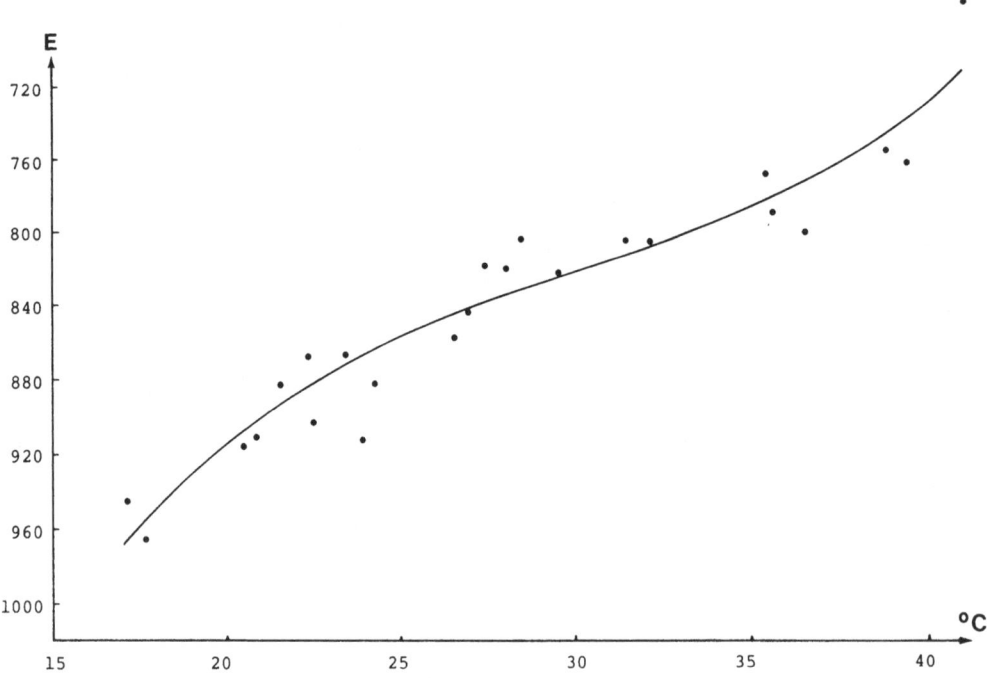

Fig. 9   Variations of angle E with respect to internal temperature of the gyro

The in situ performance of the modified GAK 1 varies from 11 to 18 cc (1 σ). After a warming-up period, a ZP - MOTOR - ZP - MOTOR - ZP sequence is observed at each station, for less than two hours. No significant improvement is gained when repeating this sequence. Each observation in the tunnel is always framed by two outdoor calibrations, one before and one after. The optimal working length has been fixed at 79 m, i.e. one pillar in two.

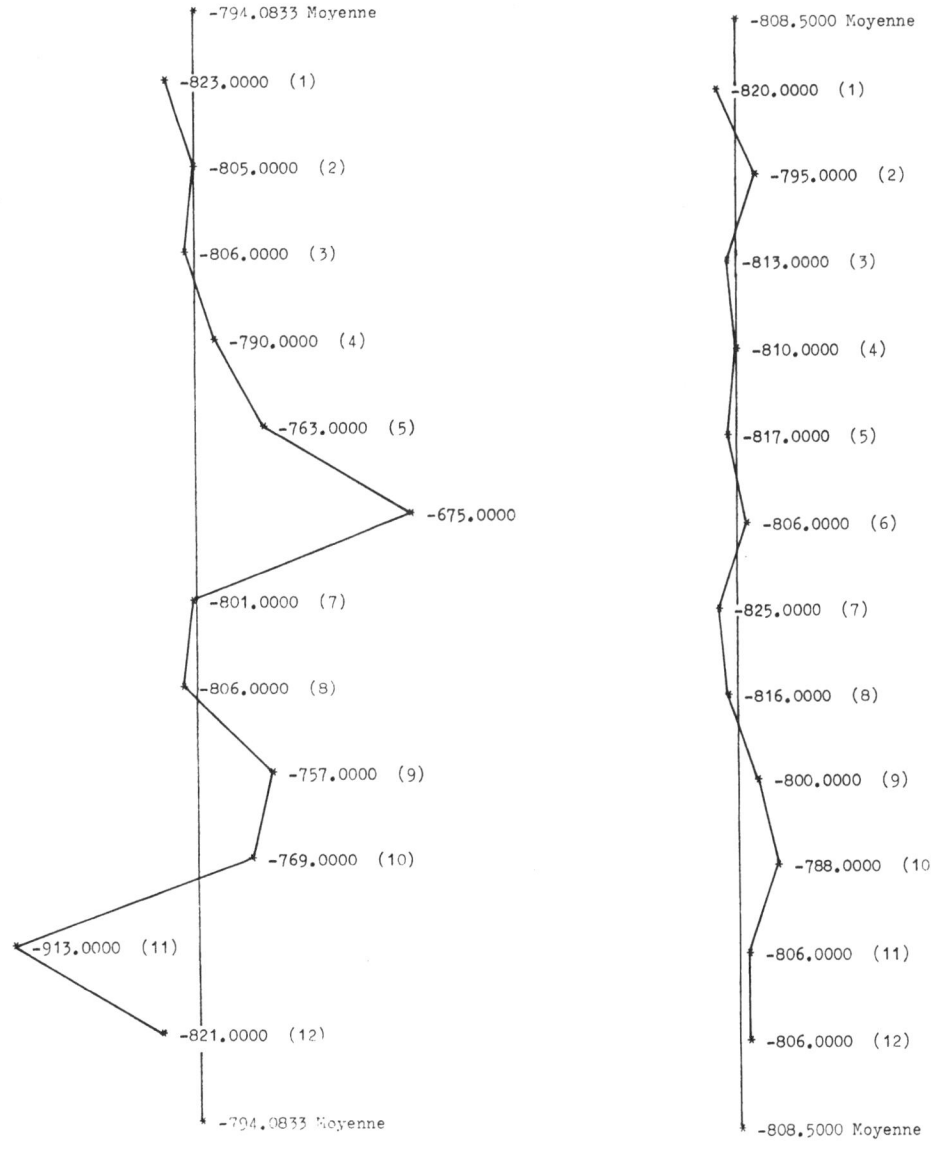

Fig. 10   Calibration of a GAK1 gyro without temperature correction

Fig. 11   The same data set as Fig. 10 but after correction

3.3 The MOM Gi-B11 gyro

As the development and testing of the modified WILD gyro was not completed in time, there was no choice but to start with a commercially available one. Only two accurate and automatic instruments were on the market : the Gyromat, from the Mining Institute of Bochum, and the Gi-B11, from MOM. The general performance of these two gyros is rather similar and the MOM was chosen because of shorter delivery times.

The MOM Gi-B11 gyro-theodolite is a semi-automatic instrument, in which the precession of the spinning motor is observed by two photocells, spaced with a well calibrated interval, according to the transit method (Fig. 12). Data acquisition and processing is achieved by means of an HP 41-CV pocket calculator through an HP-IL interface. An RS 232 option is also available and allows control by any portable computer.

The instrument has two external units, for power supply and control. It is designed to be set-up on a tripod and a special centring device has been introduced for CERN pillars. The reliability of the Gi-B11 has lead to the choice of the "accuracy option" in a long sequence of measurements, with sights of 158 m, i.e. one pillar in four. To eliminate the excentricities of the theodolite and of the centring device, observations are made in four positions and two complete cycles (Motor + ZP) are measured at each position. These eight independent determinations of North take 3½ h. They are also framed by two outdoor calibrations.

The MOM gyro is bigger than the WILD, hence a more favourable k ratio allows a simple and direct calibration procedure. If no temperature effects have been noted, a regular and unexplained linear drift is nevertheless observed and corrected (Figs. 13 and 14).

The final accuracy of this instrument when used according to the above procedures is $\sigma \simeq 7$ cc.

3.4 Processing and results

Before computation, gyro measurements are corrected for convergence of meridians and for the effects of vertical deflections, using the parameters of the local geoid previously determined and modelled [11, 12].

All measurements of the control traverses are processed in the CERN 2-D or 3-D adjustment programs [5] :

- corrected gyro observations,

- short- and long-side angular measurements,

- long-range EDM measurements,

- invar distances between pillars,

- direct and indirect levelling data.

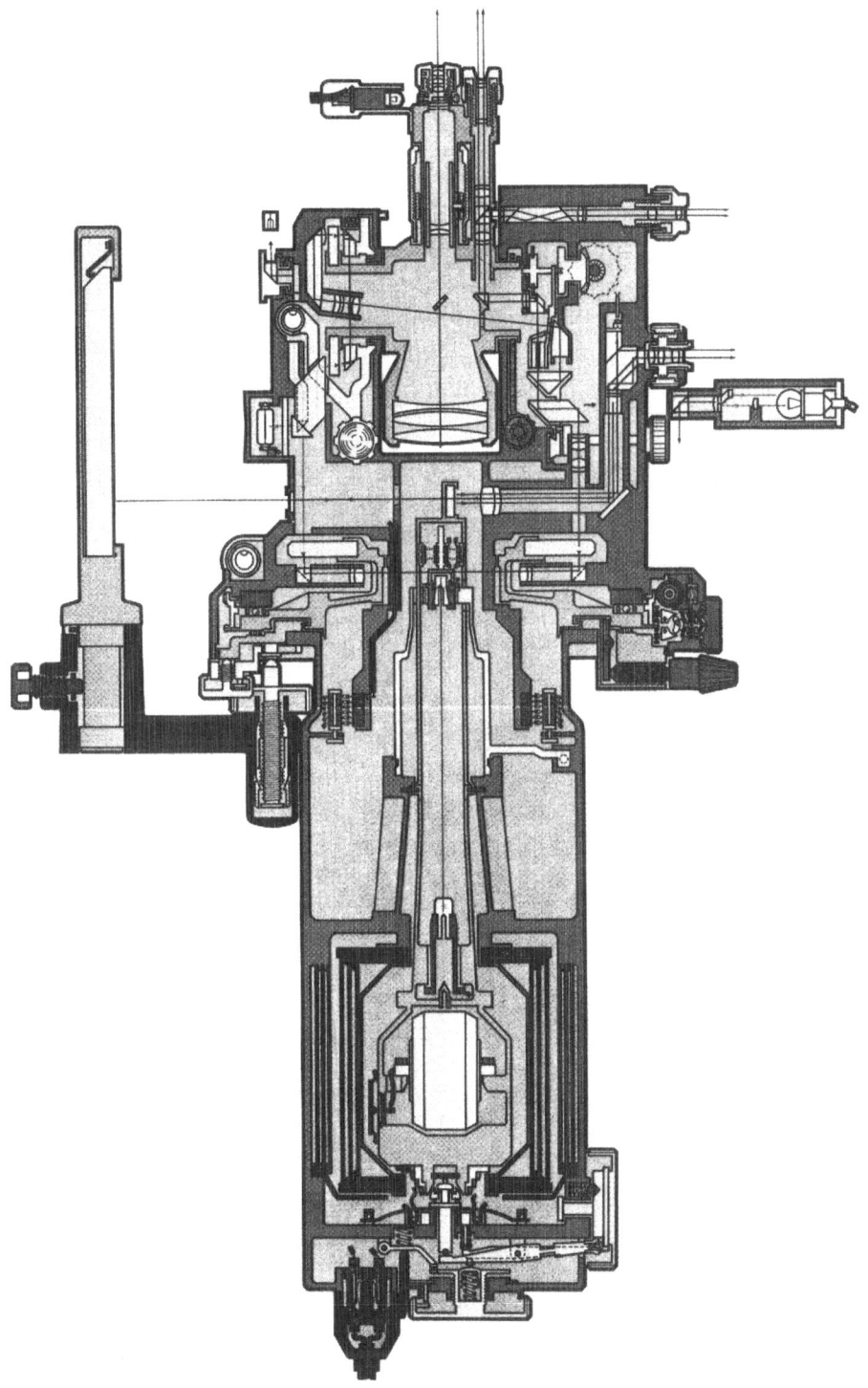

Fig. 12   Layout of the MOM Gi-B11 gyro

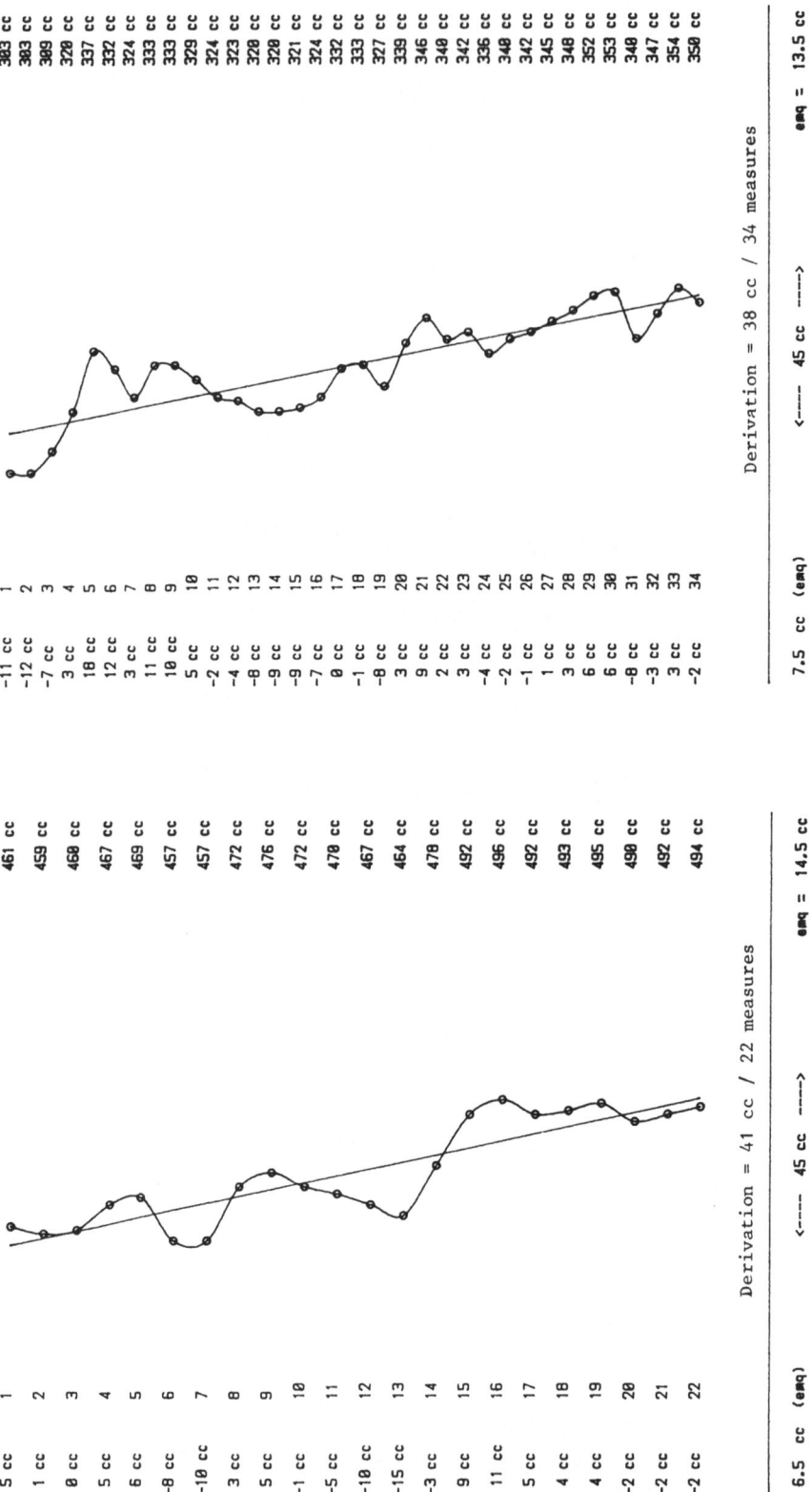

Derivation = 41 cc / 22 measures

6.5 cc (emq)   <—— 45 cc ——>   emq = 14.5 cc

Fig. 13   Calibration for piloting pits P1 - P2

Derivation = 38 cc / 34 measures

7.5 cc (emq)   <—— 45 cc ——>   emq = 13.5 cc

Fig. 14   MOM 2 calibration pits P8 - P7

When making successive least-squares adjustments as the boring advances, the progressive addition of overlapping data causes some slight changes in the computed coordinates. However, these millimetric variations have no practical consequences for the guidance of the boring machines.

Until now, three octants (3330 m) of the LEP tunnel have been completed and the radial misclosures of the control traverses have been 4 mm at P2 (Fig. 15), 22 mm at P7 (Fig. 16) and recently 14 mm at P5.

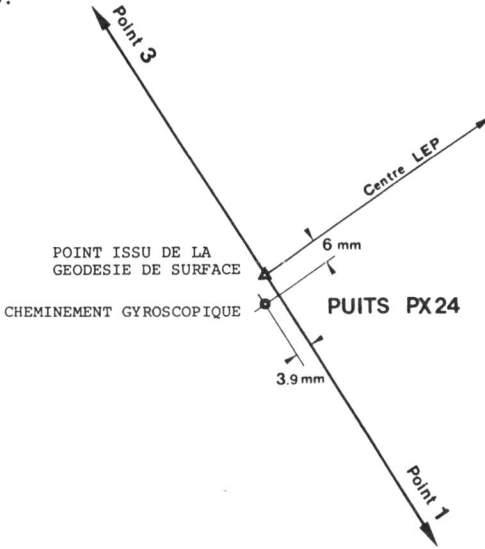

Fig. 15   Misclosure at control point P2

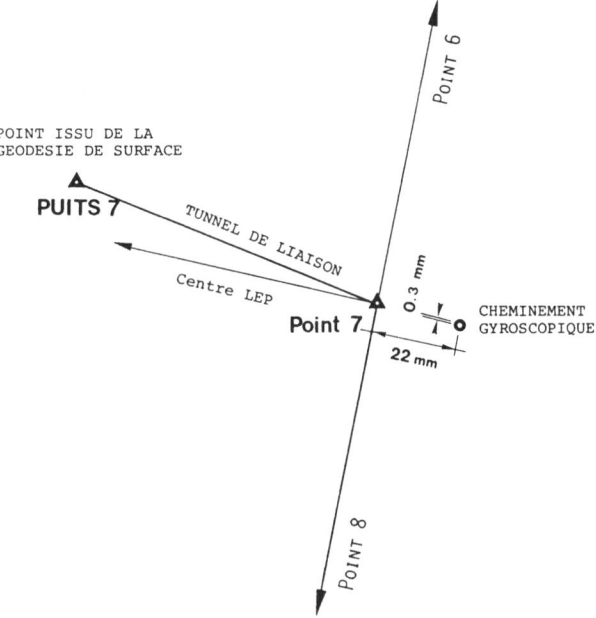

Fig. 16   Misclosure at control point P7

4. TRANSVERSE GEOMETRICAL CONTROLS OF UNDERGROUND WORKS

4.1 General remarks

An essential point in the General Specifications issued to contractors is that the thickness of concrete in vaults, tunnels or caverns, must not be less than the specified value. In order to make sure that this minimum thickness is respected, tight controls have been undertaken at the semi-finished stage of the construction.

These controls had to take account of several criteria and constraints :

- in an underground environment, measurements must be fast and reliable,

- the instrumentation must be simple, robust and easy to operate,

- remembering the 27 km length of tunnelling, the collecting and processing of the data must be very efficient,

- the final accuracy of the controls must be close to 1 cm,

- the instruments and methods used must be applicable to samll or large galleries (from 4.4 m diameter to 25 m high) and be adapted to the state of surfaces which may be rough when made with such techniques as explosives, punchers, or road-headers, or smooth when bored with a full-face machine.

Two methods have been retained : the photo-profiling method which permits very fast data acquisition, and the profile scanner (EDM measurements without reflector) for the control of large caverns.

4.2 Photo-profiling method

4.2.1 Qualities of mono-photogrammetry

The main advantages of photo-profiles are that operations in the field are fast and that the original data (the photograph) gives a view of the entire section without having to make, in situ, the difficult choice of characteristic points. The accuracy depends on the performance of the instruments, the size of the reference frame with respect to that of the measured section, and the observation and computation methods used to process the data. A compromise can always be found among these parameters to produce the required accuracy.

The only inconveniences are the intermediate steps required between acquisition and observation of the data such as processing of the films, enlargement and printing. Nevertheless, a photograph constitutes a permanent and incontestable proof of the data, which can be useful in case of dispute.

4.2.2 <u>Instrumentation</u>

As the method was mainly designed for the profiling of small galleries (from ⌀ 4.40 to ⌀ 6.00 m), it was decided beforehand to operate with non-metric 6 x 6 cameras. These are cheap and light, their distortion is well within the accuracy required in civil engineering and they can be equipped with a large changing-box in order to gain time. We have selected the Hasselblad SWC/M camera as presenting the best optical and mechanical qualities for our application since it has a wide-angle with low distortion, fixed-lenses (more robust), an external view-finder and a spherical bubble.

Profiles are illuminated with the "Photosect 40" flash-device from Rockset AB – Sweden (Fig. 17). The original equipment has been slightly modified in order to provide a reference frame of five known points instead of two. For this purpose, four rods with reflecting tips are mounted around the cental plate of the Photosect and a fifth target is attached to the centre of the plate, together with a flat retro-reflector for the direct measurement of distances.

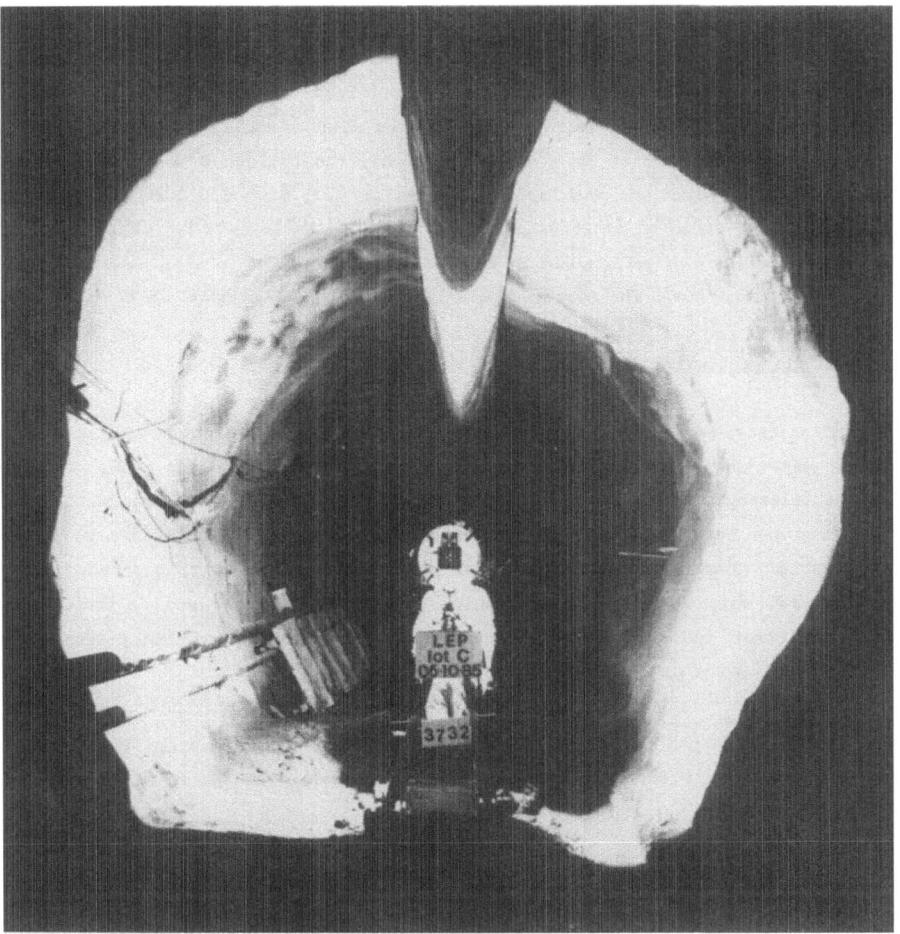

Fig. 17   A photo-profile in limestone section of the tunnel

The camera and the Photosect are each set up on special wheeled tripods which can be easily drawn along the tunnel rails. These two carriages are linked with a rope to ensure the correct framing distance. The positions of the profiles to be measured are first marked on the tunnel walls and the reference plane of the photosect is visually orientated using these marks. The analytical method used for computation of the profiles does not require rigorous perpendicularity of the optical axis of the camera with respect to the transverse section. This simplifies the operating procedure since the camera need only be roughly pointed towards the central target, using the external view-finder.

Absolute position in the tunnel is determined by means of a tacheometric station AGA 140, set up on a known reference pillar of the gyroscopic traverse used for guiding the boring machine. The angles and distance to each station of the photosect are recorded in the "Geodat" solid-state memory.

To simplify measurement of the photographs, films are enlarged by a factor three; positive prints being made on special paper with a plastic core for stability and rigidity, its flatness being ensured by suction plate. The orthoscopy of the enlarger has been verified with a test-grid and proved to be good, strict parallelism between the film-holder and the printing table not being critical since pictures are analytically rectified. Unfortunately, it was not possible to find a small electrostatic digitising table with a real accuracy of $1\sigma = 0.1mm$. Instead, the Digicon B polar digitiser CORADI was adopted. With polar and angular resolutions of 0.01 mm and 0.0040 gon respectively, the maximum r.m.s. error on coordinates is 0.03 mm. Although more accurate than initially needed, this device is well adapted to the observation of photographs and gives an additional safety margin. The measured data is collected on an EPSON HX 20 micro-computer.

4.2.3. Data processing

For efficiency as well as for accuracy, it was decided to process the data in an analytical way. The analogical methods are in fact too heavy and too slow; they require the use of telescopes to ensure rigorous orthogonality of the optical axis with respect to the object plane, and the scale of the enlargement must be visually controlled on the printing table. Two methods of analytical processing have been considered. One is by space resection which computes the position, angular parameters and calibration of the camera, the other being by "rectification" of the photograph, which expresses the projective transform between the object plane and the image plane.

Due to the fact that the object is a plane in space and the camera is non-metric, it was not advantageous to choose the space resection method. In the relation between image and object coordinates :

$$\begin{pmatrix} x - x_s \\ y - y_s \\ c \end{pmatrix} = k.R. \begin{pmatrix} X - X_s \\ Z - Z_s \\ D \end{pmatrix} \tag{5}$$

where k is a scale factor, R an orthogonal 3 x 3 rotation matrix and s the perspective centre, the unknown focal length C cannnot be determined if the distance D between the camera and one point of the reference frame is not measured.

It was more convenient – again for efficiency – to formulate the problem as a projective transform :

$$x_i = \frac{A\ X_i + B\ Z_i + C}{G\ X_i + H\ Z_i + 1} \qquad\qquad (6)$$

$$y_i = \frac{D\ X_i + E\ Z_i + F}{G\ X_i + H\ Z_i + 1}\ . \qquad\qquad (7)$$

In this way, with one more free parameter in the equations, no special care other than framing is necessary when setting up the camera. Any position of the image plane is related to the object. With eight unknowns, four known points in the object reference frame would be sufficient to obtain a solution. In order to improve the solution and to check the computation, the fifth control point (central plate of the Photosect) is used and also gives redundancy. The projective equations are linearised and solved by the least-squares method.

These computations are made with a HP 9845 desk computer. Coordinates of the gyroscopic traverse and tacheometric data are taken into account for the absolute positionning of the measured profiles. A comparison with theoretical profiles is then made and the final results are drawn on a BENSON plotter (Fig. 18).

4.2.4 Performance

All error sources of the photo-profile method have been investigated including mechanical errors of the rods making up the reference frame, misalignment of the Photosect device due to the inaccuracy of the bubble, conicity of the light-flux and random observation errors. All these factors give a total predicted error on coordinates of $\sigma < 1$ cm for $\emptyset$ 4.40 m profiles, including linkage to the gyroscopic traverse.

An experimental check has been made on the 4 m x 5 m frame of a hall door. Eight targets were placed on the frame and then measured with an accurate micro-triangulation. Eleven photographs were taken in the same manner as those in the tunnels. A statistical study of the residuals left with respect to the theoretical test network gave $\sigma x = 6.7$ mm and $\sigma z = 6.5$ mm, in good agreement with the previous predicted errors.

To give an idea of the efficiency of the method, with which more than a thousand measured photo-profiles have been obtained, the operation times – excluding travel to the work area – are the following :

- 2 minutes for the full data acquisition of one Photosect station,
- 18 minutes per photo-profile for observation, computation and plotting.

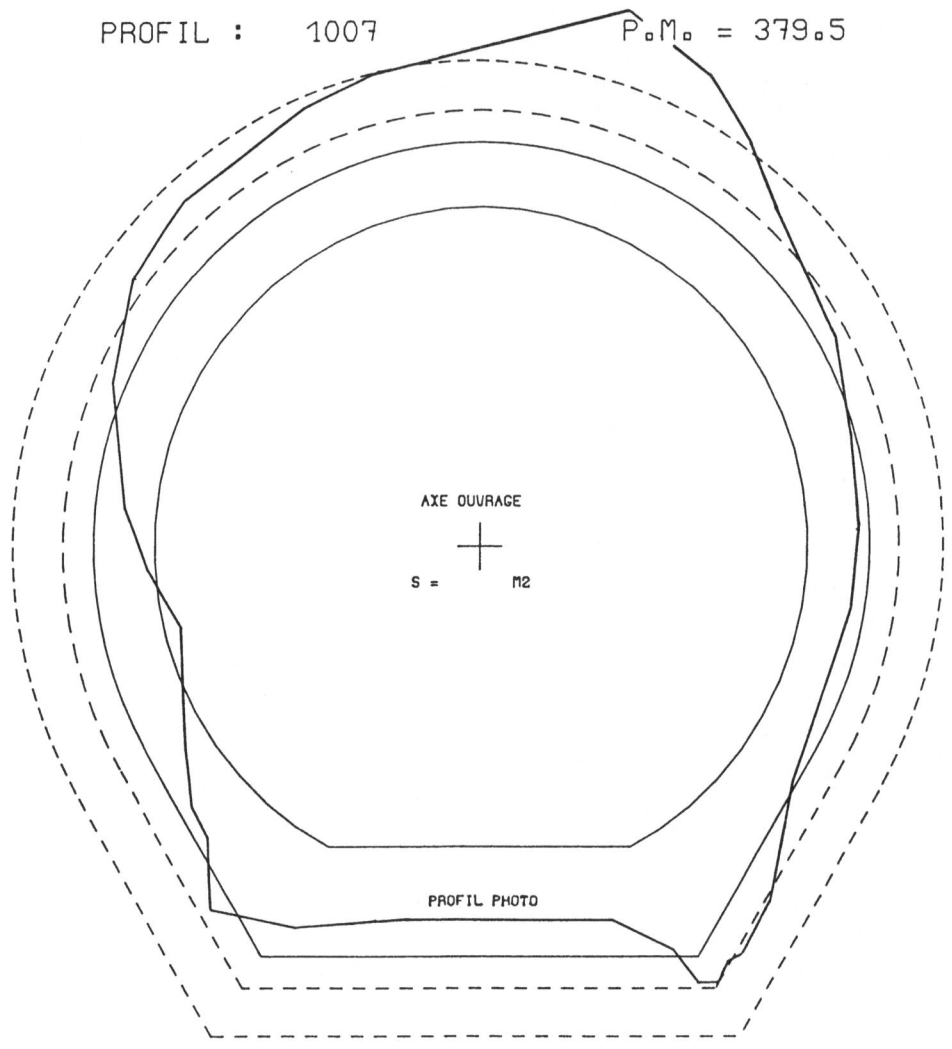

Fig. 18   Photo-profile plot

Initially developed to check the limestone section of the tunnel excavated with explosives, the method proved to be so fast, easy and accurate that it has been extended to the whole tunnel. In the molasse, excavated with full-face boring machines, the tunnel is protected by a concrete segmental lining. The transverse controls here are also made with photo-profiles every 9.80 m (coffering length) to facilitate the positioning of the concrete segments (Fig. 19). This was the simplest and quickest way to make a preliminary check, to provide useful data to the civil-engineering technicians and to guarantee that the tunnel will have the right dimensions and correct location.

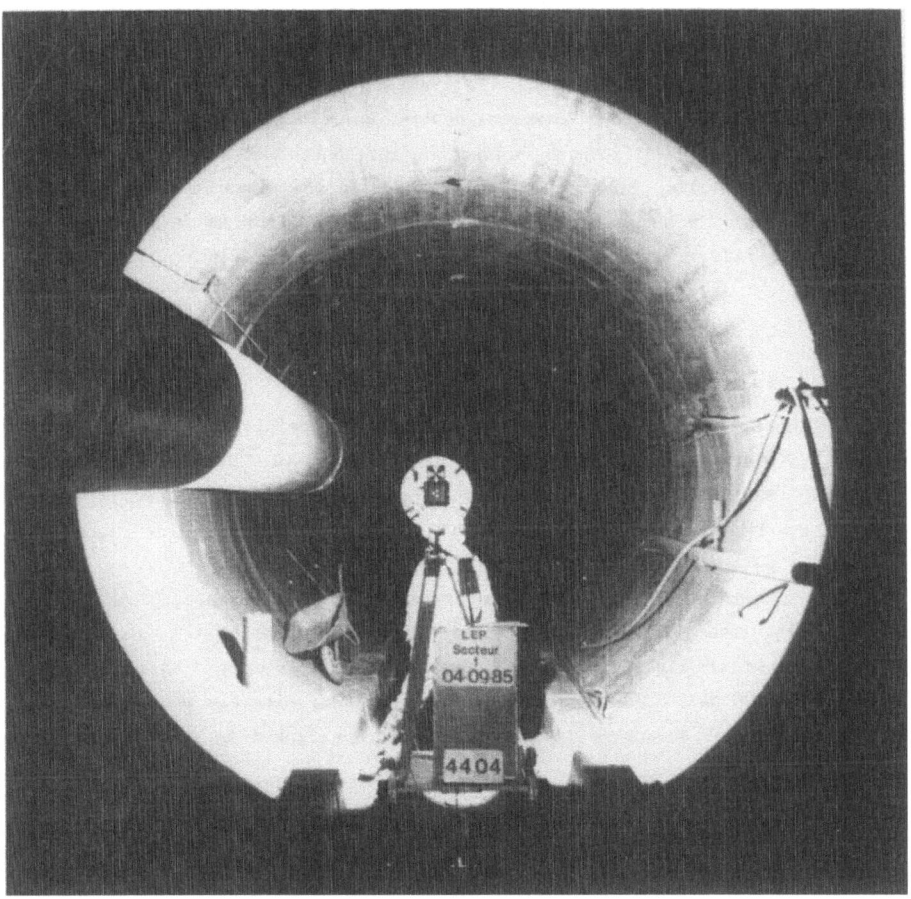

Fig. 19   A photo-profile of the semi-finished tunnel in the molasse

4.3 <u>Electro-optical profiles with a profile scanner</u>

4.3.1 <u>Necessity of another method</u>

As already stated, the accuracy of photo-profiling is dependent on the ratio between the reference frame and tunnel dimensions.   One other limitation, mainly with flash systems, comes from the difficulty to illuminate the vault.   For the large caverns which will house the huge experimental equipment of LEP, it was necessary to find a profiling method able to measure up to 25 m high with again an accurcy of about 1 cm.   The equipment of the physics experiments fits into these caverns so   tightly that construction tolerances have   to   be   very   strict   and   photo-profiling   is   no   longer   suitable. Triangulation would be a long and costly way to make the measurements, with many targeting problems.   The solution is to use an electro-optical distancemeter which can work without a retro-reflector.   Such an instrument exists and is based on the measurement of the transit time of a pulsed laser light, directly reflected by the object being measured.

4.3.2 The profile scanner Geo-Fennel FET 2

This instrument is a "total station", i.e. an electronic tacheometer, in which the distancemeter is different from the normal devices used in geodesy and surveying. Instead of measuring the phase difference between a reference beam and the return beam of modulated light travelling forwards and backwards over the distance to be determined, the instrument measures the transit time of the pulses emitted by an infra-red (λ = 905 nm) laser diode at a rate of 10^5 Hz :

$$D = \frac{1}{2} v.t .$$ (8)

The accuracy of such a measurement depends on the precision of the oscillator and on the degree to which the speed of light v can be assessed. The manufacturer does not give details about the technology and the components of this instrument, but claims an accuracy of $\sigma = \pm 1$ cm for distances up to 2500 m.

The main advantage of this difference principle is that the monochromaticity and the coherence of the laser light gives a sufficient identification of the return signal peaks even in the case of imperfect reflections of the light beam. This property allows measurements to be made without retro-reflector prisms, directly onto rock or concrete walls. An additional advantage is that cyclic errors linked to phase measurements are suppressed.

The other functions of the instrument are that of an electronic theodolite : measurement of azimuthal and zenithal angles (with automatic collimation) using incremental encoders. To facilitate profile scanning, the vertical motion is fitted with a rest disk allowing for stepwise measurements every 5 gon. The instrument is also equipped with a built-in data memory with a storage capacity of 1000 full measurements.

4.3.3 Calibration of the distancemeter

Several tests have been made at the CERN calibration base in order to verify the performance of the instrument [13]. The first calibration consisted of 20 measurements every meter, over distances ranging from 1 to 30 m, on an aluminium plate used as a reflector. In the resulting calibration curve (Fig. 20), a singular shift can be seen at about 7 m, but despite this fact, the discrepancies remain within less than ± 1 cm at the two mean levels of the curve. After having chosen another "zero" constant arising from the low level of the previous curve, a second set of calibration measurements has produced the curve of Fig. 21. Here measurements over 7 m are all contained in an error band of ± 5 mm. Errors on short distances seem to have a linear drift from + 30 mm to zero.

Other experiments have been carried out to study the influence of the reflectance of the materials. Using white or dark paper as a reflector, approximately the same shape of calibration curves have again been found but with shifts between them. Using a plate of dark rough concrete a similar curve has also been produced between 4 and 18 m. Here the discrepancies are more spread out, showing the influence of the surface irregularities.

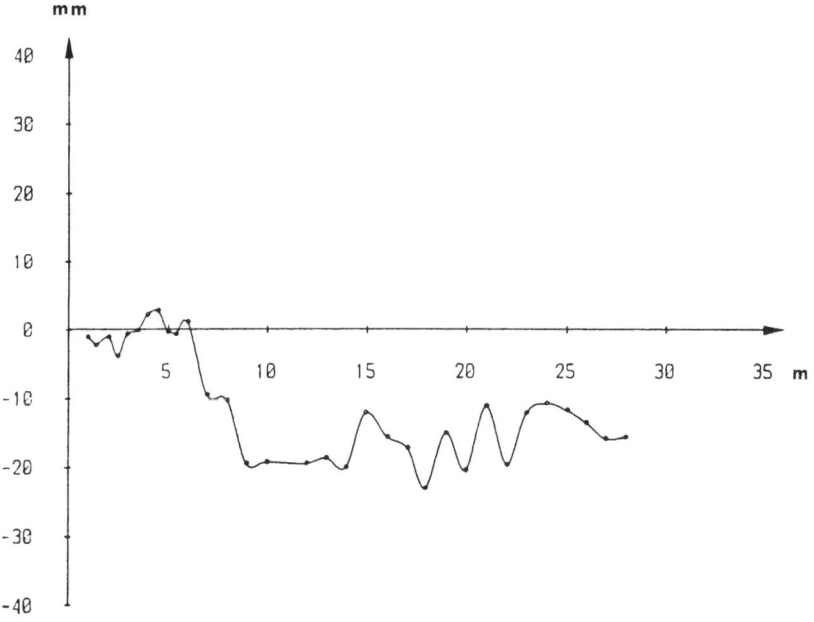

Fig. 20   Calibration curve of the FENNEL distancemeter for short distances

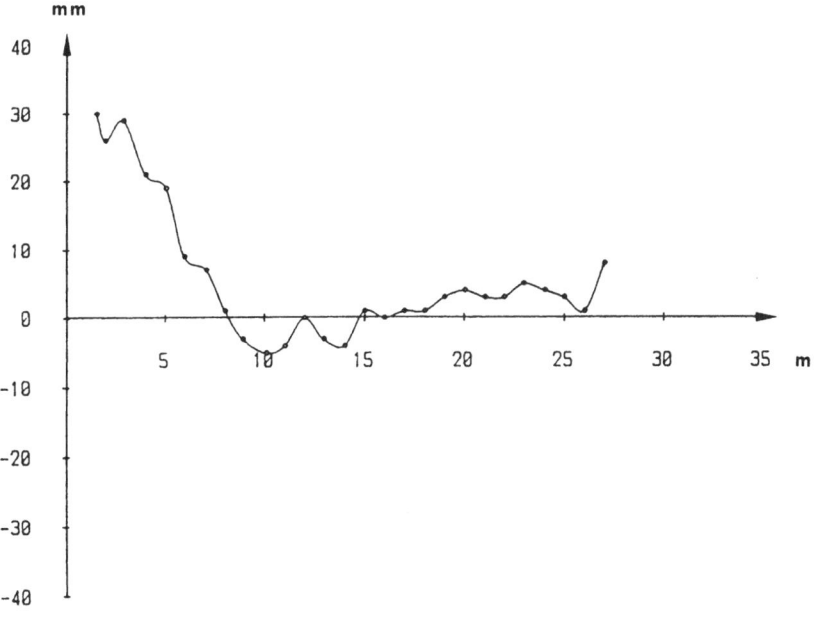

Fig. 21   Calibration curve of the FENNEL distancemeter for long distances

The conclusion of these experiments is that the measurements are strongly influenced by the nature of the materials. The best way to cancel these singular shifts is to make in situ calibrations of the distancemeter. Several distances must be measured against a graduated ribbon between chosen points and the instrument, in order to fix an appropriate constant before making further measurements. After this preliminary operation, it can be expected that the accuracy will remain at about ± 1 cm r.m.s.

4.3.4 Profile measurements in tunnels

Profil 1007   P.M. = 379.5

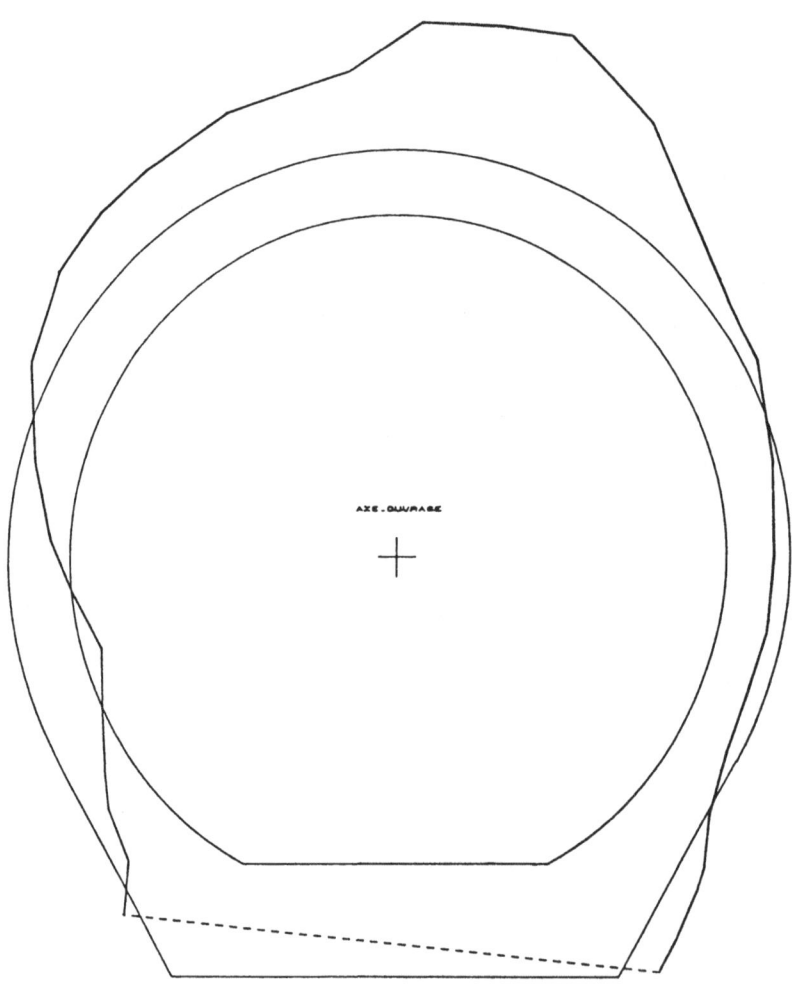

Fig. 22   FET 2 profile plot

The standard handle of the FET 2 instrument has been replaced by a special bridge which permits the force-centring of CERN device combining a light target and a retro-reflector prism. In this way, the profile station can be accurately linked to a reference pillar of the gyro traverse, using the AGA 140 tacheometric station. The local orientation of the instrument can be estimated or determined by angular measurements referring to the bearing determined with the AGA station.

After a control of the calibration, the direct survey of the profile is processed stepwise by means of a ratchet mecanism on the vertical motion. However, special attention must be paid to characteristic points, in order not to miss them.

A comparison on six profiles with the photo-profiling method has given satisfactory results. Figure 22 shows the same profile as that processed by photoprofile and shown in Fig. 18 but this time using the FET2; the two results being almost identical.

5. CONCLUSION

The geometric control of underground construction is a difficult and critical task due to specific technical problems, environmental constraints and the heavy responsability. To avoid complaints and conflicts, the credibility of these controls must be impeccable. Consequently, a high level of accuracy and reliability have been sought.

A major consideration was to ensure that the methods adopted were highly efficient, i.e. that measurements could be made rapidly and in a limited number of operations. This is extremely important since not only is access to the boring area difficult and extremely confined, but also the survey measurements cannot be allowed to unduly perturb the excavation work.

It is hoped that in the near future the technology of optical gyroscopes will provide a faster, simpler, but nevertheless accurate way to oriente tunnels.

*   *   *

REFERENCES

1) J. Gervaise, Applied geodesy for CERN accelerators, Part I, Seminar on High Precision Geodetic Measurements, Facolta di Ingeneria di Bologna, (1984).

2) M. Mayoud, Applied geodesy for CERN accelerators, Part II, Ibid.

3) M. Hayotte, Tolérances, implantation et contrôle de la géométrie, Annexe 9 de l'Appel d'Offres CERN I-1061/SB-LEP.

4) J. Gervaise, First results of the geodetic measurements carried out with the terrameter, two-wavelength electronic distance measurement instrument, Paper presented at the Geodätisches Seminar über Electrooptische Präzisionsstreckenmessung, Müunchen, (1983).

5) J. Gervaise, J. Olsfors , The LEP trilateration network, these proceedings.

6) J. Iliffe, Three-dimensional adjustment in a local reference system, Ibid.

7) G. Bain et al., Automation by microcomputer of a geodetic distance measuring instrument : the Distinvar, CERN 85-16, (1985)

8) G. Bain et al., Le Gyroscope asservi - automation par micro-ordinateur, to be published.

9) J. Iliffe, M. Mayoud, Computation of the free oscillation of the GAK 1 CERN gyro by means of numerical integration, CERN, LEP-SU/int. 84-01, (1984)

10) J.-C. Fischer, Contribution à l'automatisation du Gyroscope de Théodolite WILD GAK1 en vue de son utilisation au CERN lors de l'implantation du LEP, Mémoire de Soutenance de Diplôme d'Ingénieur Géomètre, ENSAIS, (1986).

11) M. Mayoud, Géomètrie Théorique du LEP, CERN LEP Note 456, (1983).

12) S. Burki, W. Gurtner, Deviation of the vertical, these proceedings.

13) H. Lançon, Le Profilomètre GEO-FENNEL FET2 et son exploitation dans le cadre du contrôle des ouvrages souterrains du LEP - comparaison avec la méthode du photoprofil, Mémoire de Soutenance de Diplôme d'Ingénieur Géomètre, ENSAIS, (1986).

III. Applied Geodesy for Particle Accelerators

HIGH PRECISION GEODESY APPLIED TO CERN ACCELERATORS

J. Gervaise and E.J.N. Wilson
CERN, Geneva, Switzerland

ABSTRACT

After examining the whys and wherefores of applied geodesy in high energy particle accelerator construction, the authors show how the transverse beam size and hence the dimension of the magnets are related to positional tolerances and review the methods and instruments developed for this purpose. Three accelerators have successively been built at CERN, the 28 GeV Proton Synchrotron (PS), The Intersecting Storage Rings (ISR), the Super Proton Synchrotron (SPS), and a fourth one is under construction, the Large Electron Positron Collider (LEP). Since 1954, date of the PS construction, many difficulties have been encountered and the authors show how applied geodesy has been able to overcome them. From the traditional geodetic methods available in the 1950's, used to build a 200-m diameter accelerator, up to the 1980's, many methods and instruments ranging from invar wire to Terrameter, have been developed. These have improved the reliability, accuracy and speed of the geodetic measurements and have given a philosophy and a strategy for the Applied Geodesy of a 27-km circumference collider such as LEP.

1. INTRODUCTION

Over the last three decades, particle accelerator energies have grown from several tens of GeV to several hundred GeV, and unavoidably the diameters of circular accelerators have similarly increased from two hundred metres to ten kilometres. Despite this escalation, it has proved possible to achieve tolerances on magnet positioning which are not only as tight as in the past, but even more precise due to the survey experience already gained.

The sequence of large projects undertaken by CERN, the 28 GeV Proton Synchrotron (PS, 1954-1959), the Intersecting Storage Rings (ISR, 1966-1971), the 400 GeV Super Proton Synchrotron (SPS, 1971-1976) and the Large Electron Positron Collider (LEP, 1981-), has ensured continuity in the work of the CERN Applied Geodesy Group, and each project has been in itself a major challenge. The Group has thus been constantly searching to improve reliability and speed of measurement without sacrificing in any way the accuracy essential to proper accelerator operation.

A circular accelerator consists basically of magnets arranged in series according to a certain configuration. The magnets are separated by straight sections where there is no magnetic field, and form a lattice having a certain periodicity P. To an observer standing outside, the magnetic structure resembles an underground train in a circular tunnel. However, for the accelerated proton, which sees things from the inside, the picture is quite different. It has the sensation that it is running round a magnetic channel, and this channel must be as smooth as possible to ensure that no oscillations are induced which would broaden the particle beam to which it belongs. The protons are sensitive to any bumps occurring in the channel; these bumps may be due to three causes, each independent of the other, but which have the same effect on the magnetic field that guides the beam around the machine. They are :

- slight movements of the foundation rock or of the concrete structures supporting the magnets,
- random errors in magnet alignment,
- errors in the beam guidance field itself.

These errors perturb the ideal closed orbit which passes down the axis of the magnets. The geodesist measures angles, distances, offsets and tilts. After least squares adjustment, he provides the beam orbit specialists with the following values :

- dr : the radial deviation of the synchrotron proton orbit,
- dz : the vertical deviation from the mean plane selected for the closed orbit along which the accelerated protons circulate,
- dt : the deviation in the transverse tilt of the magnets, which introduces coupling between the horizontal and vertical planes and makes orbit corrections difficult.

As an illustration of the effect of these errors and their correction, Fig. 1 shows the strong correlation between the survey results and the local orbit corrections in the SPS ring. Also shown is the effect of a correction procedure by a deliberate movement of a few quadrupoles by a few tenths of millimetres, calculated from orbit measurements made with beam position sensors. The amplitude of the closed orbit can thus be reduced by a factor 2 or 3.

It is a feature of machines such as the SPS that the relative errors of quadrupole focusing magnets are the most critical : the radial and vertical variations from one quadrupole to the two adjacent quadrupoles must remain as close as possible to the ideal value given by the theoretical geometry. Next, the curve formed by the successive positions of the quadrupoles has to be made as smooth as possible along the entire perimeter of the machine. Thus, the two basic concerns in accelerator metrology are, first to carry out an overall survey that is close to the theoretical geometry and then to smooth the curve in order to minimize the errors locally. We shall discuss this aspect in later sections.

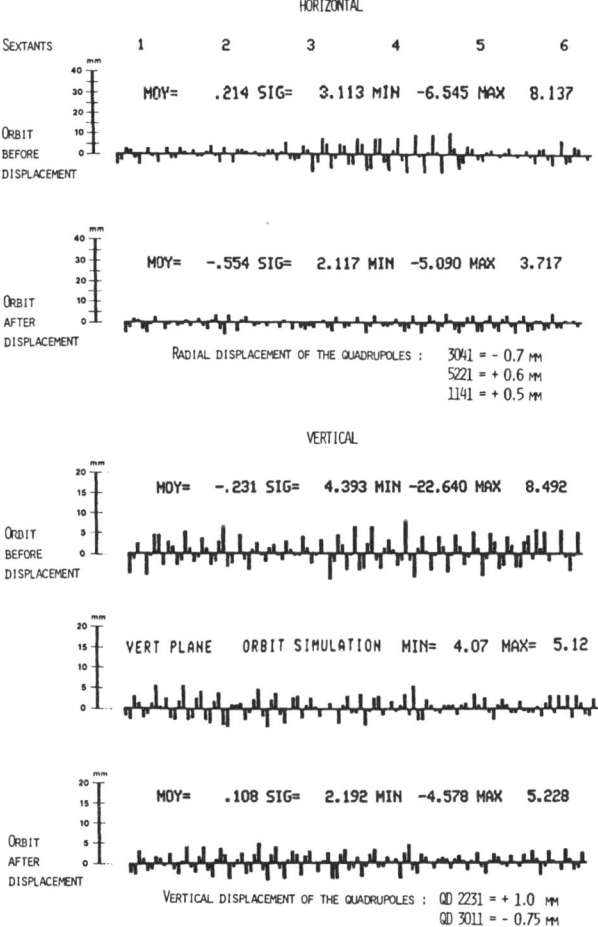

Fig. 1   Beam position measurements before and after correction

A large fraction of the considerable investment in capital and technological ingenuity necessary to build a modern synchrotron finds its way into the magnetic fields which guide and focus the beam in its circular path. Their alignment is crucial to the performance of the machine and stretches survey techniques to the limit. In the following sections, we first show how the transverse beam size and hence the dimensions of the magnets is related to positioned tolerances and go on to show how the tolerances may be realised.

2. TRANSVERSE COORDINATES

The bending fields of a synchrotron are usually vertical directed, causing the particle to follow a curved path in the horizontal plane (Fig. 2). The force acting on the particle is horizontal and is given by :

$$\bar{F} = e \ \bar{v} \ x \ \bar{B} \tag{1}$$

where : v is the velocity of the charged particle in the direction tangential to its path,
B is the magnetic guide field.

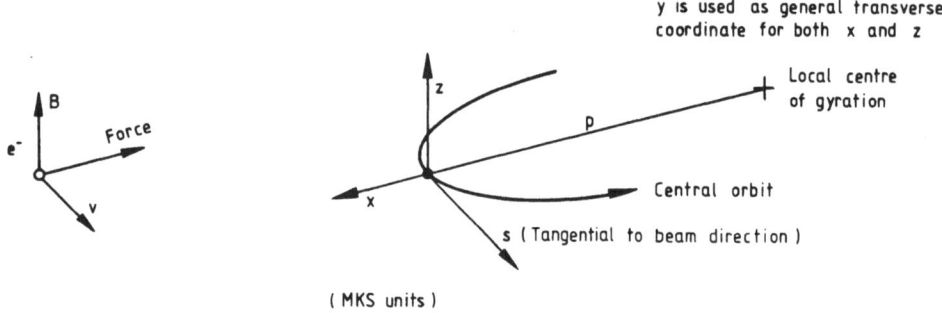

(MKS units)

Fig. 2   Charged particle orbit in a magnetic field

If the guide field is uniform, the ideal motion of the particle is simply a circle of radius of curvature ρ. We can also define a local radius of curvature ρ(s), to describe motion in a non-uniform field. We shall suppose that it is possible to find an orbit or curved path for the particle which closes on itself. We define this as the equilibrium orbit or closed orbit. The machine is usually designed with this orbit at the centre of its vacuum chamber.

Of course, a beam of particles enters the machine as a bundle of trajectories which are centred on the ideal orbit. At any instant a particle may be displaced horizontally or vertically from the ideal position. The transverse displacements are x and z, respectively. The particles may also have a divergence angle :

$$x' = dx/ds \quad \text{or} \quad z' = dz/ds \tag{2}$$

This would cause them to leave the vacuum pipe if it were not for the carefully shaped field which turns them back towards the beam centre so that they oscillate about the ideal orbit. The design of the restoring fields determines the transverse excursions of the beam and the size of the cross section of the bending magnets. The restoring or focusing magnets have four poles and are usually embedded in the lattice of bending magnets in an alternating pattern. Half focus the beam while, the other half defocus. In Fig. 3 we see an example of such a magnet pattern, which is one cell, or about 1% of the circumference of the 400 GeV SPS at CERN. This focusing structure is called FODO. For the purpose of this paper, it is sufficient to grasp the idea of a beam of particles oscillating from side to side of the centre of the machine in the focusing channel of quadrupoles. The envelope of these oscillations follows the function β(s) which has waists near each defocusing magnet and a maximum at the centres of F quadrupoles. Since F quadrupoles in the horizontal plane are D quadrupoles vertically, and vice versa, the two functions $\beta_H(s)$ and $\beta_V(s)$ are one half-cell out of register in the two transverse planes.

For formal reasons β has the dimensions of length but the units bear no relation at this stage to physical beam size.

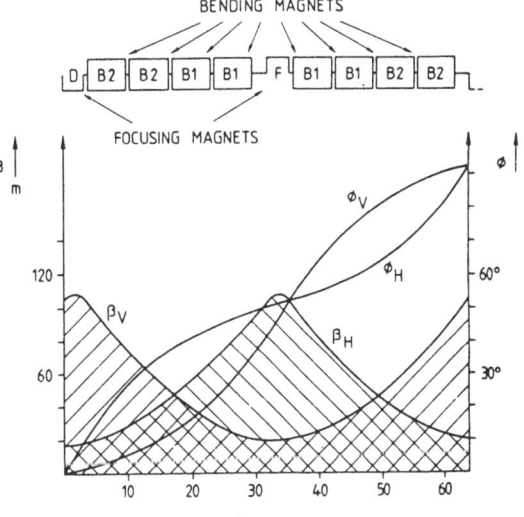

Fig. 3 One cell of the CERN 400 GeV Super Proton Synchrotron representing 1/108 of the circumference. The pattern of dipole and quadrupole magnets (F and D) is shown. Beam particles make betatron oscillations within the shaded envelopes.

Particles do not follow the β(s) curves but oscillate within them in a form of modified sinusoidal motion whose phase advance is φ. The phase change per cell in the example shown is close to π/2 but the phase advance is modulated throughout the cell.

Later, we shall discuss how the functions β and φ are related to the focusing pattern and to beam sizes.

3. <u>BENDING MAGNETS AND MAGNETIC RIGIDITY</u>

Most of the ring of a synchrotron consists of a guide field of bending magnets.

Suppose the particle has a relativistic momentum vector p and travels perpendicular to B into the plane of the diagram (Fig. 4). After time dt it has followed a curved path of radius ρ whose length is ds and its new momentum is p + dp. Since we may equate the force and rate of change of momentum :

$$e \ \bar{v} \ x \ \bar{B} = \frac{dp}{dt}$$

(3)

and we see from resolution of momenta that :

$$\frac{dp}{dt} = |p| \ \frac{d\theta}{dt} = \frac{|p|}{\rho} \ \frac{ds}{dt} \ .$$

(4)

On the other hand, if the field and plane of motion are normal, the magnitude of the force may be written :

$$e \; |\bar{v} \times \bar{B}| \; = \; e \; |B| \; \frac{ds}{dt} \; . \tag{5}$$

Equating (4) and (5) we find we can define a quantity known as magnetic rigidity :

$$(B\rho) \; = \; \frac{p}{e}. \tag{6}$$

Strictly we should use the units Newton.second for p and express e in coulombs to give $(B\rho)$ in tesla.metres. But in charged particle physics a more common convention is to quote pc in units of electronvolts. Whereupon :

$$(B\rho)[T.m] \; = \; \frac{pc[eV]}{C[m.s^{-1}]} \; = \; 3.3356(pc)[GeV] \; . \tag{7}$$

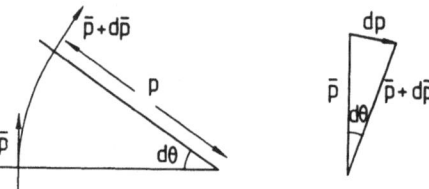

Fig. 4   Vector diagram showing differential changes in momentum for a particle trajectory

Figure 5 shows the trajectory of a particle in a bending magnet or dipole of length ℓ. Usually the magnet is placed symmetrically about the arc which is the particle's path. One may see immediately that :

$$\sin \; (\theta/2) \; = \; \frac{\ell}{2\rho} \; = \; \frac{\ell B}{2(B\rho)}, \tag{8}$$

and if $\theta \ll \pi/2$,

$$\theta \; \approx \; \frac{\ell B}{(B\ell)} \; . \tag{9}$$

The bending magnet aperture must be wide enough to contain the sagitta of the beam which is :

$$\pm \; \frac{\rho}{2} \; \left[1 \; - \; \cos(\theta/2) \right] \; \approx \; \pm \; \frac{\rho\theta^2}{16} \; \approx \; \frac{\ell\theta}{16} \; . \tag{10}$$

The ends of bending magnets are often parallel but in some machines are designed to be normal to the beam. There is a focusing effect at the end which depends on the angle of these faces.

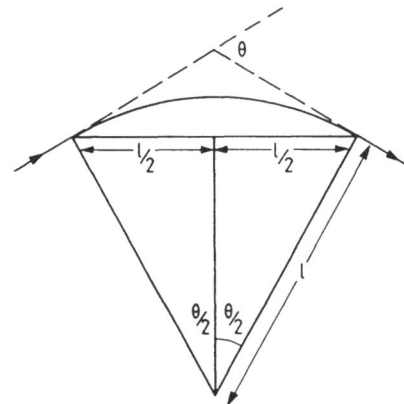

Fig. 5   Geometry of a particle trajectory in a bending magnet of length ℓ

4. QUADRUPOLE MAGNETS AND FOCUSING

The principal focusing element in a modern synchrotron is a quadrupole magnet. The poles of a quadrupole alternate and the field shape (Fig. 6) is such that it is zero on the axis of the device but its strength rises linearly with distance from the axis. This can be seen from a superficial examination of Fig. 6, remembering the product of field and length of a field line joining the poles is a constant. Symmetry tells us that the field is vertical in the median plane (and purely horizontal in the vertical plane of asymmetry). The field must be downwards on the left of the axis if it is upwards on the right.

This last observation ensures that the horizontal focusing force, evB_z, has an inward direction on both sides and, like the restoring force of a spring, rises linearly with displacement, x. The strength of the quadrupole is characterized by its gradient dB_z/dx, normalized with respect to magnetic rigidity :

$$k = \frac{1}{(B\rho)} \frac{dB_z}{dx} \quad .$$

(11)

The angular deflection given to a particle passing through a short quadrupole of length ℓ and strength k at a displacement x is therefore :

$$\Delta x' = \ell k \cdot x \cdot$$

(12)

If we compare this with a converging lens,

$$\Delta x' = x/f \quad ,$$

(13)

we see that the focal length of such a quadrupole is :

$$f = 1/(k\ell) \ . \tag{14}$$

The particular polarity of quadrupole shown in Fig. 6 would focus positive particles coming out of the paper, or negative particles going into the paper in the horizontal plane. A closer examination reveals that such a quadrupole is defocusing in the vertical plane and deflects particles with a vertical displacement away from the axis. We shall see later that in spite of this seemingly damning characteristic a FODO pattern of alternating gradient has a net focusing effect in both planes.

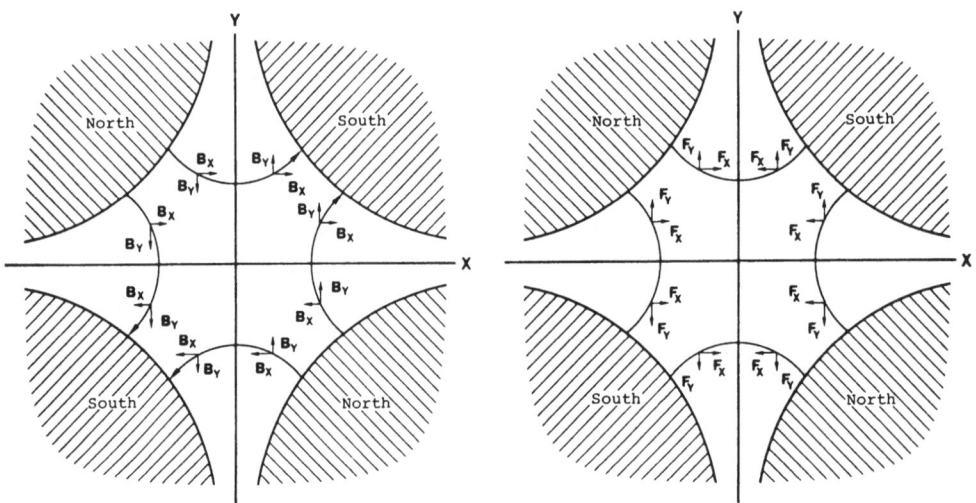

Fig. 6   Components of field and force in a magnetic quadrupole.
Positive ions approach the reader on paths parallel to the y axis.

5. THE GUTTER ANALOGY OF FOCUSING

It is important to start with an almost tangible concept of focusing. If we are prepared to ignore vertical defocusing for a moment and consider an infinitely long quadrupole we find that a particle oscillates within it exactly like a small sphere rolling down a slightly inclined gutter with constant speed. Figure 7 shows two views of this motion and from the right hand view we recognize the motion as a sine wave : note too that the sphere makes four complete oscillations along the gutter. In accelerator terms its motion has a wave number Q = 4.

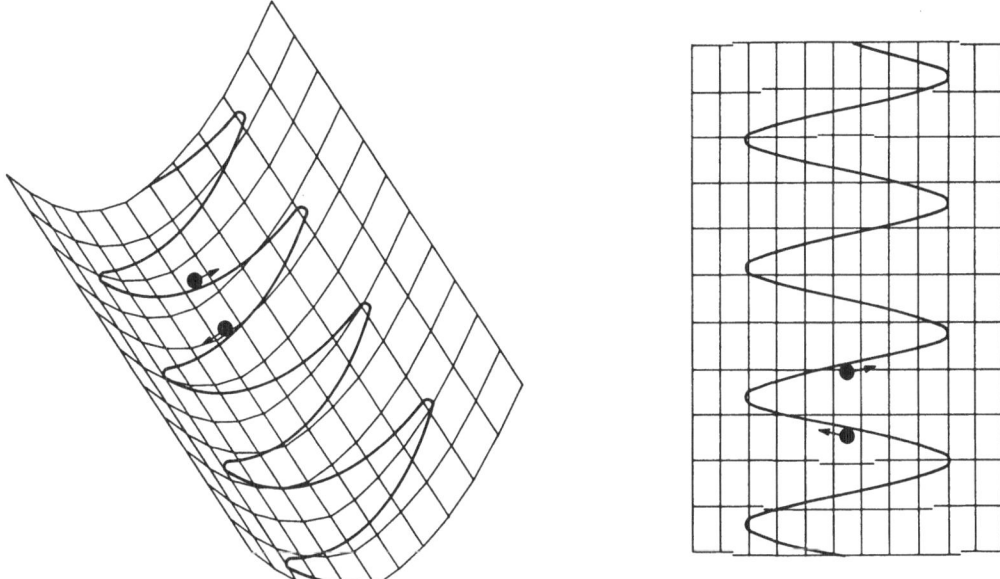

Figure 7   Two views of a sphere rolling down a gutter as it is focused by the walls.

Now extend this analogy by bending the gutter into a circle rather like the brim of a hat. Suppose we provide the necessary instrumentation to measure the displacement of the sphere from the centre of the gutter each time it passes a given mark on the brim. We also have a means to measure its transverse velocity which, with the aid of a microcomputer, we might convert into its divergence angle :

$$x' = \frac{dx}{ds} = \frac{v_\perp}{v_\parallel} \; .$$

Suppose we make the hat out of a slightly different length of gutter than shown so that Q is not an integer. If we plot a diagram of x' against x we can plot a point for each arrival of the sphere. This is called a phase space diagram. The sphere may have a large transverse velocity as it crosses the axis of the gutter or it might have almost zero transverse velocity as it reaches its maximum displacement.

The locus will be an ellipse (Fig. 8) and the betatron phase will advance by Q revolutions each time the particle returns. Of course, only the fractional part of Q may be deduced.

The time has come to define some of the beam dynamical quantities more rigorously. The area of the ellipse is a measure of how far away from an ideal trajectory the particle is :

$$\text{Area} = \pi\epsilon \; [\text{mm.mrad}] . \tag{15}$$

The maximum amplitude, or major axis, of the ellipse is defined :

$$x = \sqrt{\epsilon\beta},\qquad\qquad\qquad (16)$$

so that to satisfy equation 15 :

$$x' = \epsilon/\beta.$$

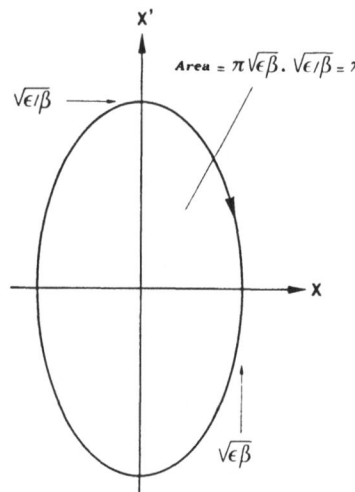

Fig. 8   The elliptical locus of a particle's history in phase space as it circulates in a synchrotron.

The quantity β is a property of the focusing system, not the beam, and is the function plotted in Fig. 2. In an alternating gradient focusing system such as Fig. 1 the brim of the hat will vary its width and curvature around the crown and β will follow this variation in some way. Note that the aspect ratio of the ellipse is just β.

6.   ALTERNATING GRADIENT FOCUSING

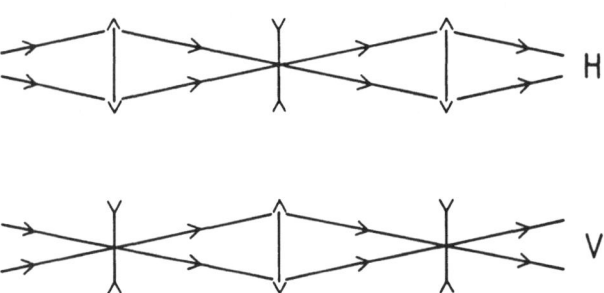

Fig. 9   Ray diagrams to show the existence of a contained trajectory in an alternating gradient optical system. The upper diagram shows the horizontal motion and lower shows the vertical.

The principle is best illustrated in Fig. 9 which shows an optical system in which each lens is concave in one plane while convex in the other. It is possible, even with lenses of equal strength to find a ray which is always on axis at the D lenses in the horizontal plane and therefore only sees the F lenses. The spacing of the lenses would have to be 2f. If the ray is also central in the lenses which are vertically defocusing the same condition will apply simultaneously in the vertical plane. At least for one particular particle or ray the trajectory will be contained.

Now let us return to the FODO cell structure. We shall see that by suitable choice of lens strength, the function β(s), which follows the peak excursion of a beam, can be made to be periodic in such a way that it is large at all F quadrupoles and small at all D quadrupoles. Symmetry will ensure this is true also in the vertical plane. Particles oscillating within this envelope will always tend to be further off axis in F quadrupoles than in D quadrupoles and there will therefore be a net focusing action. Earlier we said that β is the aspect ratio of the phase space ellipse. At F quadrupoles the ellipse will be squat and at D quadrupoles it will be tall.

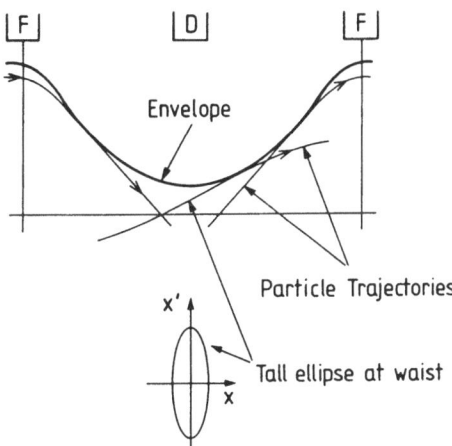

Fig. 10   The paths of particles within a FODO lattice are within the envelope of betatron motion and, like the rays of Fig. 8, are always closer in the D quadrupoles so they receive a net focusing effect. The phase space ellipse is tall and narrow at the D lens where the beam has a large divergence spread.

7. THE EQUATION OF MOTION

The motion of the particle in the transverse direction is described by Hill's Equation. In the vertical plane, this is written :

$$z'' + k(s)z = 0, \qquad (17)$$

$$k(s) = \frac{1}{(b\rho)} \frac{dB_z(s)}{dx}. \qquad (18)$$

This is the famous Hill's Equation, a linear equation with periodic coefficient, k(s), which is the distribution of focusing strength around the ring.

In the horizontal plane, we must include a term for the curvature ρ:

$$x'' + \left[\frac{1}{\rho(s)_2} - k(s) \right] \ x = 0. \tag{19}$$

The equation is reminiscent of simple harmonic motion but with a restoring constant $k(s)$ which varies around the accelerator. We assume that $k(s)$ is periodic on the scale of one turn of the ring or some smaller unit, a cell or period, from which the ring is built.

The solution of Hill's Equation is not unlike simple harmonic motion :

$$y = \sqrt{\epsilon} \ . \ \sqrt{\beta(s)} \cos \ [\varphi(s)+\lambda], \tag{20}$$

where y is a general transverse coordinate (either x or z),
 $\sqrt{\epsilon}$ is the beam's contribution to the amplitude (ϵ = emittance),
 $\sqrt{\beta(s)}$ is a periodic function which describes the modulation of the amplitude due to the structure of the ring.

Note that the envelope of the motion of a beam of particles with random starting phase, λ, will be the function :

$$\sqrt{\epsilon\beta(s)}.$$

The term $\cos[\varphi(s)+\lambda]$ describes the oscillation of an individual particle within this envelope. The phase must satisfy the condition :

$$\varphi(s) = \int \frac{ds}{\beta(s)}. \tag{21}$$

or Eq. (17) will not satisfy the equation of motion, Eq. (19).

8. FLOQUET COORDINATES

In this section we shall consider imperfections and to do this we introduce a transformation into Floquet coordinates. Earlier we derived Hill's Equation :

$$\frac{d^2y}{ds^2} + k(s)y = \frac{\Delta B(s)}{(B\rho)} = F(s) \tag{22}$$

Here we have included a driving term on the right hand side which represents a perturbation due to a field error.

The solutions of Hill's Equation are :

$$y \quad = \quad \sqrt{\epsilon \beta(s)} \; \sin(\varphi(s) + \lambda),$$

$$y' \quad = \quad \sqrt{\epsilon/\beta(s)} \; [\cos(\varphi(s) + \lambda) - \alpha \sin(\varphi(s) + \lambda)], \tag{23}$$

$$\varphi(s) \quad = \quad \int ds/\beta;$$

which form an invariant :

$$\Upsilon \; y^2 + 2 \; \alpha \; yy' + \beta \; y'^2 = \epsilon. \tag{24}$$

The equation and its solutions are so reminiscent of a forced harmonic oscillator that it is tempting to seek a transformation which removes the s dependence of $k(s)$, $\beta(s)$ and $\varphi(s)$. Such a transformation, named after Floquet, involves a change to new variables :

$$\eta = \frac{x}{\sqrt{\beta(s)}}, \quad \psi = \int \frac{ds}{Q \; \beta(s)}. \tag{25}$$

Substitution in Eq. (22) yields :

$$\frac{d^2\eta}{d\psi^2} + Q^2\eta = g(\psi) = Q^2\beta^{3/2}(s)F/(s) . \tag{26}$$

We have reduced the problem to that of a forced harmonic oscillator at the expense of a little physical reality. Transverse displacements are scaled by the local beta function and ψ does not correspond to azimuth. However, ψ, like azimuth, θ, advances 2π per turn and is equal to azimuth at the quadrupoles of a FODO lattice. We shall make use of these new Floquet coordinates in the next section where the effect of errors is often treated as a forcing term. Note that errors are weighted with $\beta(s)^{3/2}$ on the right hand side.

9. ORBIT DISTORTION FROM ONE KICK

The size of the vacuum chamber and magnet aperture can become uneconomically large if field imperfections which can come from misalignments are allowed to distort the perfection of the central synchronous orbit which does not hit the walls. In extreme cases there may be no single closed orbit. To be able to calculate and correct this effect of orbit distortion we must first understand the effect of adding a short dipole of integrated field δ $(B\ell)$ somewhere in the machine. Note that in the jargon of the accelerator designer the effect of such a dipole is a "kick" though the dipole deflects by the same angle on each turn. A "bump" might be a better word but this is reserved for another use.

The effect of the kick is best visualized in the Floquet coordinates of a machine unwound as in Fig. 11. We know the orbit must be a solution of :

$$\frac{d^2\eta}{d\psi^2} + Q^2\eta = 0 \tag{27}$$

and if we place the kick at $\psi = \pi$, symmetry tells us the solution must start as :

$$\eta = \eta_0 \cos Q\psi \tag{28}$$

with a maximum diametrically opposite the kick.

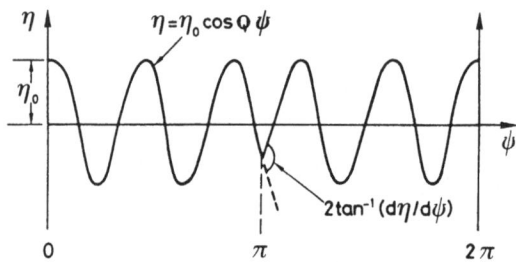

Fig. 11   A closed orbit for one turn of a synchrotron plotted in Floquet coordinates.
The deformation is due to a single dipole at the centre of the diagram.

At the location of the kick the slope of the orbit :

$$- \eta_0 Q \sin\pi Q \tag{29}$$

changes sign and this change must therefore equal half the kick angle :

$$\delta(B\ell)/B\rho \tag{30}$$

transformed into $d\eta/d\psi$ coordinates.

To perform this transformation, we remember :

$$\frac{d\psi}{ds} = \frac{1}{QB(s)}, \quad \frac{dx}{ds} = \beta^{\frac{1}{2}}(s) \frac{d\eta}{ds} . \tag{31}$$

and can deduce that at the kick :

$$x' = \frac{dx}{ds} = \beta_\kappa^{\frac{1}{2}} \frac{d\eta}{d\psi} \cdot \frac{d\psi}{ds} = - \frac{\eta_0}{\sqrt{\beta_\kappa}} \sin \pi Q = \frac{\delta(B\ell)}{2B\rho} . \tag{32}$$

Thus, the amplitude of the distortion is :

$$\eta_0 = \frac{\sqrt{\beta}\kappa}{2|\sin\pi Q|} \frac{\Delta(B\ell)}{(B\rho)} \tag{33}$$

and elsewhere the physical distortion scales with $\beta(s)$ modulated as $\cos\upsilon \, \psi(s)$:

$$x = \left[\frac{\sqrt{\beta(s)}\beta_\kappa}{2|\sin\pi Q|} \cdot \frac{\delta(B\ell)}{(B\rho)} \right] \cos Q \, \psi(s) \; . \tag{34}$$

10. SOURCES OF CLOSED ORBIT DISTORTION

Of course, no one in their right mind would place such a dipole magnet in an otherwise perfect machine. In practice there are a large number of uncorrelated kicks due to misalignments and errors in the magnets of the ring.

Table 1

Causes of Distortion

Type of element	Source of kick	rms value	$(\Delta B\ell/B\rho) = \delta y'$	Plane
Gradient magnet	Displacement	$\langle\Delta y\rangle$	$\kappa\ell\langle\Delta y\rangle$	x,z
Bending magnet (Bending angle = θ)	Tilt	$\langle\Delta\theta_e\rangle$	$\theta\langle\Delta\theta_e\rangle$	z
Bending magnet	Field error	$\langle\Delta B/B\rangle$	$\theta\langle\Delta B/B\rangle$	a
Straight sections	Stray field	$\langle\Delta B_s\rangle$	$d\,[\langle\Delta B_s\rangle/(\rho B)]$	x,z

Table 1 shows some of the common sources of such kicks. Apart from obvious field errors from geometry and remanent field variations among the magnets, there are two sources of distortion related to alignment; the displacement of a quadrupole produces a dipole kick for a particle on the central orbit, and a tilted bending magnet will deflect vertically. The fourth column of Table 1 gives the angular deflection from which Eq. (35) gives :

$$y_i = \frac{\sqrt{\beta(s)}\beta_i}{2\,|\sin\pi Q|} \cdot \delta y_i \cdot \cos\upsilon(\varphi(s)+\lambda_i) \tag{35}$$

where λ is a random phase.

Taking the rms of all these uncorrelated distortions (including the phase λ) gives :

$$y_{rms} = \frac{\beta^{\frac{1}{2}}(s)}{2\sqrt{2}|\sin\pi Q|} \left[\sum \beta_i(\delta y_i)^2 \right]^{\frac{1}{2}} \; . \tag{36}$$

In making allowances for the fact that chance may not be on the side of the accelerator designer, it is traditional to quote twice this value as an allowance which ensures that in more than 90% cases a large aperture allowance is often considered unnecessary luxury. Proven techniques exist to apply correction to obtain the first turn and to the circulating beam orbit.

11. FOURIER ANALYSIS OF ORBIT DISTORTION

Certain frequencies in the azimuthal pattern of errors contribute most to the distortion. This response of the accelerator to these "resonant" frequencies is an important concept which reappears in the theory of gradient errors and later in treating nonlinear effects. Again, we use Floquet's transformed betatron motion to see the driving effect of the error pattern, $F(s)$. If :

$$F(s) = \frac{\Delta B(s)}{B\rho} \tag{37}$$

then the motion is described by the differential equation :

$$\frac{d^2\eta}{d\psi^2} + Q^2\eta = g(\psi) = Q^2\beta^{3/2}F(s) \ . \tag{38}$$

We can think of $g(\psi)$ as the new driving term and Fourier analyse it :

$$g(\psi) = Q^2 \sum_{k'} f_k e^{ik\psi} \tag{39}$$

where

$$f_k = \frac{1}{2\pi} \int_0^{2\pi} B^{3/2}(\psi)F(\psi)e^{ik\psi} \, d\psi \ . \tag{40}$$

since

$$d\psi = \frac{ds}{Q\beta} \tag{41}$$

$$f_k = \frac{1}{2\pi Q} \int_0^{2\pi R} \beta^{1/2}(s)F(s)e^{ik\psi(s)}ds \ . \tag{42}$$

The differential equation now becomes :

$$\frac{d^2\eta}{d\psi^2} + Q^2\eta = \sum_{k=1}^{\infty} \int Q^2 f_k e^{ik\psi} \tag{43}$$

and has the solution :

$$\eta = \sum_{k=1}^{\infty} \frac{v^2}{Q^2 - k^2} f_k e^{ik\psi} \; . \tag{44}$$

We see that the machine will be particularly sensitive to Fourier components where k is close to the Q value of the ring. An example of this is to be seen in Fig. 12 which shows an uncorrected closed orbit in the FNAL main ring. Each bar is the position with respect to the centre line of one of the 100 or so position pick-ups around the ring. the ring has been uncoiled into a straight line. By counting the peaks one can see that the response is rich in frequencies 19 and 20. This is not surprising since Q = 19.25. Even if the error spectrum has the characteristics of white noise (all f_k's equal) the denominator of Eq. (44) will be much smaller for k = 19 or 20.

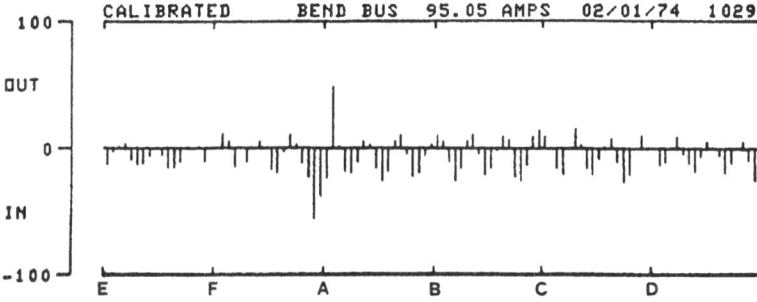

Fig. 12 Beam position monitors at each quadrupole of the FNAL Main Ring display the shape of the orbit before correction (200 units is the width of the vacuum chamber)

We could turn this to our advantage. We could correct such an orbit error by installing 38 small dipoles of alternating sign around the ring to cancel the 19th harmonic. A more economic way would be to simply install one of each sign at opposite ends of a diameter. These dipoles, effective and function perturbations, generate all odd harmonics in the driving function, with equal strength but only the 19th will have a major effect on the orbit distortion and can be used to cancel the 19th harmonic pattern in the distortion (Fig. 13).

Fig. 13 Diagram showing the sign of correction dipoles necessary to excite or compensate even or odd Fourier components of distortion around the ring of a synchrotron

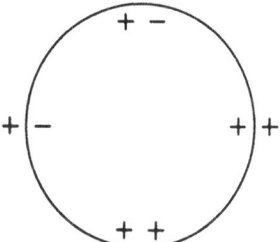

There are few machines where one would be satisfied just to correct these leading harmonics since the sum of the residue is still about 30% of the distortion before correction. Nevertheless, this principle of harmonic correction is sometimes used to correct a machine without realigning all the magnets. Displacing a quadrupole produces a dipole field on the central orbit. Displacing four dipoles at quadrants will produce the desired pattern.

12. LOCAL ORBIT CORRECTION

It is often necessary to steer the beam from side to side or up and down in the vacuum chamber, either to centre it or, to move it close to some injection or extraction device. Clearly one wants to localize this distortion as much as possible.

The most convenient way of making such a bump, the half-wave bump, is shown in real space in Fig. 14 and in phase space in Fig. 15 where β_0 is the β at the peak and β_κ is the β at the kick.

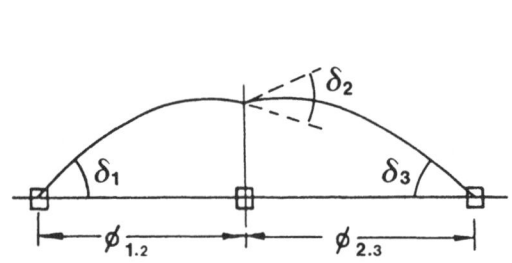
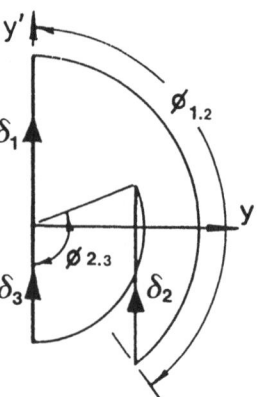

Fig. 14   A triad bump of dipoles producing a local orbit distortion.

Fig. 15   Triad bump seen in phase space.

The change in divergence at C is produced by small DC bending magnets which bend through an angle δ_1, δ_3. If one is fortunate enough to find two places separated by a betatron phase advance of π the bump will be a simple sine wave of amplitude :

$$\Delta y = \beta_0 \beta_k \, \delta_1 \, . \tag{45}$$

If both dipoles are at the same beta value their strength can be equal.

This bump has the characteristic that it leaves the beam centred elsewhere in the machine. However, if we are forced to some phase difference other than π the bump must contain a third element to make a cusp and satisfy the boundary conditions $(x, x') = 0$ at entrance and exit. Such a bump is called a triad.

Triads may be superimposed, each calculated to restore the local measurement of distortion (the bar in Fig. 11) to zero. They will overlap, but when added algebraically and applied to a set of dipoles as numerous as the measurements in Fig. 11, they can correct distortion to a small percentage of its initial value. Such a scheme is frequently used at injection in a synchrotron where the strength of dipoles needed is rather small and where the beam is large. In some machines triad bumps are made by displacing quadrupoles from their ideal position.

13. CERN PROTON SYNCHROTRON (CPS) [1954-1959]

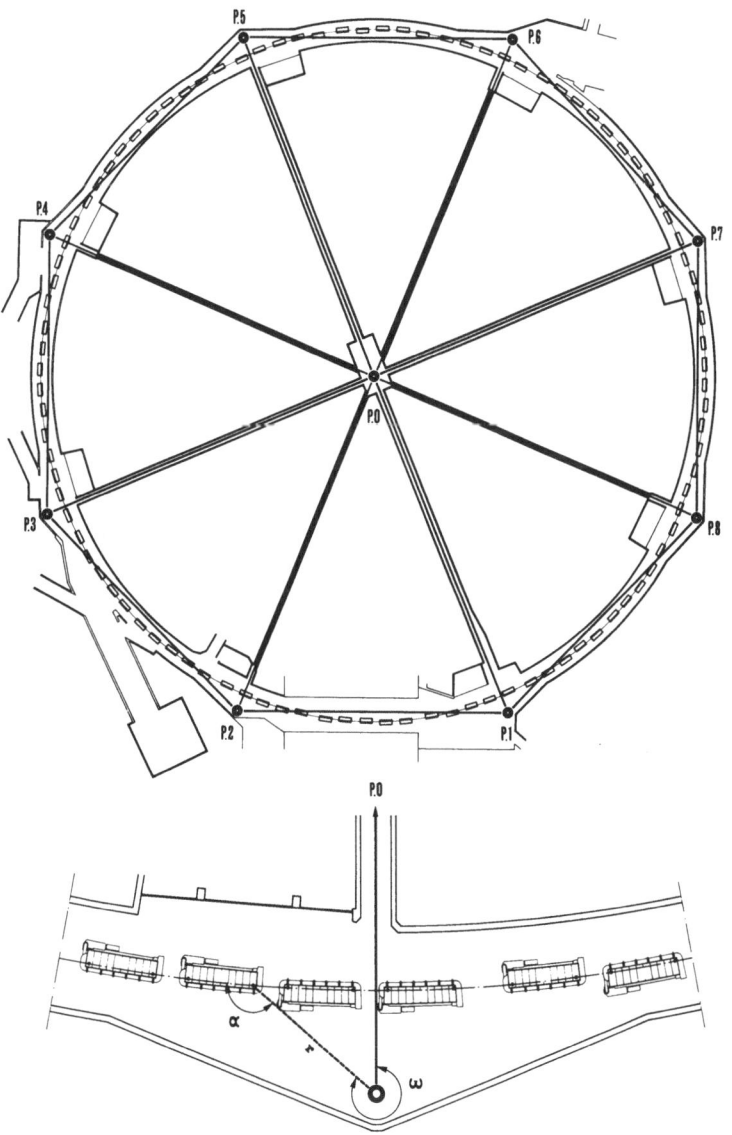

Fig. 16 : Basic survey configuration for the 28 GeV Proton Synchrotron

The problem of aligning the magnets of a circular particle accelerator was first encountered in 1954, following the decision to build a 28 GeV proton synchrotron in Switzerland, at Meyrin. The selection of survey instruments, and therefore the techniques available for meeting the tolerances required were fairly limited. At that time, the logical consequence was to use traditional geodetic methods in order to work out a possible approach to this problem. The estimated values of the achievable tolerances were introduced in the computations for the closed orbit, in order to define the size of the magnet aperture. This aperture determined the dimensions of the magnets and therefore the dimensions of the tunnel and, in part, the overall cost of the accelerator.

What tools did the surveyor have at his disposal in 1954? Computers had only just made an appearance and least squares adjustment had still to be made on calculating machines. The first direct consequence was that the geodetic reference figure had to be as simple as possible to avoid major adjustments which were always time-consuming with calculating machines. Figure 16 illustrates the octagon (P_1-P_8) with a central point (P_0), which was the basic network of the synchrotron's geodetic system.

It also shows the elements (r, ω, α) had to be measured so that the magnets were properly aligned on the calculated orbit. The second consequence was that, as calculations had to remain simple, maximum accuracy had to be achieved in the measurements.

At that time, geodesy relied essentially on angle measurement using a theodolite (Fig. 17). A better accuracy could be obtained by using forced centring when locating the various instruments, and by building the sighting targets with the same precision as the reference holes of the forced centring system.

Distance measurement was the second method available. In 1954, length measurement by electromagnetic methods was just making its appearance, with instruments such as the geodimeter and tellurometer. The original model of the geodimeter was tested in 1947 with light modulated at 10 MHz. The standard deviation of this instrument (5 mm + 10^{-6} of the distance) makes it unsuitable for very short distances. The same also applies to the tellurometer, which is based on the use of transmitters and receivers operating on very short wavelengths (3000 MHz and above). The only available tool was, therefore, the invar wire (or tape) used for measuring the bases at the end of primary triangulations (Fig. 18).

The third of the three tools was the Wild N3 bubble level used with Taylor spheres placed on the bevel edges of the sockets. This method offers a standard deviation of 0.3 mm/km.

To correct the position of the magnets according to the results of measurement, they were installed on three jacks. This enabled movement in three directions (translation) and movements through three angles (rotation), corresponding to the six degrees of freedom of all non-deformable bodies.

Fig. 17   Invar wire reading system   Fig. 18   Knife-edge pulley
          (microscope, theodolite &
          target on the same axis)

Over a ten-year period, repeated measurements of the monument positions in the PS reference figure showed the long-term consistency taking into account movement of the molasse, movement of the monuments, and random errors in measurements to be 0.1 mm per 100 m per year. A measurement carried out independently of any triangulation confirmed this figure. From 24 August 1965 to 13 February 1968 a pair of Marussi horizontal pendulums were mounted on the central pillar P_o anchored in the molasse 10 m below ground level. These instruments measure the variations of their support in relation to the direction of the vertical, and therefore of the movement of the vertical axis of the 10 m pillar. Fig. 19 shows that the over-all movement of the molasse, and of the monument itself, scarcely exceeded 0.15 mm in a North/South direction over a period of two and a half years.

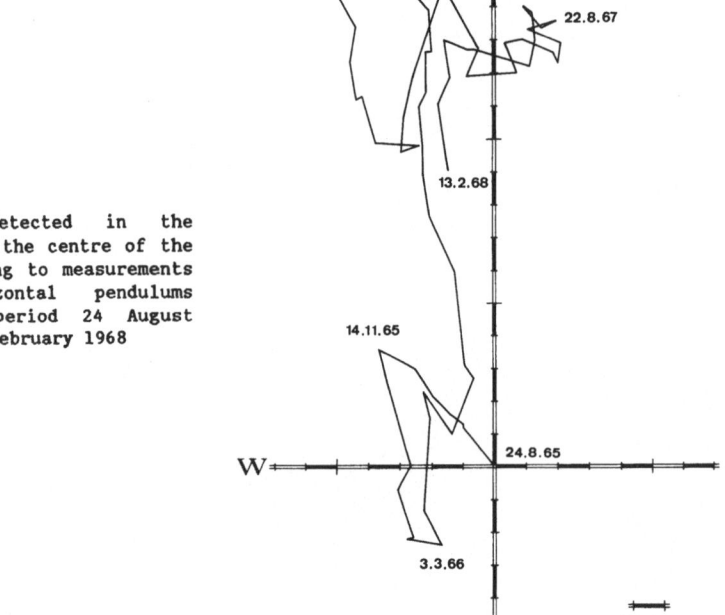

Fig. 19   Movement   detected   in   the
monument at the centre of the
PS, according to measurements
with   horizontal   pendulums
over   the   period   24 August
1965 to 13 February 1968

The requirements of the CERN Survey Group have been such that more than 10 km of invar
wire have been used to date. When the PS was being built, wires of 81 and 105 m were
used. It was necessary to use frictionless pulleys, because an excess tension of 1 g on
wires of such length resulted in variations dl of the order of 10 μm. The development
of pulleys in which the ball-bearings were replaced by balanced knife-edges enabled the
excess tension to be reduced to 0.002 g. During microscopic measurements of the scales
fixed at each end of the wires, it was noted that they were constantly becoming elongated
under a tension of 196.18 N (Newton). As we were the first to use wires longer than 100 m
with pulleys that were virtually frictionless, we were naturally in a position to detect
non-elastic elongations. The Bureau International des Poids et Mesures in Sèvres (Paris)
was immediately informed, and as a result carried out a series of tests on a 24 m invar
wire which subsequently confirmed our results.

Figure 20 shows the elongation of invar wire and tape when subjected to tension for a
prolonged period (Proceedings of the 1960, 1962, 1963 Meetings of the Bureau International
des Poids et Mesures, Paris). Figure 21 shows the elongation of the invar wires as a
function of time, Fig. 22 as a function of length.

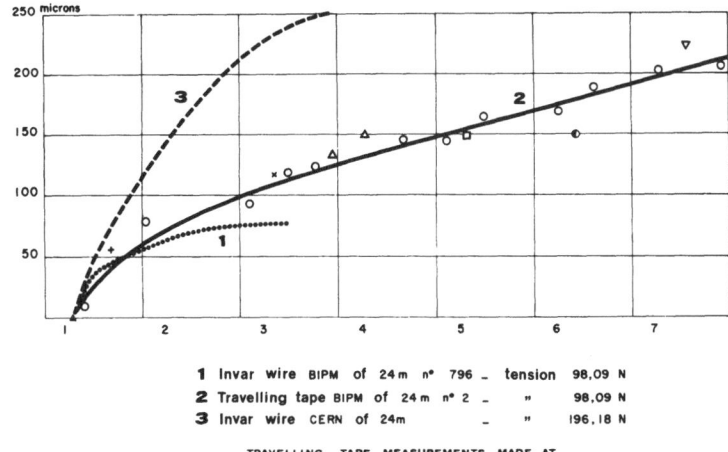

1 Invar wire BIPM of 24 m n° 796 _ tension 98,09 N
2 Travelling tape BIPM of 24 m n° 2 _ " 98,09 N
3 Invar wire CERN of 24 m _ " 196,18 N

TRAVELLING TAPE MEASUREMENTS MADE AT

O Bureau International des Poids et Mesures, Sèvres. △ Institut Central de Recherches Scientifiques de Géodésie, Moscou
▲ National Bureau of Standards, Washington. □ Physikalisch - Technische Bundesanstalt, Braunschweig.
+ National Physical Laboratory, Teddington. ◉ Institut Géodésique de Finlande, Helsinki.
× National Standards Laboratory, Chippendale. ▽ Geographical Survey Institute, Tokyo.

Fig. 20   Elongation of invar wire and tape as a function of the number of years of use

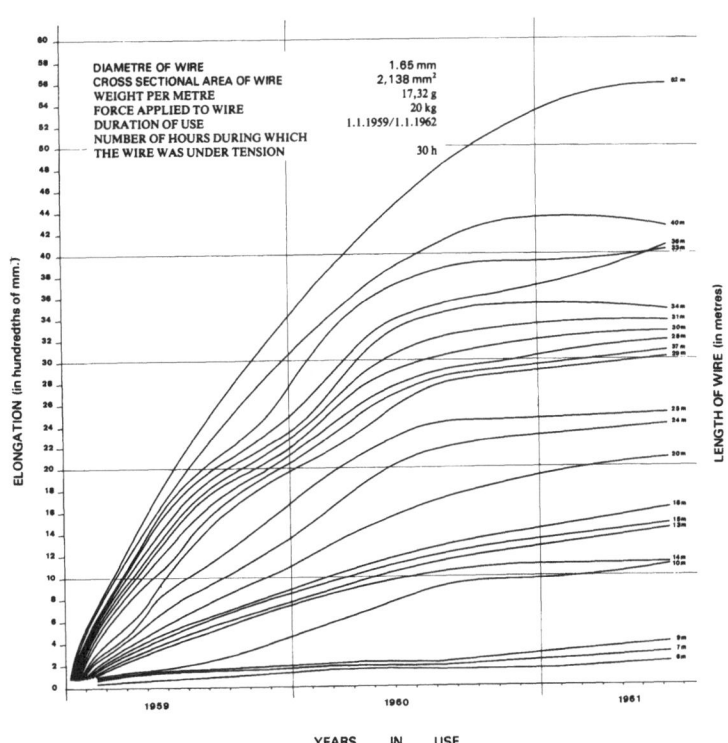

DIAMETRE OF WIRE	1.65 mm
CROSS SECTIONAL AREA OF WIRE	2,138 mm²
WEIGHT PER METRE	17,32 g
FORCE APPLIED TO WIRE	20 kg
DURATION OF USE	1.1.1959/1.1.1962
NUMBER OF HOURS DURING WHICH THE WIRE WAS UNDER TENSION	30 h

Fig. 21   Elongation of invar wire as a function of time

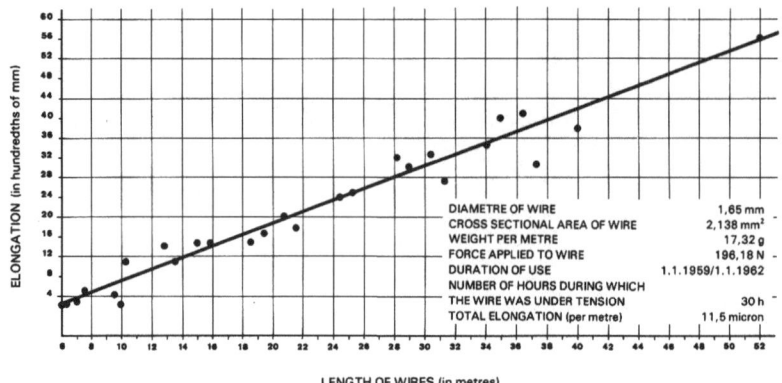

DIAMETRE OF WIRE — 1,65 mm
CROSS SECTIONAL AREA OF WIRE — 2,138 mm²
WEIGHT PER METRE — 17,32 g
FORCE APPLIED TO WIRE — 196,18 N
DURATION OF USE — 1.1.1959/1.1.1962
NUMBER OF HOURS DURING WHICH
THE WIRE WAS UNDER TENSION — 30 h
TOTAL ELONGATION (per metre) — 11,5 micron

Fig. 22   Elongation of invar wire as a function of its length

These elongations were detected because CERN posesses, for standardization of its wires, a 64 m bench constructed by the Société Genevoise d'Instruments de Physique and located, between 1959 and 1969, in one of the radial tunnels of the CPS. After 1969 (Fig. 23) the bench was installed in a specially equipped tunnel near the ISR. Only the 4 m rule and the microscopes were retained. The rest of the equipment was modernized. Until the advent of laser interferometers (Fig. 24 & 25), this standardization bench was the only one in the world capable of calibrating invar wires of any length between 0.40 and 64 m.

Fig. 23   Calibration bench

Fig. 24   Interferometer and display

Fig. 25   Corner cube reflector mounted at the end of a 4 m rule,
with microscope to read the rule

14. INTERSECTING STORAGE RING (ISR) [1966-1971]

The use of computers for geodetic calculations brought about a fundamental change in the philosophy of measurement and, to a certain extent, in accelerator design. The PS and the machines whose design was based on it were the last to have radial tunnels. The studies carried out at that time for the ISR and the SPS already illustrated a basic survey system based on a chain of braced quadrilaterals. In the case of the PS, the centre is real and physically exists, whereas in the case of the ISR and SPS the centre is virtual. It was therefore necessary to find a way of superimposing two sets of successive measurements. Furthermore, the elimination of all angle measurement and the sole use of distance measurement modified the observation equations in the computations, and obliged us to reconsider the entire problem of trilateration adjustments. The studies were carried out simultaneously for the ISR and the SPS.

The ISR lattice consists of 48 magnet periods divided into four super-periods, whereas the complete geodetic figure comprises 32 quadrilaterals. Figure 26 shows one of the octants of the overall framework.

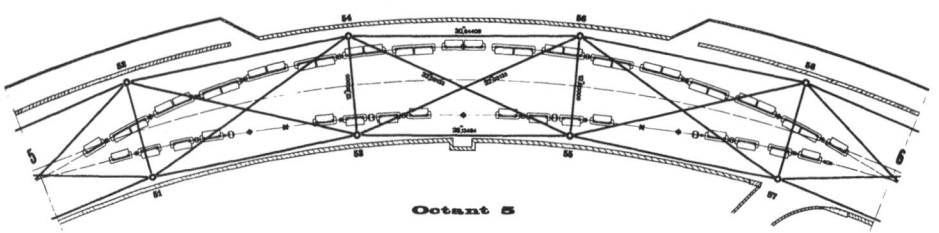

Fig. 26   Braced quadrilateral survey system of one octant of the ISR

Starting with one monument as reference, the error ellipses were calculated for each monument up to those diametrically opposite (Fig. 27). In this way, it is possible to judge the distortion of the reference figure. The value of the semi-major axis of the largest error ellipses is 0.2 mm for a probability of 0.40, and 0.6 mm for a probability of 0.99.

This means that from one set of alignments to the next, the variation dr in the radius of the orbit is of the order of 0.08 mm in relation to the theoretical orbit, but its position in space has a 99% chance of being inside a circle of errors of 1.2 mm in diameter. The precision in the relative position of magnets in relation to another one, inside two adjacent quadrilaterals, is better than 0.1 mm.

To achieve this result, it was decided to eliminate systematically all angle measurement. The distances were measured with invar wire, using the "distinvar" for the reference geodetic figure and for positioning the magnets. Only two days were needed to measure the 32 quadrilaterals of the reference figure and to calibrate the invar wires before and after measurement in the ring.

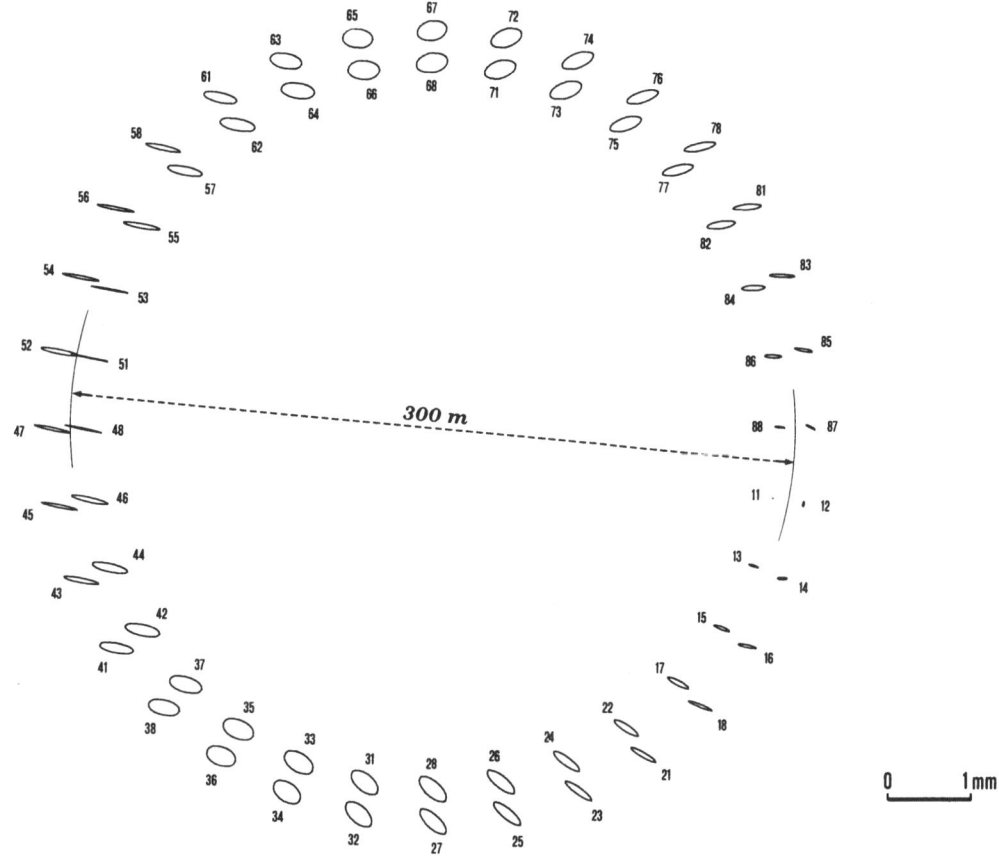

Scale of error ellipses

Fig. 27 Ellipses of errors computed when taking monument 11 as the starting point and direction 11-51 as orientation vector. They show the way in which errors accumulate around the Storage Rings' circumference

15. SUPER PROTON SYNCHROTRON (SPS) [1971-1976]

On 27 January 1971, beam collisions were produced for the first time in the ISR. On 19 February the construction of the SPS was approved by the majority of Member States and immediately there was a new challenge for the CERN Applied Geodesy Group. The PS and the ISR were surface structures built by the "cut and fill" method with prefabricated concrete tunnels. Unlike the NAL site (Batavia, USA), which is horizontal, flat and free from vegetation, woods or forest, the SPS tunnel could only be located underground in the Chattian molasse plateau of the Geneva Basin and be bored with a tunnelling machine.

Three major problems were successively encountered during construction of the SPS machine :

- the gyroscopic traverse for guiding the boring machine,
- the installation of magnet supports before the final completion of the tunnel, i.e. without a complete underground geodetic figure,
- the final installation of quadrupole and dipole magnets and auxiliary components within the required tolerances.

Each of these three problems was a challenge to the geodesists, and could be successfully solved only by making a series of extremely precise measurements, based on a highly accurate geodetic network on the surface.

The special feature of the SPS tunnel was that the work started by the digging of six shafts located at regular intervals around the 6,911 m circumference. The main tunnel was thus divided into sextants representing the six super-periods of the accelerator. When making the underground traverse to guide the boring machine, it was not possible to measure a geodetic bearing at each end of one sextant of the accelerator. The surface coordinates could be transferred down each shaft to the tunnel, but not accurate bearings.

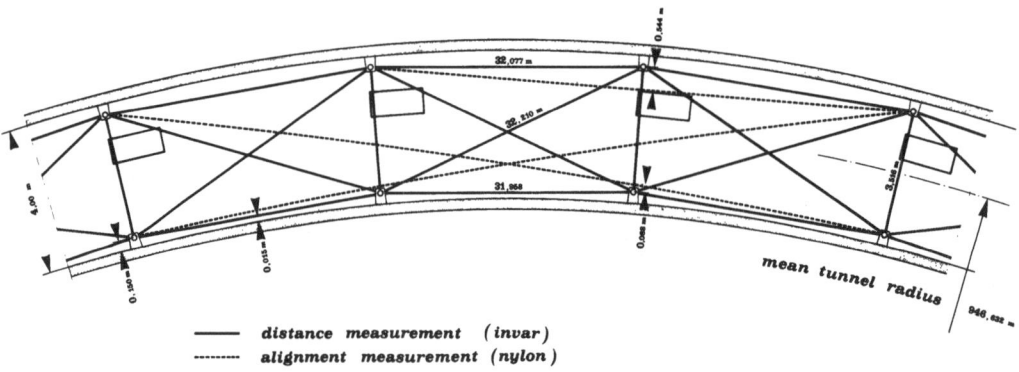

——— distance measurement (invar)
-------- alignment measurement (nylon)

Fig. 28   Geodetic reference figure of the SPS

The geodetic reference figure of the SPS (Fig. 28) is a modified version of that of the ISR. In the latter, because of the relatively favourable ratio between the width and length of a braced quadrilateral (1 to 2.5) distance measurements alone were used for the ISR framework consisting of 32 braced quadrilaterals. In the SPS, this ratio is approximately 1 to 10. To keep the deviation of the radius within the tolerances, the reference figure had to be stiffened by measuring the offsets of the monuments of three consecutive quadrilaterals with nylon wire equipment. As the SPS is a separated-function machine, the components which must be accurately positioned are the quadrupoles situated at one end of each semi-period of the accelerator's lattice, every 32 m. In this case, therefore, it was necessary to adapt the metrology system to the periodicity of the quadrupoles. Opposite each quadrupole, two brackets were fixed, one on the outer wall, the other on the inner wall. In addition to the distance measurement of each braced quadrilateral, nylon wire alignment inside three consecutive quadrilaterals ensured the precise shape of the geodetic framework in each sextant.

To make an a posteriori precision estimate it is worthwhile to recall the notion of the redundancy coefficient :

$$\tau = \frac{N \text{ observation}}{n \text{ unknowns}}$$

The value of τ for the ISR is 1.28, and 1.93 for the SPS.

In the ISR, the statistical study of the residuals after adjustment gives a standard deviation $\sigma = 18$ μm, which with $\tau = 1.28$ corresponds to an estimated accuracy of 34 μm.

In the SPS, the same study of length measurements gave a standard deviation $\sigma = 33$ μm, which, with $\tau = 1.93$, provides also an estimated accuracy of 34 μm.

These results, obtained from tens of thousands of measurements made with the distinvar over 15 years, show the very great homogeneity and reliability of this process in CERN's specific operating conditions. The above results obtained with this type of metrology are leading to a limit of precision of between 30 and 40 μm for distances ranging up to 50 m.

On the 2 January 1976, the coordinates of the brackets used for the underground survey chain were calculated. Without having to move any support already fixed on the foundation raft, it was possible to perform the fine adjustment of all the magnets and beam instrumentation, and on the 3 May 1976, protons were circulating in the SPS for the first time.

The fact that the protons were able to circulate round the machine without any correction to the orbit showed that the degree of accuracy required from the geodesists at the design stage had been kept. In fact, we already knew this from a detailed analysis of the survey measurements and radial smoothing results. This analysis was based on the rms sum of all the random deviations on the radial position of the quadrupoles.

For the six points obtained from the surface network, the radial standard deviation was 1.3 mm. With regard to the geodetic network of the brackets, which used these points as reference, an examination of the ellipses shows that, after overall adjustment of the circumference, the radial standard deviations are basically similar for each sextant. They steadily increase from the reference pillar, reach a maximum at the middle of the sextant, after which they decrease and finally disappear at the next reference pillar; in each sextant, the pattern of the error ellipses is that of a very long cigar. For the ultimate measurement made, this maximum represents a standard deviation of 1.2 mm.

The complete set of radial uncertainties in the surface network and underground framework provides, for each sextant, a cigar-shaped envelope. This envelope provides, among other things, a picture of the possible variations in the accelerator radius, which has an uncertainty varying from 1.3 mm at the reference points to 2.5 mm in the middle of each sextant. However such large mid-sextant values have no direct influence on the correct operation of the accelerator provided that the relative radial error dr remains within the narrow limit imposed by the machine's parameters (0.15 mm), thus compelling the curve of the successive quadrupole positions to remain continuous, with discrete increments.

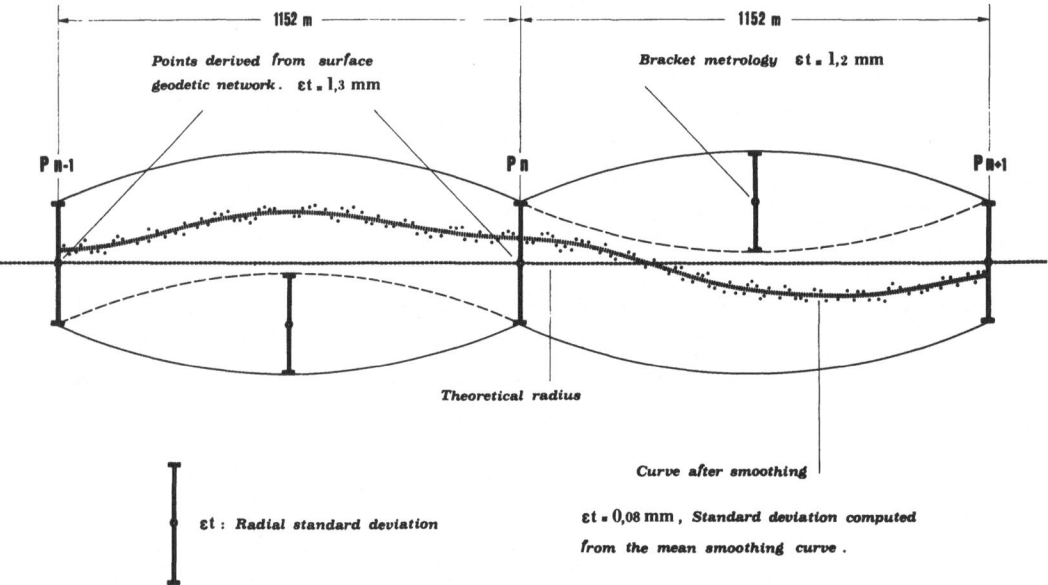

Fig. 29   Combination of random errors and curve obtained after quadrupole smoothing

Its general appearance will be that of a curve of the second or third degree. This curve is contained in a random manner inside the envelope shown in Fig. 29 and constitutes an approximation of the theoretical line among an infinite number of possible locations within this envelope.

Smoothing was done by measuring, with a nylon wire device, the local curvature of successive sets of three adjacent quadrupoles, (Fig. 30). A comparison of this curve with the theoretical curve gave local information on the positions of the quadrupoles in relation to a mean curve, for which the absolute position did not need to be known. This was done, in the least squares adjustment, by solving the observation equations with the double condition v^2 and dR^2 minimum, v being the residuals of the measurements and dR the deviation in relation to the curve.

This measurement was made for checking purposes only, but it provided a direct estimate of the relative residual errors. On the basis of the smoothing results, the calculated value of the parameter dr (the radial deviation of a quadrupole in relation to the two adjacent quadrupoles) was 0.08 mm for the overall circumference of the synchrotron.

Levelling gave rise to the usual problems. The overall accuracy obtained is 0.5 mm/km, and a relative accuracy of 0.10 mm was attained for one quadrupole in relation to the two adjacent quadrupoles.

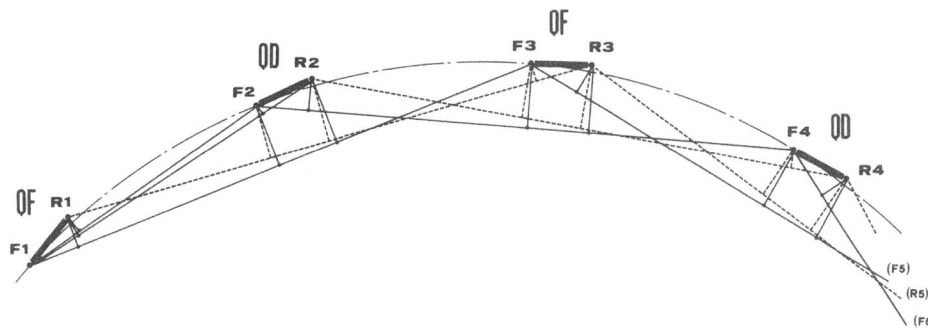

Fig. 30   Smoothing principle

16. INSTRUMENTATION

To obtain such an accuracy it is obvious that the entire geodetic system uses precision-type reference sockets for centring the measuring instruments (Fig. 31). It is possible to correct the verticality of the sockets at any time.

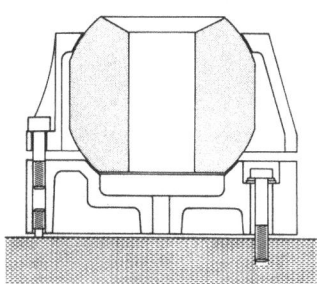

Fig. 31 : Precision-type reference socket

To facilitate and speed up the precise positioning of the quadrupoles, special jacks with polyurethane pads were developed by the CERN Survey Group (Fig. 32).

Among the new devices developed and used at CERN for applied geodesy, it seems worthwhile to describe four of the most typical belonging to the special CERN Survey instrumentation.

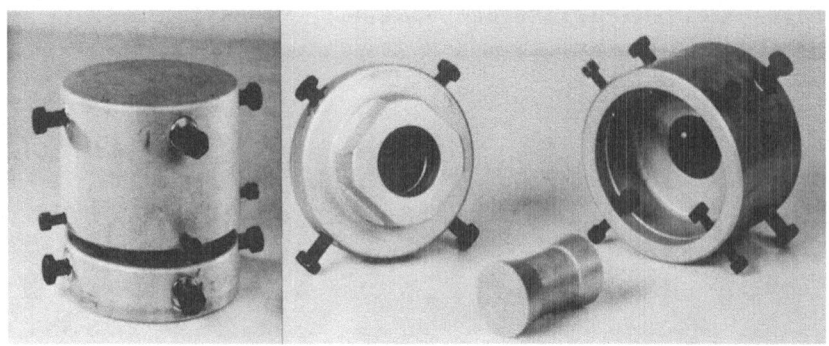

Fig. 32   Polyurethane jacks

17. LENGTH MEASUREMENT DEVICES

The self-aligning reflector makes it possible to carry out distance measurements with a laser interferometer without using mechanical guidance. In good conditions, measurements can be made up to 60 m with an error less than 0.01 mm (Fig. 33).

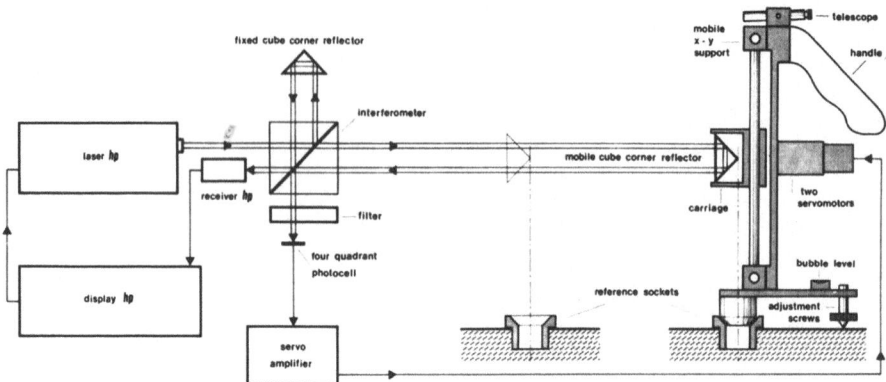

Fig. 33   Laser interferometer

The beam from the laser is split in two by the interferometer prism, one part is reflected back to the prism by the fixed reflector and the other by the mobile reflector. The two beams are recombined and reach the receiver, where the phase difference between them is detected. This difference is proportional to the distance over which the mobile reflector has been moved. The receiver signal is fed to the display unit, which shows the measured distance in mm, 1/4 wavelengths or inches.

The reflector is mounted on a carriage, which can be moved in the horizontal and vertical directions by two servo-motors. The servo-motors are controlled by a servo-amplifier, which receives the signal from the four-quadrant photocell. This forms a servo-loop, which keeps the returned laser beam centred on the photocell by moving the reflector. For the vertical adjustment of the XY-support at the reference sockets, there is a bubble level and two adjustment screws.

The Distinvar (Fig. 34), developed by the CERN Applied Geodesy Group in 1962, has been used for thousands of measurements.

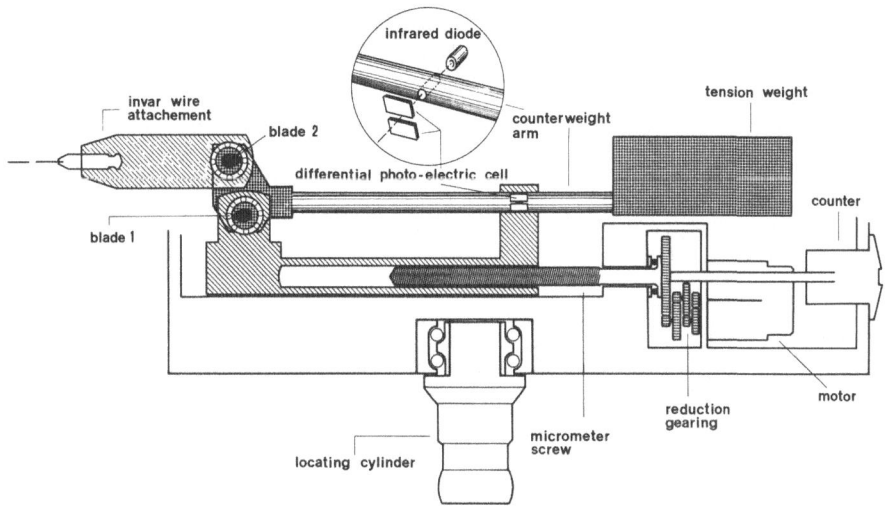

Fig. 34   Distinvar

The Distinvar consists of three parts : the 1.65 mm geodetic invar wire with its plug-in fittings, the fixed reference point and the measuring head. The latter two components have a male cylinder for precise centring inside the 30 mm diameter sockets. These sockets must be adjusted for verticality and serve as the origins for the length measurements.

The equipment is a precision balance fitted on a carriage. The balance serves to apply a constant horizontal tension on the invar wire via a system of levers attached to a fixed weight (1.5 kg). The carriage seeks the centre of equilibrium of the instrument. This centre of equilibrium is determined by a system for detecting the position of the balance-beam. According to the information provided by the detection system, the carriage is automatically moved by a reversible motor via a micrometer screw until it reaches the centre of equilibrium. A mechanical counter integral with the screw displays the longitudinal position of the carriage. The length measurement is obtained by subtracting the dial reading from that obtained during the calibration of the wire.

The total travel of the carriage is 50 mm. Depending on the invar wires used, the Distinvar can measure lengths between 0.40 and 50 m. Repeatability of the reading is 0.01 mm. Measurement accuracy is not a function of the length of the wire but of the sensitivity of the balance. Thousands of measurements made with the Distinvar over many years show the standard deviation to be 0.03 mm.

18. OFFSET MEASUREMENT DEVICES

The laser alignment system with tracking receiver is an active target system using a laser reference beam. It is used at CERN for offset measurements which arise when establishing the geodetic framework of accelerators. This system is particularly useful in turbulent air conditions where the stretched wire method described later is unsuitable. It can measure up to 600 mm over a distance of 100 m with a resolution of 0.01 mm.

The apparatus consists of a laser source, with its special optical system, and a differential photo-cell receiver.

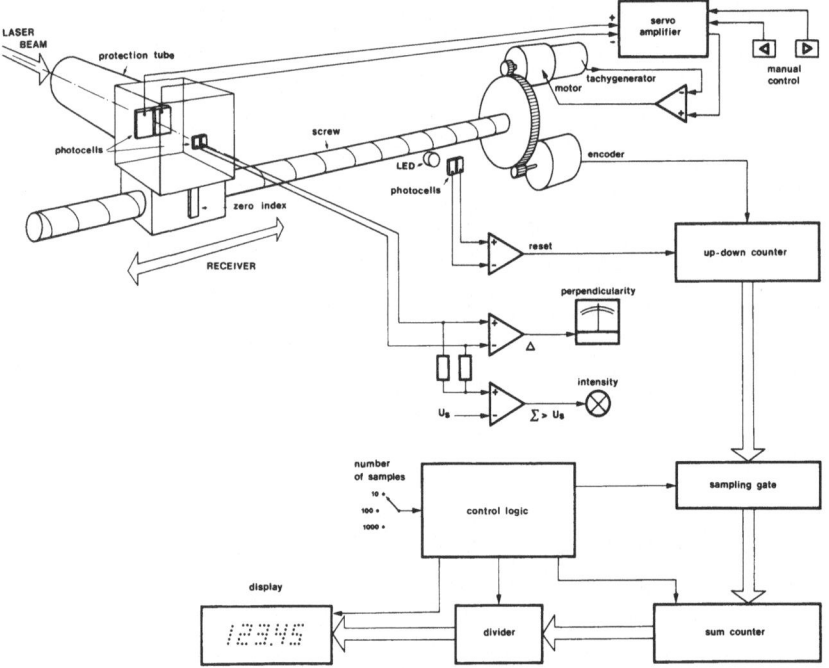

Fig. 35   Laser alignment system with tracking receiver

The laser is a commercial He-Ne unit of 1 mW beam power, followed by an optical system which reduces the divergence by a factor of ten and increases the beam diameter by the same factor. The laser beam increases from 11 mm at the output of the optical system to

only 14 mm at 100 m. A 25 mm diameter diaphragm is located at the focus of the system.
The diaphragm serves as a spatial filter and cleans up the beam; it eliminates
non-parallel rays and assures a Gaussian distribution of light intensity across the beam
diameter. The transmitter is mounted on a adjustable plate which allows the beam to be
steered onto the receiver. Once adjusted, the plate can be locked (Fig. 35).

The carriage of the photo-electric receiver is driven by a precision screw
perpendicular to the laser beam, and is servo-controlled to remain centred on the beam.
The position of the carriage along the screw is measured by an encoder which feeds 100
pulses per millimetre to an up-down counter. This counter stores the instantaneous
position of the carriage relative to a "zero", established by means of a reference cursor
attached to the carriage and passing between a fixed LED and a photo-cell pair, resetting
the counter to zero at the "zero" position. Instantaneous positions are read 10 times per
second via a sampling gate and added together in a summing counter. A manual selector
allows the choice of 10, 100 or 1000 samples.

The nylon wire alignment system is designed to replace angular observations by precise
measurement of the shortest distance (offset) from one point to a straight line used as a
reference. This line consists of a taut nylon wire. The vectors thus obtained are
introduced into the planimetric computations with other measurements.

Once the wire has been stretched between the extreme points, the measurement system is
installed on the intermediate point to be determined. The offset vector to the nylon wire
is then measured. The mobile detector passes a reference point, thus resetting the
distance counter, and then continues its travel until it locates the wire. The detector
then centres under the wire and the value of the measured offset is displayed (Fig. 36).

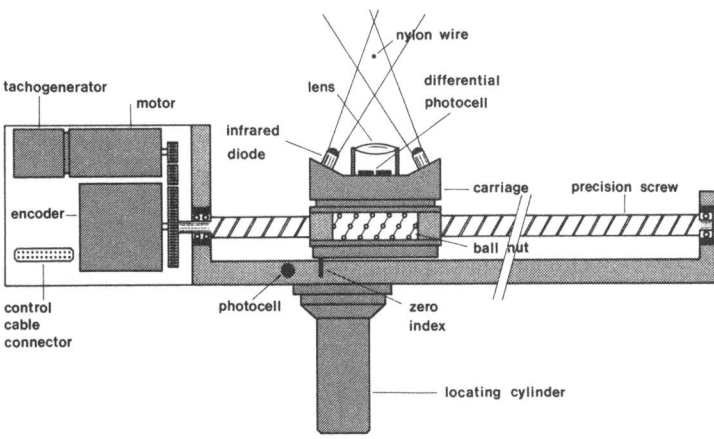

Fig. 36 Nylon wire alignment system

The method of operation consists of measuring the various offsets at least twice with
different origins, in order to obtain redundant and independent observations. The
calibration of the instrument is carried out by turning the detector system through

180°. The maximum measurable offset is 520 mm, the accuracy of the detector position is 0.01 mm. The automatic system has now been used intensively with a standard deviation of 0.05 mm over 100 m long chords. The observation conditions must be stabilized to ensure that even very slight draughts do not introduce systematic errors into the measurements. In such conditions, better results are obtained by laser alignment.

19. CONCLUSION

By way of conclusion, it is useful to summarize the evolution of the Applied Geodesy Group from 1954 to 1982 and to emphasize the mile stones which have been passed.

For the PS, in 1954, the improvement of the first order geodetic instruments and the use of primitive computing machines then available inevitably had a strong influence on the methods used for the positioning of the magnets. The development of these devices, the understanding between geodesists and mechanics and the good integration of the Applied Geodesy Group in the CERN structures made it possible to carry out an absolute metrology.

In 1961, the quite fortuitous invention of the Distinvar completely transformed the measurement methods. This permitted the elimination of the construction of radial tunnels in accelerator buildings and the angle measurements in geodesy. The appearance of more sophisticated computers opportunely added to these developments and led to the success of the ISR metrology. Collaboration between geodesists and specialists in mechanics and electronics became more and more essential.

Since 1972, the surface geodesy required by the SPS dimensions has called for the use of E.D.M. instruments the accuracy of which was about 10^{-6} of the measured distance. The appearance of the laser interferometers made the calibrations easier. For the underground metrology, the development at CERN of the laser and nylon wire alignment measuring devices allowed the measurement of small angles to be replaced by offset measurements. Here, too, the success of work was shared equally between geodesists and mechanical and electronic engineers.

For LEP, the automation of the measuring devices will again modify the practices of the technicians. The positioning of the magnetic components will result more from the smoothing of the quadrupole curve than from the computation of their coordinates. The use of the interferometer with a self-aligning reflector will avoid repeated visits to the calibration base during the positioning operations.

In spite of all these successive developments, including the automation of the instruments, which increases the speed and reliability of the measurement, the role of the individual people remains preponderant. The geodesist must still remain very close to the reality of the quantity being measured. Whether automatic or not, the most important and time consuming factor in the use of the distinvar will always be the correct rolling and unrolling of the wire. To achieve a precision of 10^{-7} with the Terrameter, i.e. 0.1 mm/km, will never be an easy task. The Terrameter is a Formula 1 car with its team of mechanical and electronic technicians. It is clear that the technology and equipment developed to meet the unusual CERN requirements could have a fruitful spin-off in all the fields where accuracy and speed of measurement are essential to the success of the work being carried out. At the human level, one of these spin-offs has been that many young surveyors and engineers have spent up to two years at CERN gaining a knowledge of, and experience in, this new technology with which they can profitably return to their home countries.

*   *   *

BIBLIOGRAPHY

PARAGRAPHS 2 TO 13

Bruck H., 1968, Accélérateurs circulaires de particules , PUF, Paris.

Courant E. & Snyder H., 1970, An. Phys. 3, I.

Livingood J., 1961, Principles of Cyclic Accelerators , Van Nostrand.

Livingood M. and Blewett J., 1962, Particle Accelerators (McGraw-Hill, N.Y., 1962)

Steffen K.G., 1965, High Energy Beam Optics , Wiley.

Wilson E.J.N., 1977, Proton Synchrotron Accelerator Theory, CERN 77-07.

PARAGRAPHS 1 & 14 to 19

Gervaise J., 1970, Positioning of the CERN Intersecting Storage Rings : the geodetic Approach , CERN 70-18, ISR Division, Geneva.

Gervaose J., 1974, High Precision Survey and Alignment Techniques in Accelerator Construction. Proceedings of the Meeting on Technology arising from High Energy Physics, CERN 74-9, Geneva.

Gervaise J., 1976, Geodesy and Metrology at CERN : a Source of Economy for the SPS Programme , CERN 76-19, SPS Division, Geneva.

Gervaise J., 1977, Precision Survey and Closed Orbits in Circular Accelerators , Processing of Special Large Matrices for Geodesy, Interpretation of the Results, CERN SPS-SU/78-4, Workshop 1-4 March 1977, Geneva.

Gervaise J., 1978, Invar Wire Measurement - A Necessary Evil , Proceedings of the Symposium on High Precision Geodetic Length Measurements, Helsinki.

Gervaise J., 1981, High Precision Instrumentation Developed for CERN , Technical papers, The American Congress on Surveying and Mapping, Washington D.C.

Gervaise J., 1983, Applied Geodesy for CERN Accelerators , Chartered Land Surveyor / Chartered Minerals Surveyor, Vol 4, No 4, RICS Journal Limited (p. 10 - 36), London.

Gervaise J., 1984, Applied Geodesy for CERN Accelerators , Seminar on High Precision Geodetic Measurements, Facolta di Ingeneria di Bologna, 16-17 October 1984.

THREE DIMENSIONAL ADJUSTMENTS IN A LOCAL REFERENCE SYSTEM

J. Iliffe *)
University College London, GB

ABSTRACT

This paper presents the way in which a general adjustment program
was written for three dimensional networks at CERN. It describes
the solutions which were adopted in developing an adjustment
procedure based on the local geoid, and the constraints which
existed regarding the generality of the program's application.
The eleven types of observation currently permissible are given,
together with a short description of some of the more "exotic"
ones, which are peculiar to CERN. The input procedure, with its
free-format convention and the system of key-words, is described,
as is the information presented to the program user in the form of
output.

1. INTRODUCTION

During the three decades and more of CERN's existence, the use of computers has played
a central, and increasingly important, role in all areas. In the Applied Geodesy Group,
as the techniques used to survey the larger and larger accelerators become increasingly
sophisticated and complex, so the methods by which the surveying and geodetic networks are
adjusted have had to develop apace. Each new accelerator has presented a new challenge,
and has led to the development of calculating techniques and computer software to meet
this challenge.

The latest project at CERN, the Large Electron Positron Collider (LEP), is almost four
times as large as its immediate predecessor and has, in its turn, brought a new set of
problems to be overcome. For the accuracy required in its construction, all computations
relating to the adjustment must take account both of the ellipsoid and the geoid in the
region of CERN. A spherical earth model is no longer sufficient.

In addition to the considerations of accuracy, the sheer volume of work, the profusion
of data to be handled in such a project has entailed the writing of a computer program
which would reduce the work done by the users to a minimum. This applies to the tedious
process of preparing the data on computer input files; it also means that the program must
wherever possible accept raw observations without the need for the prior reduction of data.

---

*) Former CERN technical student 1984-1985

Thus, the twin dictates of accuracy and ease of use already define the broad outlines of the program which is needed for the adjustment of the LEP geodetic network. Perhaps the next most important point was that the program, although designed specifically for the LEP, should also be appliable to all other surveying and geodetic networks at CERN. This follows naturally from a need for standardisation, and from a belief that the program could be made sufficiently attractive to the user for it to be prefered to all previously existing programs. So, although by default the calculations are refered to the ellipsoid and the geoid, the user may request that these be replaced by, for example, a spherical earth model with an unknown geoid. Eventually, it is intended that the program should incorporate functions which had previously been performed by complementary programs (such as the calculation of approximate coordinates and two- or three-dimensional adaptation).

The program which has been written is called the Logiciel Général de Compensation (General Adjustment Program), or LGC for short. This paper describes the way in which it was written to conform to the principles outlined in this section. Other, more particular, constraints existed, but these will be described in context.

2.   THE GEODESY OF LEP

2.1 The CERN coordinate system

The adjustment of the surveying and geodetic networks of the LEP is performed by LGC, the adjustment program, by the method of the variation of coordinates, with a least-squares minimisation of the weighted residuals. The coordinates which are varied in the adjustment are in the CERN XYZ cartesian system which, although not necessarily being linked to any physical quantity over the whole of the LEP area, nonetheless allows the easiest solutions mathematically. It is first necessary to define the coordinate system.

The origin, Po, is the pillar at the centre of the Proton Synchrotron, and the Z axis of the CERN system is deemed to be along the vertical at this point. The XY plane is thus tangential to the ellipsoid there, and the Y axis is defined as having an azimuth of 37.77864 gons. The system is a right-handed one, and hence the three axes are uniquely defined. For historical and practical reasons the coordinates of the origin, Po, are defined as follows :

    Xo = 2000.00000 m
    Yo = 2097.79265 m
    Zo = 2000.00079 m
    Ho = 433.65921 m

As has already been stated in the introduction to this paper, a spherical earth is no longer a sufficient model to use for the geodesy of the LEP. Such an approximation would have entailed an error of up to 22.5 mm in the transformation between measured altitudes and Z coordinates [1]. The ellipsoid used instead was that chosen in 1980 by the Union Internationale de Géodésie et Géophysique, and of which the parameters are :

$$a = 6378137 \text{ m}$$
$$e^2 = 0.0066943800229$$
$$e^{12} = 0.0067394967755$$

It can be seen, then, that the Z axis is only parallel to the vertical at the origin, Po, and becomes less so as the distance from the origin increases. In general, a point's XYZ coordinates are only a convention without physical meaning, and yet there are rare occasions when it may be assumed that the XY plane represents the local horizon at any point. As this is not usually justifiable, it will be specifically stated when this assumption has been made.

There are several combinations of the coordinates which may be held fixed at any particular point in a network. Obviously, there is a need both for points which are completely free and for ones which are fixed in all three dimensions. In addition there are points which are fixed in X and Y but free to move in Z (in a levelling network for example) and ones which are fixed in Z but free to move in X and Y. An important variant on the latter is when it is not Z, but H which is held fixed; this cannot easily be expressed directly in the adjustment, but it is sufficient to recalculate Z from the new X and Y coordinates after each iteration (Fig. 1).

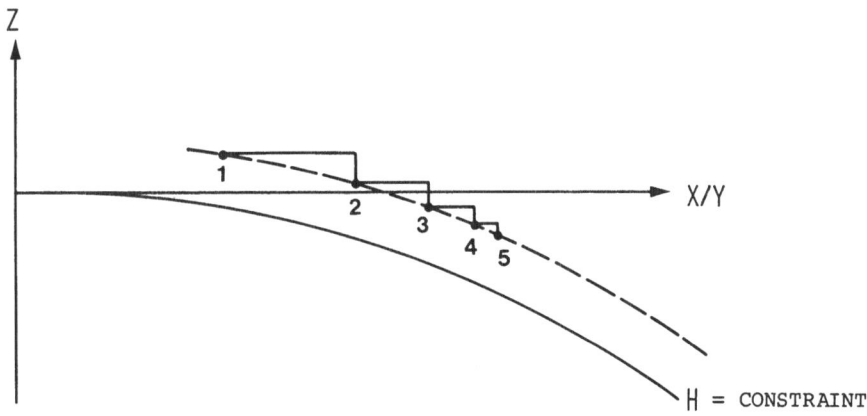

Fig. 1   Constraint onto the horizon

2.2 The Geoid

The geoid in the area of the LEP has been computed by calculating the contribution of mass elements over the whole of Switzerland and surrounding areas; the results obtained were checked by comparison with astro-geodetic measurements. The geoid in this area is defined as being tangential to the ellipsoid at the CERN origin, Po, and the deviation of the vertical and the geoidal heights are defined relative to the reference ellipsoid. It should be noted here that the maximum deviation found was 15 seconds of arc, and the maximum separation was 200 mm [2].

Having computed the position of the geoid at 121 points on a 10 km x 10 km grid over the whole of the LEP area, the results were parameterised to obtain a simple algorithm by which the geoid (and its inclination) could be computed for any point in the area. This was done for two levels : at level zero and at the level of the LEP tunnel. For each one a "best fit" paraboloid was fitted to the known data, each one having certain dimensions and orientation. This information was included in LGC, the adjustment program, and hence the deviation of the vertical and the geoidal height could be calculated for any point at any stage in the program by calling a simple subroutine, XIETAN.

2.3 Heights and Z coordinates

The geoidal heights are used to convert a height relative to the geoid to one relative to the reference ellipsoid. This is done according to the formula :

$$H_E = H_G + N \tag{1}$$

where H_E is height relative to the ellipsoid, H_G is height relative to the geoid, and N is the geoidal height.

A height relative to the geoid is, of course, the one which has a physical meaning; the geoid is an equipotential surface and H_G can be measured. The ellipsoidal height, H_E, is a necessary quantity when converting from a measured altitude to the Z coordinate, as can be seen from Fig. 2.

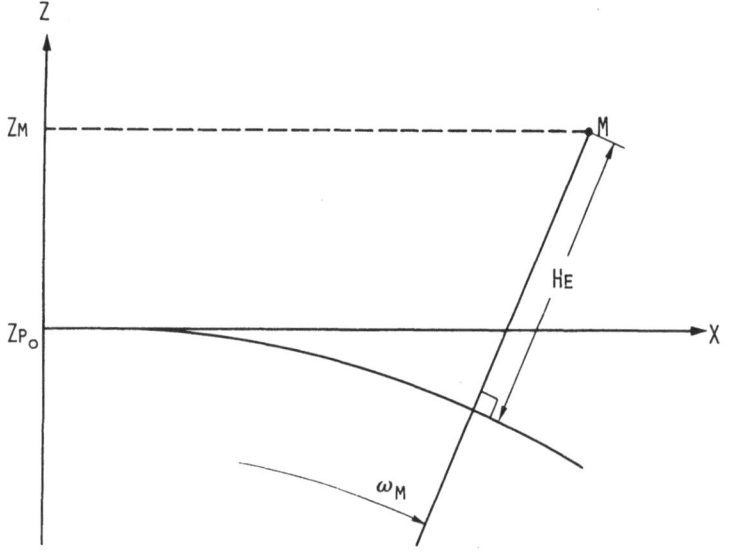

Fig. 2   Geometrical height above the reference ellipsoïd

This conversion is performed (in both directions) several times in the course of the adjustment, especially in those cases where H is held fixed and Z is updated after each iteration. For this reason, there are two subroutines (TRANZH and TRANHZ) which perform the transformation from geoidal heights to Z coordinates, and vice versa.

2.4 The rotation matrix

When considering observations which are made with respect to a local vertical or in the local horizon (eg. zenithal distances or theodolite observations) the solution process is assisted greatly if the observation equation is first derived with respect to a coordinate system based on the local vertical. Horizontal angles, for example, do not provide any information on the distance between two points in the direction of the vertical at the stationed point; but away from the origin, Po, it may have some bearing on the difference in Z.

A local reference system needs to be defined. The local z axis is, naturally, colinear to the vertical. The y axis is then defined as being in the plane of the z and Z (CERN) axes, pointing away from the origin, Po. The x axis forms a right handed system with the y and z axes (Fig. 3).

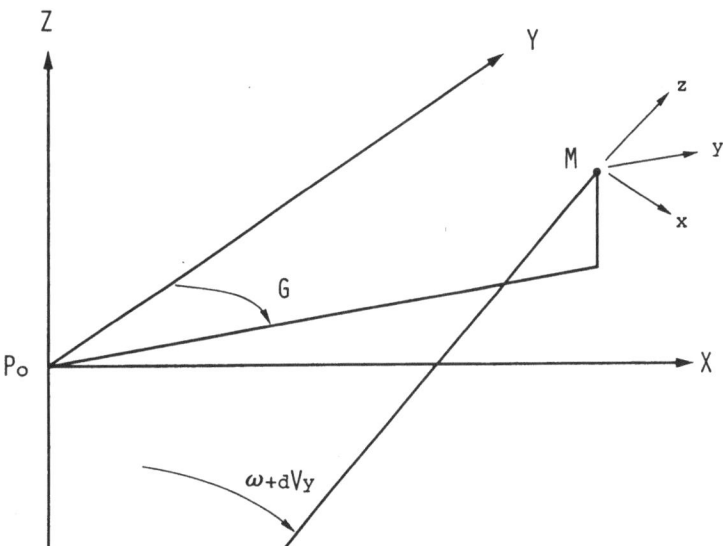

Fig. 3   Local parameters of an observation station

Having a standard definition of the local coordinate axes means that a general rotation matrix can be derived for converting from one coordinate system to the other. This rotation matrix expresses the compound action of three fundamental stages :

1. Rotation G about the Z axis,
2. Rotation ω plus dVy (deviation of the vertical at M in the direction PoM) about the new X axis,
3. Rotation dVx (deviation perpendicular to PoM) about the y axis.

The resultant rotation matrix, R, is given by :

$$R = \begin{pmatrix} \cos G - dVx \sin (w+dVy) \sin G & -\sin G - dVx \sin (w+dVy) \cos G & -dVx \cos (w + dVy) \\ \cos (w+dVy) \sin G & \cos (w+dVy) \cos G & -\sin (w+dVy) \\ dVx \cos G + \sin (w+Vy) \sin G & -dVx \sin G + \sin (w+dVy) \cos G & \cos (w+dVy) \end{pmatrix} (2)$$

where it has been assumed that $\sin dVx \to dVx$ and $\cos dVx \to 1$.

In the first instance this matrix is used to derive the difference in xyz coordinates between two points from the difference between the tabulated approximate XYZ coordinates. These differences (Δx, Δy, Δz) are then used to derive the observation equation in terms of the incremental displacements in x, y, and z at each point. The rotation matrix is then used, in the opposite sense from the above, to express the equation in terms of dX, dY, and dZ, the unknowns in the adjustment. That is, since :

$$\begin{pmatrix} \Delta x \\ \Delta y \\ \Delta z \end{pmatrix} = R \begin{pmatrix} \Delta X \\ \Delta Y \\ \Delta Z \end{pmatrix} \qquad (3)$$

and if the observation equation in the local horizon is expressed as :

$$C \begin{pmatrix} dx_M \\ dy_M \\ dz_M \end{pmatrix} - C \begin{pmatrix} dx_N \\ dy_N \\ dz_N \end{pmatrix} = C \begin{pmatrix} d\Delta x \\ d\Delta y \\ d\Delta z \end{pmatrix} = (\text{Obs.} - \text{Comp.}) + v_i \qquad (4)$$

where C is the 1x3 matrix of coefficients,

then,

$$R^T C \begin{pmatrix} d\Delta x \\ d\Delta y \\ d\Delta z \end{pmatrix} = (\text{Obs.} - \text{Comp.}) + v_i. \qquad (5)$$

Thus, the coefficients of the observation equation can be rotated as if they were a spatial vector.

This rotation procedure has been simplified by placing it in a subroutine, ROTVEC, where it can be called from any part of the program. It is necessary to supply, as dummy arguments to this routine, the origin of the local coordinate system (the name of the stationed point), the direction of the rotation (indicated by a 0 or a 1) and the three components of the vector. The components of the rotated vector are returned in the same locations. It is not necessary to supply w, G, dVy, and dVx, the component elements of R; w and G are calculated by a subroutine (OMEGAS), called at the start of each iteration, and stored; dVx and dVy are called as necessary from XIETAN from within ROTVEC.

2.5 Special cases

Up to this point mention has only been made of the way in which LGC performs the adjustment for the LEP. Yet it was stated in the introduction that this was to be an adjustment program for all surveys carried out at CERN. Two special cases have to be provided for : the spherical earth model and the limited dimension local origin survey.

The sperical earth model is applicable to those surveys related to the previous accelerators. This is the simplest case to deal with : instead of the radius of curvature of the ellipsoid depending on the azimuth from Po to a particular point, it is simply replaced by the mean value of R = 6371 km. The geoidal corrections are not applicable in this case, either, and so the subroutine XIETAN is modified to return a geoidal height and a deviation of the vertical equal to zero.

The other special case refers to small three-dimensional networks where the curvature of the earth is of no importance. In such cases the coordinates of the points are usually grouped about a local origin with coordinates $(0, 0, 0)^T$; the CERN origin is approximately at $(2000, 2000, 2000)^T$ and so there would be problems at the local origin, where the XY plane would not be parallel to the horizon. This is corrected by applying a coordinate shift within the program which is invisible to the user.

Both of these cases are selected by writing special key words (SPHERE, and OLOC, respectively) in the input list. The necessary modifications are carried out automatically.

3. THE OBSERVATIONS

3.1 Formation of the equations

The fact that LGC is an adjustment program for many different types of network has meant that a wide variety of observations have to be dealt with; at present there are around ten, including no less than four different sorts of off-set.

The observation equations are not stored as separate entities; they are computed from the approximate coordinates at each iteration, and one by one they contribute to the formation of the normal equations. This is done first of all by creating the coefficients and placing them in the vector CO. Each element in this vector has a corresponding rank in the normal matrix, and these are found by calling the subroutine IMAT, which calculates the ranks for each point involved in the observation. These ranks are placed in the vector IR, in exactly the same order as the corresponding coefficients in CO. The last element in CO is the "observed minus computed" term, and this has no corresponding rank. Once all the coefficients have been found, the subroutine EQNORM calculates their contributions to the normal equations and the vectors CO and IR are cleared ready for the next observation.

Two types of observation have special characteristics which require further explanation. These are the horizontal angles and the zenithal distances. Rounds of horizontal angles require a correction (Vo) to be made to the horizontal circle reading; an approximation is made for each set-up before the first iteration by calling the routine VOAPP2. The fine corrections are treated as unknowns and updated after each iteration. Since these are scalar, that is non-vectorial, values they are not rotated with the other terms of the observation equations.

It was decided, when considering the case of the zenithal distances, that the most satisfactory procedure would be to leave the heights of the instruments as unknown quantities. These are set to 0.200 m initially, and updated after each iteration.

3.2 Simulations

It is possible to run the adjustment program, LGC, in simulation mode to assess the strength of a proposed network. The observation equations will be exactly the same as for an observed network, except that the observations themselves will not exist and it is necessary to create them, in order to write the "observed minus computed" terms. This is done by obtaining a random number from a sample with mean zero and standard deviation equal to the standard error of the observation, and adding it to the computed "observation". This is done for each "observation", and the network is solved using these values.

This process is repeated several times (the acutal number is specified by the user in the input file) and the result is several versions of the same network, which can be analysed to determine the spread, maximum movements of points, and so on.

3.3 Types of observation

The observations which can be handled by the adjustment program are as follows :

1) Spatial distances, completely independent of the geoid; Straightforward equation.

2) Horizontal distances, not horizontal distances in the usual sense; rather, they are the distances between points which have been reduced to level zero and then the XY plane (Fig. 4). This observation is for handling data which has already been reduced for other purposes. Obscure, but occasionally useful.

3) Vertical distances, to take account of the geoidal heights at each point.

4) Horizontal angles, whose equation is determined in the local horizon and then rotated to the CERN XYZ system.

5) Orientations, as for horizontal angles, but without the horizontal circle correction, Vo.

6) Zenithal distances, for which seven coefficients have to be calculated, including that for the height of the instrument. The program allows an extension to be specified on the targeted point. This can justifiably be assumed to be parallel to the vertical at the stationed point, and the equation is simplified considerably as a result.

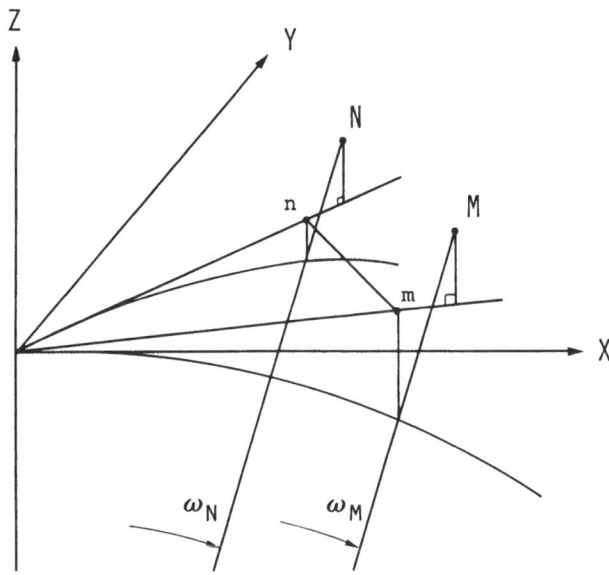

Fig. 4   Horizontal distances

7) Horizontal off-sets, perpendicular off-sets where the measurement is made in the horizontal plane regardless of the relative heights of the three points (Fig. 5). Since the three points A, B, and C are very often almost co-linear it is necessary to have the concept of "left" and "right" measurements. These are conventionally ascribed as negative and positive respectively. For this observation, the horizontal plane can, with negligible loss of accuracy, be assumed to be the same as the XY plane.

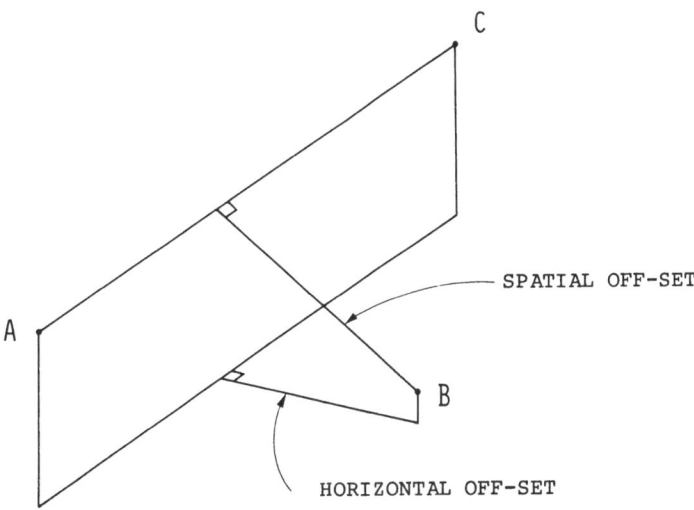

Fig. 5   Horizontal offset measurements

8) Spatial off-sets, being perpendicular off-sets where the measurement is made in the plane defined by the three points (Fig. 5). They can be made at any angle - on a wall, upside down, etc. - and for this reason the concept of positive and negative measurements had to be abandoned. It is impossible for the program to determine on which side of the plane the measurement was made.

9) Theodolite off-sets, made from a theodolite bearing, where there is no "third point". This measurement cannot be made independently, but must be included in a round of angles for the horizontal circle correction to be calculated (Fig. 6).

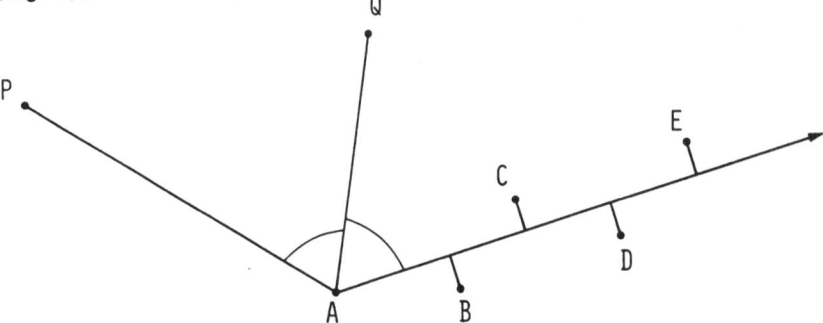

Fig. 6   Theodolite off-sets

10. Vertical off-sets, made between one point and a plumb-bob hanging off another.

11. Radial observations, correctly described as constraints and not observations, but as they are treated in a similar way it is appropriate to describe them here. The "observation" is a direction in which a point is deemed to be free to move ; movement perpendicular to this direction gives rise to a residual and is therefore restricted. This facility is useful when trying to avoid the "compression effects" obtained when both ends of a linear network are fully constrained.

4.   THE NORMAL EQUATIONS

In considering the formation of the normal equations, the first question to be addressed is how they are to be stored. Obviously, since they are symetric, it is only necessary to store the lower or upper half of the left hand sides. For larger networks, and especially the elongated variety typical in tunnels, the matrix will tend to be "banded". That is, the only non-zero elements lie in a narrow band on either side of the leading diagonal. In such a case, there will be advantages in only storing this central band, and not wasting space by occupying it with zero elements. Conversely, there are disadvantages in terms of the time taken to store and retrieve elements in such a fashion. So for LGC it was decided that it would be sufficient to store the whole upper half-matrix of the normal equations, and only introduce a banded storage pattern at a later date if it was required.

The matrix is stored in a vector of one dimension, which is composed of the successive columns of the upper triangular matrix (Fig. 7).

Hence the element in the ith row and j^{th} column is stored in the pth position, where :

$$P = \frac{j^2 - j + i}{2} \, . \qquad (6)$$

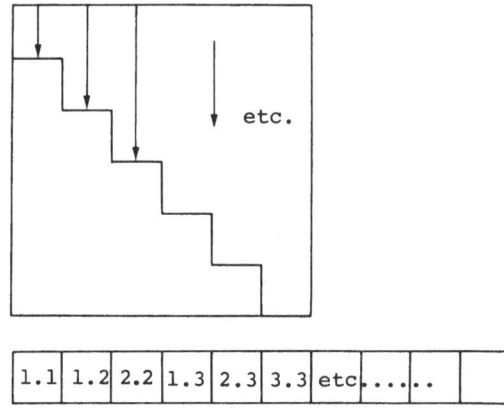

Figure 7   Columnwise storage

The elements of the normal matrix are formed by the subroutine EQNORM. This is called for each observation when the coefficients of the observation equation have been formed. The cross-product of each pair of coefficients is added to the corresponding element of the normal equations.

When the formation of the normal equations is complete, the matrix of the left hand sides, N, is decomposed as a first step to finding the solution vector. This is done in the subroutine EQSOLV. The pattern of decomposition chosen was :

$$N = U^T D U \qquad (7)$$

where U is an upper triangular matrix with all elements on the leading diagonal equal to 1 and D is a diagonal matrix. The advantage of this as opposed to a decomposition into U^T U (Choleski) is that time is saved by not having to find square roots of the diagonal elements. The elements of U and D can be stored in the same locations as those of N, the latter being overwritten as the decomposition procedes. It is not necessary to store the leading diagonal of U.

Having decomposed the matrix, the solution vector is found by forward and back substitution. If any of the unknowns are superior to the limit of 0.01 mm then another iteration is deemed to be necessary. The subroutine returns to the main program, the coordinates are augmented, and the solution re-starts with the formation of the observation and normal equations.

If, on the other hand, all the unknowns are less than the limit, then the vector solution is complete. Some elements of the inverse of the normal matrix will be needed for the analysis which follows the finding of the solution. Then, since

$$N^{-1} = U^{-1} D^{-1} U^{T^{-1}} = U^{-1} D^{-1} U^{-1^T} \qquad (8)$$

the first step is to find the inverses of U and D. These are again stored in the same locations.

The subroutine EQSOLV stops at this point. Obviously, it would be wasteful to find all the elements of N^{-1}, since in general relatively few of them will be needed. It is more efficient to calculate individual elements from U^{-1} and D^{-1} as they are needed, even though there will often be cases where elements are calculated more than once.

So, having inverted U and D, the subroutine EQSOLV returns to the main program. Here the coordinates are updated again, as for the previous iterations. But this time, instead of starting another iteration, the program calculates the residuals of the observations and from these the unit variance, σ_o, where

$$\sigma_o = \frac{\Sigma (v_i \sigma_i)^2}{n - k} \tag{9}$$

where V_i are the residuals, σ_i are the standard errors of observation, n is the number of observations, amd k is the number of unknowns.

Individual elements of the inverse of the normal matrix can subsequently be found by calling the subroutine ELMINV. This multiplies out the relevant rows and columns of U^{-1} and D^{-1}, and multiplies the result by σ_o. The result is returned as the inverse element corresponding to the rows and columns specified in the CALL statement.

5. THE INPUT PROCEDURE

5.1 Key-words

As was stated in the introduction, a principle feature of LGC is the ease with which it can be used. This is due in large part to the way in which the input routines have been structured, and an essential feature of this is the system of key-words.

A key-word is an asterisk, followed by a string of characters which define the meaning of the word. The meanings can be divided into two broad categories. Firstly there are those key-words which introduce a set, or a type, of data. For example, *ANGLES is followed by the list of horizontal angular observations, and this list is understood to terminate when the next key-word is encountered.

The second category of key-words is of those which do not precede a data set, but select particular options of the adjustment program. An example of this type is *SPHERE, which selects a spherical earth model, rather than an ellipsoidal one.

Almost any length of key-word may be written, but in general only the first four characters are checked. Hence, in the previous example, the key-word *SPHERICAL EARTH MODEL would have exactly the same effect as *SPHERE.

The complete list of key-words, together with a brief summary of their meanings, is given in the appendix.

5.2 Free format

Despite the fact that the program was written in FORTRAN, it has been done in such a way that the data on any one line does not need to correspond to any particular format. The important thing is the order of the data, and hence the lines :

	1.0	2.3	4.05
and		1.0	2.3	4.05

would both have the same effect. The only rule affecting the position of the data is that separate pieces should be separated by at least three blanks.

The way in which this is achieved is that each line is read as a character string of length 80, and then divided up into an array containing the 80 individual characters. Five subroutines may be called to "read" the data from this array, instead of having actual READ and FORMAT statements.

These fives routines are :

RINTEG - reads integers
RREALN - reads real numbers
RALPHA - reads character strings
RMOT - reads words, curtailing after four characters, and
RTITRE - reads entire lines.

These routines find the next piece of data, irrespective of its position on the line. Redundant blanks are ignored. The five routines are complemented by :

RNEXT - called to fetch the next character
RLINE - reads the next line when the current one ends
RSTART - initializes all parameters
RFAIL - called when a discrepancy is found
REND - decides whether to continue or to print failure message and stop.

As can be seen from the presence of the routine RFAIL, various checks can be made on the data which is being read. These errors occur either half way through an item of data (e.g. finding a decimal point in an integer), or as a result of the unexpected presence or absence of an item. The general principle is that an error should be noted, and the user warned, but that action should be taken to jump over the suspect data and continue reading. The program is not halted until an attempt has been made to read all the data present.

5.3 Layout of data

Without giving detailed specifications of the ways in which the various classes of data are entered, it is nevertheless instructive to show an example of the flexibility which can be attained. For example, to enter measured distances the essential data is the key-word followed by a list of the pairs of points and the distances between them :

```
* DMES
AA      BB      1234.5
CC      BB2      896.12
CDE     FGH     1066.88
BB      CDE     1492.0
```

This is an irreducable minimum for the data, and clearly the next information required is the standard error of each measurement. If, as is often the case, these are the same for all observations, then the generic standard error can be put on the same line as the key-word, signifying that it applies to each observation. For example,

```
* DMES 1.0    1.0
```

implies that each observation has a standard error of 1.0 mm + 1.0 ppm. This value can be over-ruled for individual observations, however, by placing an alternative value next to the observation in question. So, in the previous example,

```
* DMES     1.0    1.0
AA      BB      1234.5
CC      BB2      896.12    2.0    5.0
CDE     FGH     1066.88
BB      CDE     1492.0
```

All observations have a standard error of 1.0 mm + 1.0 ppm, except CC-BB2 for which it is 2.0 mm + 5.0 ppm.

If no standard error is given for the group as a whole then the program will assume a value and warn the user that it has done so.

Another feature which can be introduced is the ability to specify a constant to be added to each observation. Constants are introduced with a "/", to distinguish them from standard errors, and apply to each observation after and including the one whose line they are on, until another constant is found. Thus, in the previous example, to add 100 mm to the first observation, 200 mm to the next two, and 0 to the final one :

```
* DMES     1.0    1.0
AA      BB      1234.5    /0.100
CC      BB2      896.12    2.0    5.0    /0.200
```

```
CDE     FGH     1066.88
BB      CDE     1492.0    /0
```

This assumes that the network is observed, rather than simulated. If the latter were the case then the observations would not be written. The previous example, with the same standard errors, but without the constants would become :

```
* DMES    1.0     1.0
AA      BB
CC      BB2     2.0     5.0
CDE     FGH
BB      CDE
```

6. RELATIVE ERRORS

One option available on LGC is that which allows a list of pairs of points to be specified, between which the relative errors are to be calculated. Since these relative errors are calculated in three dimensions, it is necessary to define the individual components which are displayed in the output list. They are : radial error, σ_r; bearing error, σ_G; vertical angle error, σ_v; height error, σ_z; and length error, σ_L. These are shown in Fig. 8.

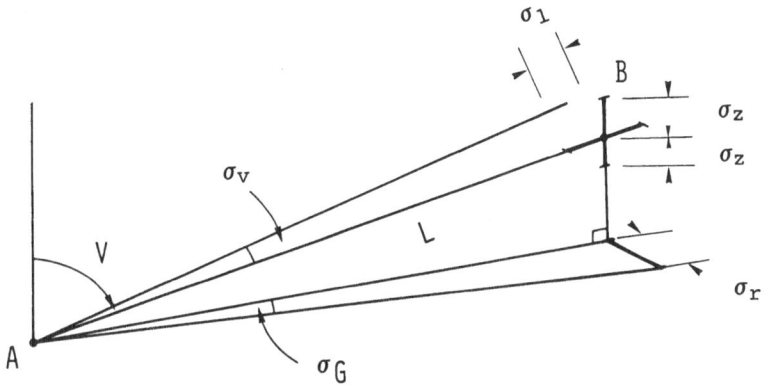

Fig. 8   Parameters for relative errors

These errors are calculated from the characteristic equation for the observations which correspond to them. The first step is to form the 6 x 6 sub-matrix, Q, of the inverse of the normal matrix. This is composed of the six rows and columns corresponding to the three unknowns at each point (Fig. 9).

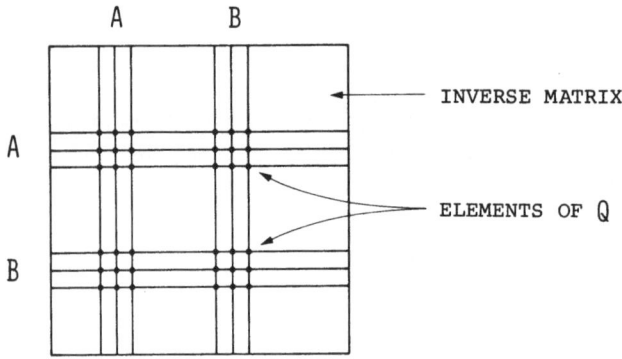

Fig. 9 Sub-matrices for relative errors

Note that if any coordinate at either point is fixed, then there will be no corresponding row and column in the inverse matrix. Since it is fixed, then it follows that its variance, and all its associated covariances, are zero. The relevant elements of Q are, accordingly, set to zero.

The relative height error is found directly from Q, and is given by the formula :

$$\sigma_Z^2 = \sigma_{Z_A}^2 + \sigma_{Z_B}^2 - 2\sigma_{Z_A Z_B}. \tag{10}$$

All the elements on the right hand side of this equation are found in Q. Having found σ_z, the angular error σ_v can be found from the relation :

$$\sigma_v = \frac{\sigma_z}{L} \cdot \sin V \tag{11}$$

To find the relative length error, σ_L, it is necessary to perform the matrix equation :

$$\sigma_L^2 = C_L Q C_L^T \tag{12}$$

where C_L is a 1 x 6 vector containing the theoretical coefficients of what would be the observation equation for a measured distance between the two points. In a similar way, the standard error of the bearing, σ_G, is found from the equation :

$$\sigma_G^2 = C_G Q C_G^T \tag{13}$$

where C_G contains the coefficients of the observation equation for the bearing from one point to the other. The relative radial error, σ_r, is found from the formula :

$$\sigma_r = \sigma_G \cdot L \sin V. \tag{14}$$

The pairs of points between which the relative errors are required are introduced by the key-word *EREL.

7. **THE RESULTS OBTAINED**

7.1 Final adjusted coordinates

Each adjustment of an observed, rather than a simulated, network leads to a set of final adjusted coordinates in the X, Y and Z dimensions. These are printed for each point, with the points grouped as fixed points, free points, and so on. In general there is also a geoidal height, or measured altitude, corresponding to these coordinates, but if this is in a network with a local origin, in which case the geoidal height is meaningless, it is printed as zero.

The coordinates are usually printed up to the fifth decimal place, or hundredths of millimetres. If such precision is unnecessary, however, it may be clearer to specify that only the tenths of millimetres need to be printed. The user can do this by writing the key-word *DIXI in the input list. (The etymology of this may be obscure ; it is from the French " dixièmes ", or " tenths ").

Together with the list of coordinates are printed the movements of the points in each dimension ; DX, DY and DZ. These are the adjusted minus the approximate values. The standard errors of the positions of the points are also printed. These three values, in the X, Y and Z directions, are extracted from the inverse of the normal matrix.

7.2 The observations

After the list of adjusted coordinates comes the list of observations. These are grouped into class (measured distances, horizontal angles, etc.) and for each observation the following information is printed :

- the points stationed and observed
- the observation
- its adjusted value
- the standard error (sigma)
- the residual
- the ratio residual / sigma

In addition, certain types of observation have items of information printed which are peculiar to them. For the measured (spatial) distances, the altimetric sensitivity is printed. This value, which is useful in assessing the contribution to determining the Z coordinate at a point of a distance bearing upon it, is the variation caused to the distance by a change of 1 cm in the height of the point. For each zenithal distance observed, any extension specified on the target point, and the height of instrument which has been calculated, are printed.

The residuals of all angular observations are expressed both in centessimal seconds (cc), and in terms of the radial displacement at the observed point (Fig. 10).

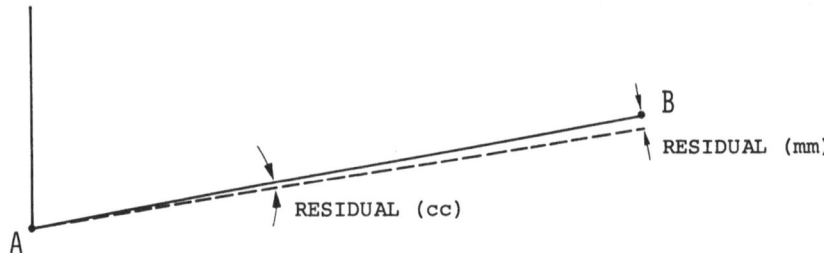

Fig. 10   Angular and radial residuals of an angle measurement

For each class of observation, the mean and standard deviation of the residuals are calculated. Denoting these as X and S' respectively, S'^2 is taken as an unbiased estimate of the true variance of the population, and the statistic :

$$\frac{X - \mu}{S' / n} \qquad (15)$$

is distributed as a Student t-distribution with (n-1) degrees of fredeom, where μ is the population mean and n is the sample size.

So, taking a 5% level of significance, the corresponding 95% confidence interval for μ is defined as :

$$\left[\bar{X} - \frac{t_{0.05} \cdot S'}{\sqrt{n}} , \quad \bar{X} + \frac{t_{0.05} \cdot S'}{\sqrt{n}} \right] \qquad (16)$$

where $\mathrm{Prob} \left\{ \left| \frac{\bar{X} - \mu}{S'/\sqrt{n}} \right| > t_{0.05} \right\} = 0.05$.

The confidence limits for the standard deviation of the residuals can be calculated as well, since the statistic :

$$\frac{(n-1) \, S'^2}{\sigma_0^2} \qquad (17)$$

is distributed as χ^2 with (n-1) degrees of freedom, where σ_0^2 is the variance of the population from which the sample is drawn. Again with a significance level of 5%, the limits χ_0^2 and χ_1^2 are defined as :

Prob. $\left\{ \chi^2 < \chi_0^2 \right\} = 0.05/2$,
Prob. $\left\{ \chi^2 < \chi_1^2 \right\} = 0.05/2$.

The 95 % confidence interval for the variance is then given by :

$$\left[\frac{(n-1) \; S'^2}{X_1^2} \quad , \quad \frac{(n-1) \; S'^2}{X_0^2} \; . \right] \tag{18}$$

There are library routines in the CERN computing system which enable the Student t-distribution and the χ^2 distribution to be used. Having the confidence limits for the mean and standard deviation of the residuals means that the user is better placed to interpret the effect of individual, or groups of, observations. Ultimately it is hoped that an option will be available where the residuals are printed as a histogram. A visual inspection can identify irregularities in the distribution which figures cannot.

7.3 Output for simulations

When LGC is running in simulation mode, several simulations are carried out on the network. It is not necessary to print the results of each simulation in full. Instead, the adjusted coordinates of all points are printed as before, but instead of a complete list of the observations a summary of the number in each class, together with the mean and standard deviation of the residuals, is printed. On the other hand, since it would be useful to have a list of the observations (but not have it repeated for each simulation) this is printed for the last simulation only.

After all the simulations have been completed, a final list of points is printed, showing for each one :

- the standard errors (1σ) of the position in the X, Y and Z directions,
- the maximum positive and negative movements in each direction over all the simulations.

7.4 Other information printed

The next item to be printed after the observations (or the summary of the simulations) is the list of relative errors. The five components defined in section 6 are printed here.

For all networks, simulated or observed, a final page is written which gives for each point :

- the X and Y coordinates at level zero
- the ellipsoidal height (non-measurable "geometric" altitude)
- the value of the geoid-ellipsoid separation at the LEP and zero levels
- the deviations of the vertical and n (North and East components)
- the deviations of the vertical in the CERN X and Y directions.

ACKNOWLEDGEMENT

I gratefully acknowledge the help and encouragement given to me by Michel Mayoud and many others during my stay at CERN

*   *   *

REFERENCES

1) M. Mayoud, Géométrie théorique du LEP, CERN LEP Note N° 456 (1983).

2) B. Bell, A simulation of the gravity field around LEP, Private communication, (1985).

APPENDIX

BREVIARY OF KEY-WORDS

*ANGL	List of horizontal angles
*CALA	List of fixed points
*DHOR	List of distances reduced to level zero and to the X,Y plane
*DIXI	Specify tenths of a millimeter rather than hundredths
*DMES	List of measured distances
*DVER	List of vertical distances
*ECHO	List of offsets in the X, Y plane
*ECSP	List of offsets in a three-dimensional plane
*ECTH	List of offsets from a theodolite direction
*ECVE	List of offsets from a plumb-bob line
*END	Last key-word
*EREL	List of pairs of points for relative errors
*FIN	Last key-word
*LEP	Work on the LEP level
*OLOC	Local origin
*ORIE	List of orientation
*POIN	List of points variable in X, Y, z
*PDOR	Introduce the orientation point
*PUNC	Specify list of adjusted coordinates at end of output
*RADI	List of radial observations
*SIMU	For a simulation
*SPHE	Spherical earth model
*TITR	Introduction of title
*VXY	List of points variable in X and Y
*VZ	List of points variable in Z only
*ZENH	List of zenital distances - instrument height known
*ZENI	List of zenital distances

COMPUTER AIDED GEODESY FOR LEP INSTALLATION

Part I : THE INSTALLATION PROCEDURE

M. Hublin
CERN, Geneva, Switzerland.

ABSTRACT:

In this first part of the paper, we shall review the problems encountered by the geodesist to place the LEP machine in the right position. The absolute positioning of a new machine (LEP) and the relative positioning of its quadrupoles are familiar procedures to us; the new problem with LEP is the increase of the number of elements to be aligned and the decrease in the duration of the installation. Then we shall describe the transfer of the coordinates down the shafts, the underground network, the alignment of the quadrupoles, dipoles and elements of the straight sections.

1. INTRODUCTION: SOME INFORMATION FROM THE SURVEY POINT OF VIEW

Name :	L.E.P. = Large Electron-Positron Storage Ring
Definition :	Circular accelerator including 4500 elements to be aligned
Location :	At the end of the following chain (Fig. 1):
	LIL = Lep Injector Linacs
	EPA = Electron Positron Accumulation Ring
	PS = Proton Synchrotron
	SPS = Super Proton Synchrotron
Length :	26.7 km
Specification :	Relative alignment accuracy of the quadrupoles: 0.1 mm r.m.s.
Duration of the installation :	From June 1986 to December 1988

2. LEP, WHAT DOES IT IMPLY FOR US ?

2.1 Well known problems

2.1.1 The absolute positioning

Being the last one of the chain including LIL, EPA, PS, SPS, LEP must be perfectly aligned with respect to them, i.e. at its theoretical position. It is the homogeneity of the geodetic network which makes this constraint possible.

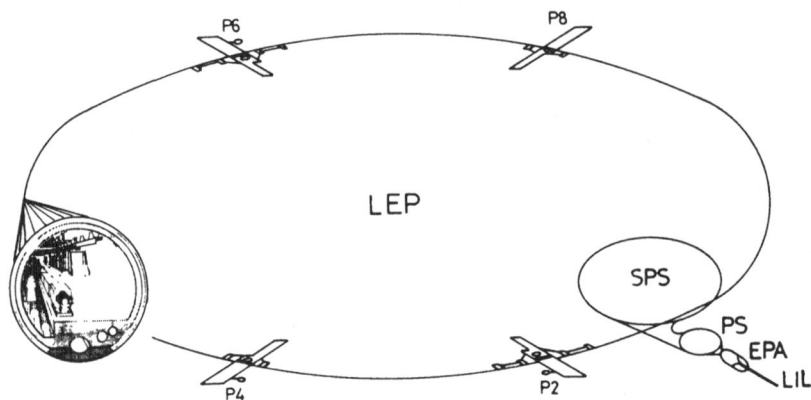

Fig. 1   The chain of CERN accelerators

2.1.2 The relative positioning of the quadrupoles

Machine physicists demand from us a relative alignment accuracy for the quadrupoles of
0.1 mm r.m.s; i.e. that the position of each quadrupole with respect to the preceding and
following ones is good to 0.1 mm, in both the vertical and radial directions. We have
already achieved these specifications in the SPS machine and we shall see in the next
section the methods and instrumentation we will use for the LEP.

2.2 New problems with LEP

	SPS	LEP
Number of elements to be aligned	1600	4500
Number of measurements	14000	32000
Duration of the installation	22 months	30 months

The table indicates the differences between LEP and SPS, the preceding big machine at
CERN. The ratio in each of the first two rows is about 2.5, in the last row it is only
1.4. This shows the problem of LEP for the geodesist : a significant increase in the
number of measurements and an appreciable decrease in the duration of the installation.

3.   THE INSTALLATION PROCEDURE

About 4500 elements have to be aligned in the LEP tunnel, including 800 quadrupoles,
3200 dipoles and some 500 devices located in the eight straight sections.

3.1 Transfer of the coordinates down the shafts

A concrete pillar of 40 cm diameter is built at the bottom of each of nine shafts.
Vertically above each pillar, on the surface, a metal bracket assures a link to the

geodetic pillar of the zone and also to another geodetic pillar further away. The distances between these geodetic points will generally be measured by the Terrameter (Fig. 2).

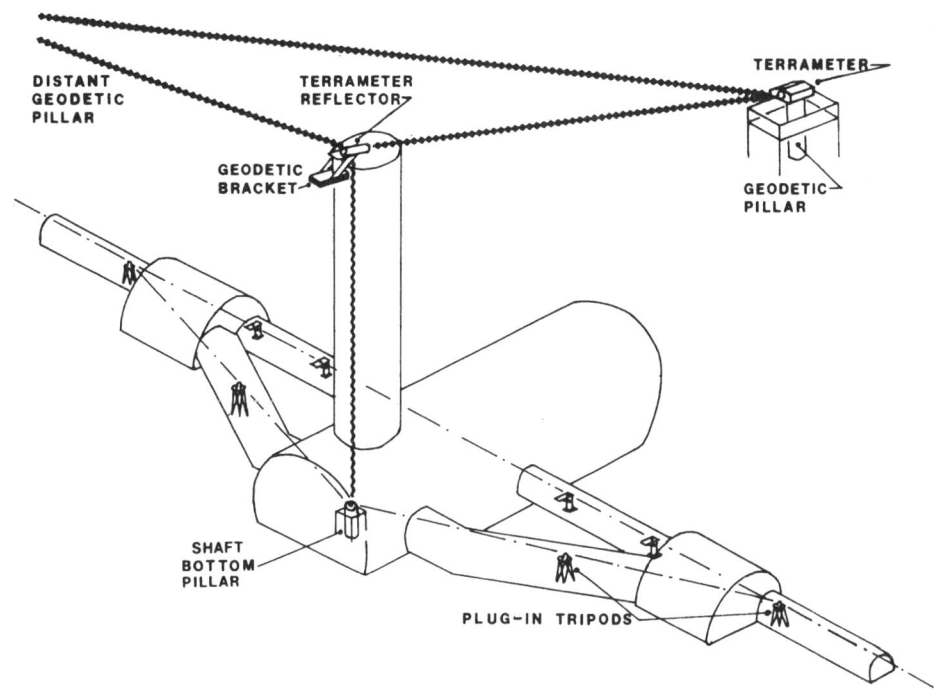

Fig. 2   Coordinate transfer arrangements

Two types of instrument will be used to measure the offset distance between the reference marks of the bracket and the pillar. The first one is a zenith telescope giving a relative accuracy of 1/200000, or half a millimetre per 100 metres depth. In the second one, an electronic theodolite coupled with an EDM is used; the instrument is placed 15 centimetres from the reference mark of the pillar at the bottom of the shaft, and aimed at a target placed on the bracket at the top of the shaft.

3.2 The underground network

We thought previously of using the reference points of the quadrupoles for the control points, but this conflicted with the installation schedule, so we replaced them with plug-in tripods equipped with a 3.5 inch diameter sphere with a centring socket of 30 mm bore. The approximate coordinates of the tripods are given by the civil engineering reference network.

About ninety tripods will be installed between the pillars built at the bottom of two consecutive shafts, which are at this time the only points of which the coordinates are known. The coordinates X, Y and H of the polygonal traverse defined by the tripods will be determined by three types of measurement.

3.2.1 The offset measurement

In place of conventional angle measurements, an offset will be measured to each tripod from a line defined between the two neighbouring tripods. In order to obtain the necessary redundancy for a least squares computation, the offsets are also measured on each group of four tripods, as illustrated in Fig. 3. In calm conditions, a nylon wire device is used to measure these offsets, but in a draught a laser device is necessary (Fig. 4).

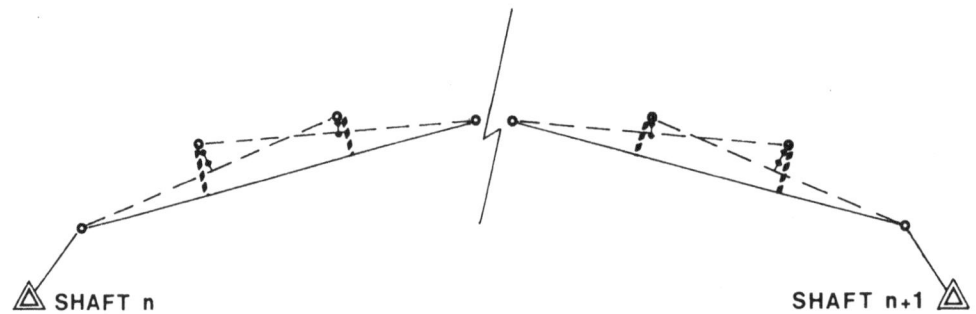

SHAFT n   SHAFT n+1

Fig. 3   Offset measuring arrangement

3.2.2 The distances between consecutive tripods will be measured using an invar wire stretched between the instrument called the "distinvar", (now fully automated) and its fixed point, both fit into the afore mentioned sockets on the tops of the tripods (Fig. 5). One measure in ten will be accomplished with the portable-laser interferometer, providing a calibration distance for the distinvar.

3.2.3 The difference in level between the tripods will be measured by Nivella (a levelling instrument in the process of development).

The principle is the same as that of the nylon wire device : a photo-diode detector travels up and down a threaded rod.

Two consecutive tripods are equipped with such an instrument, and a laser installed on a mobile tripod is positioned between the two and directed alternately at the two threaded rods, the laser-beam defining a horizontal reference line. The photo-diodes search for the instersection of the laser with the vertical rod, giving the height of the beam above the tripod reference point (Fig. 6).

The offset and distance measurements will be treated by a least squares adjustment to determine the X, Y coordinates of the tripods. The levelling will be calculated separately.

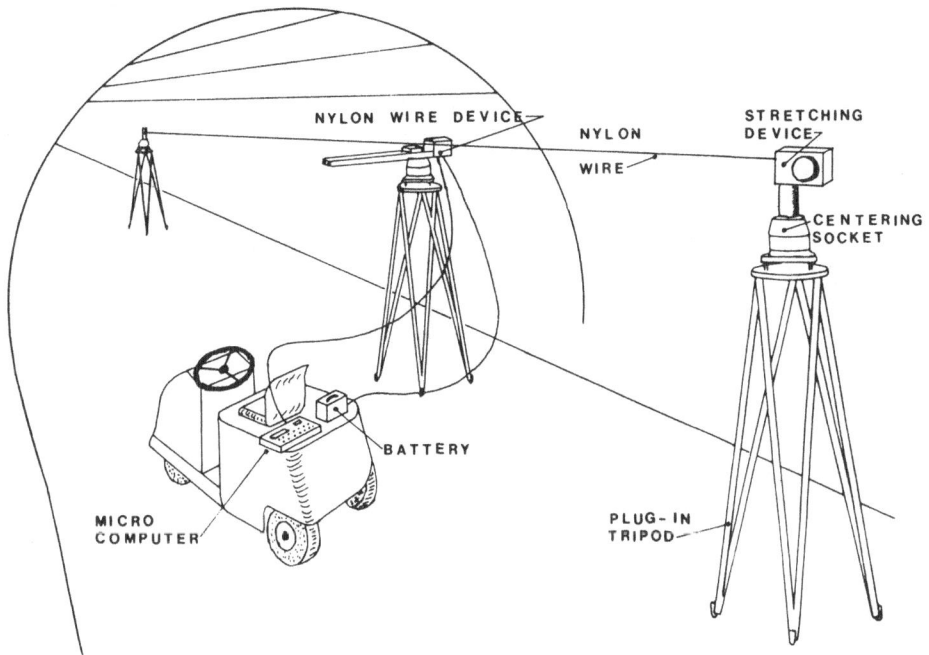

Fig. 4　Offset measurements by nylon wire

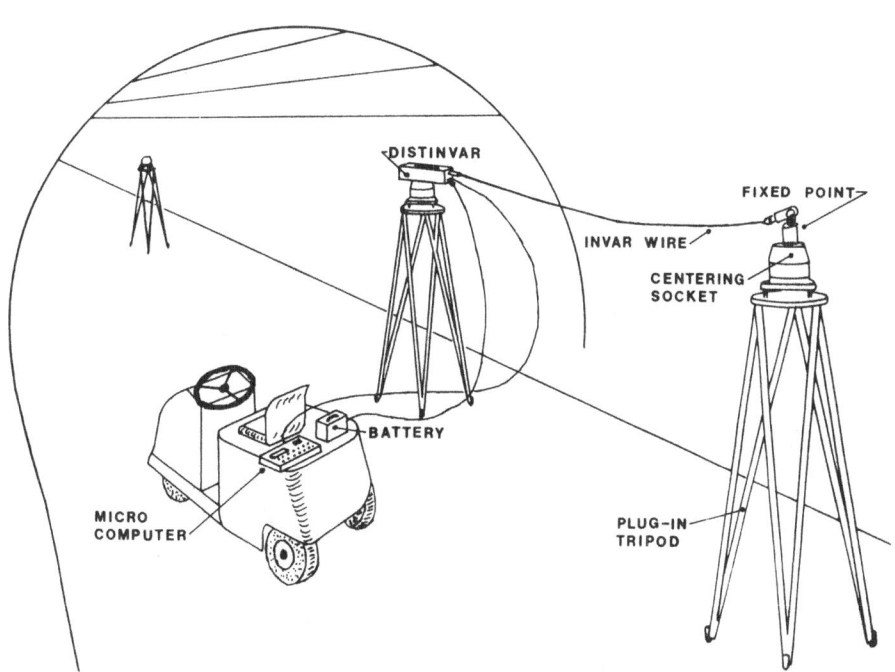

Fig. 5　Distinvar inter-tripod distance measurements

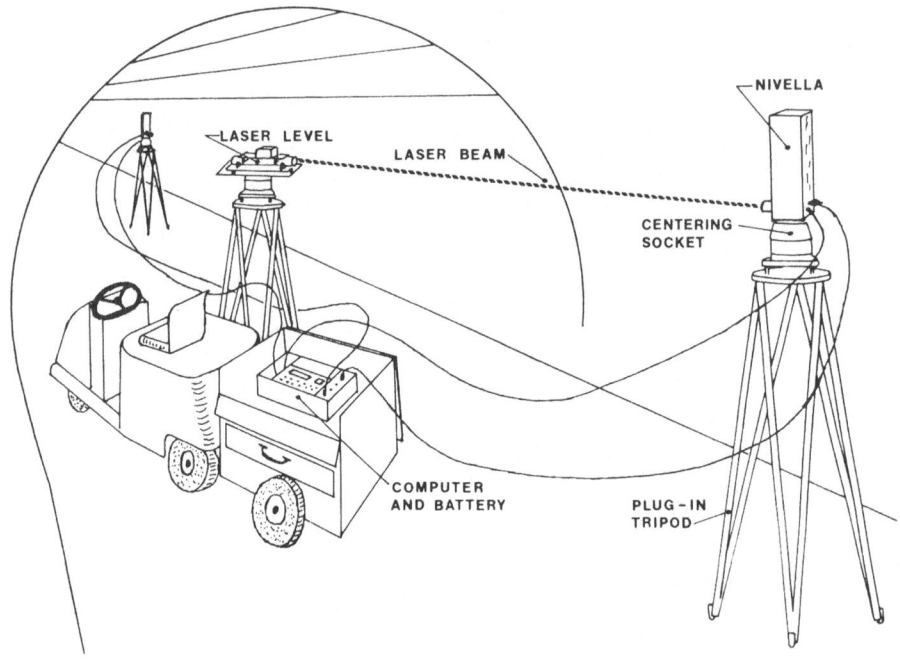

Fig. 6   Nivella height difference measurement

3.3 The alignment of the quadrupoles

The task of the quadrupoles is to focus the beam of particles. Some distortion of the closed orbit can occur if the relative position of the quadrupoles is not good to 0.1 mm r.m.s.

Every quadrupole bears two 3.5 inch diameter spheres with centring sockets of 30 mm bore and two reference marks for the positioning of a precision spirit level. The straight line passing through the centres of the two spheres is parallel to the magnetic axis of the quadrupole; the accuracy required by the mechanics to install the reference marks is 0.1 mm r.m.s. in the radial and vertical directions.

To align a quadrupole it is necessary to calculate, and then to effect the transverse, vertical and longitudinal offsets of the reference marks of the quadrupole with respect to the two adjacent tripods. The instruments used are the same as those used for the network of tripods : distinvar, nivella and nylon wire device used concurrently (Fig. 7).

To move the quadrupoles to their theoretical position, the three jacks of the girder supporting the quadrupole are used. The force necessary to move the jack in the vertical direction is about 10 daN; in order to accomplish this, an auxiliary hydraulic device (in the process of development) will be used.

Once each quadrupole is in its theoretical position, its reference sockets become the control points for alignment of the following elements.

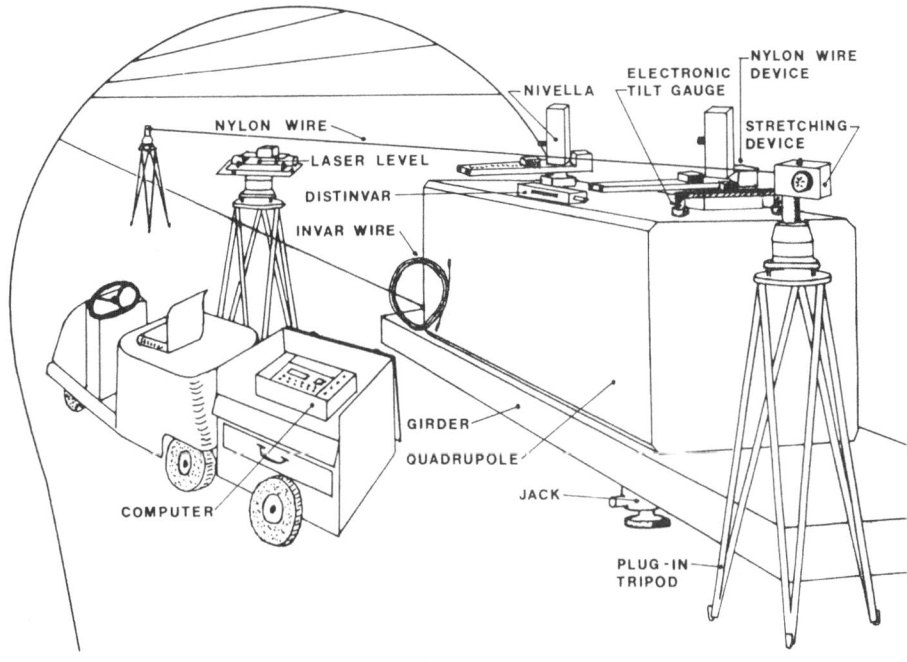

Fig. 7   Quadrupole alignment arrangement

3.4 The alignment of the dipoles

Maintaining the particles in their sub-circular orbit is the function of the dipoles. The required accuracy for the alignment is of the order of 0.3 mm r.m.s.; the operation itself is more difficult than for the quadrupoles because the 5.75 m long dipoles are joined in pairs and attached by a vacuum chamber and the excitation bars (Fig. 8).

The provisional positioning will be accomplished by two laser beams detected in the vertical and radial directions by photocells mounted on a measuring-rod placed at the extremities of the pair, in the centring sockets (Fig. 9). At the same time, a tiltmeter will give the correct inclination of the magnet.

3.5 The straight-section elements

These attain the injection, detection, dumping and orbit correction of the particle beam (Fig. 10). The principle of their alignment is the same as that of quadrupoles and dipoles.

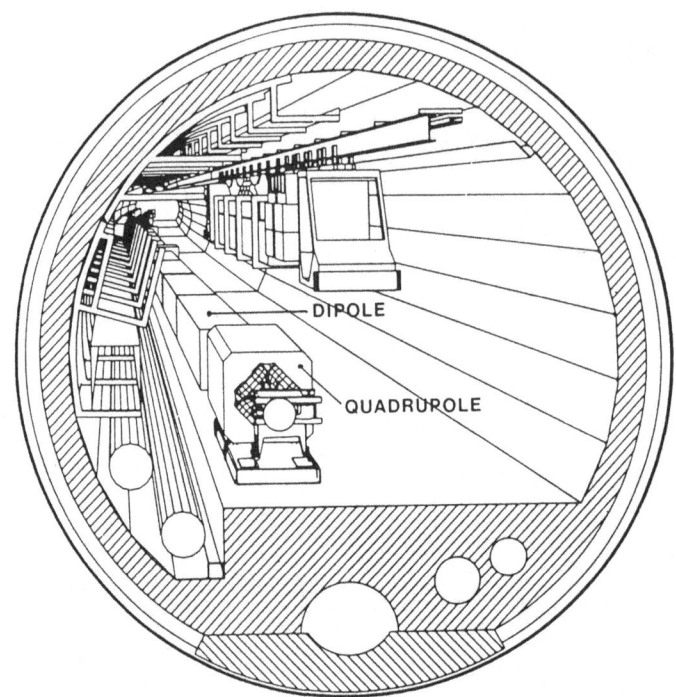

Fig. 8   Quadrupoles and dipoles installed in LEP

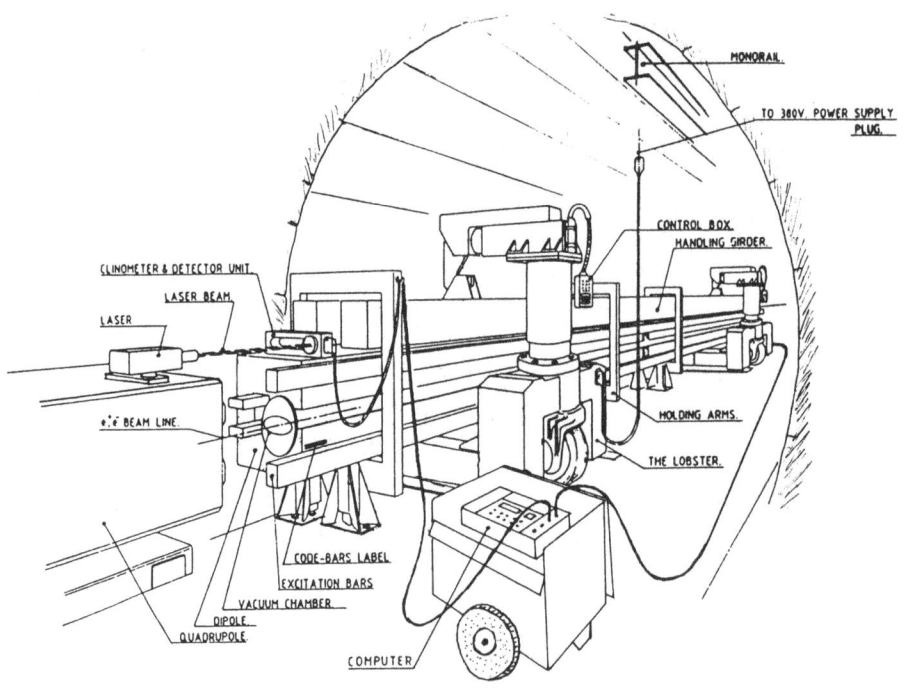

Fig. 9   Dipole alignment arrangement

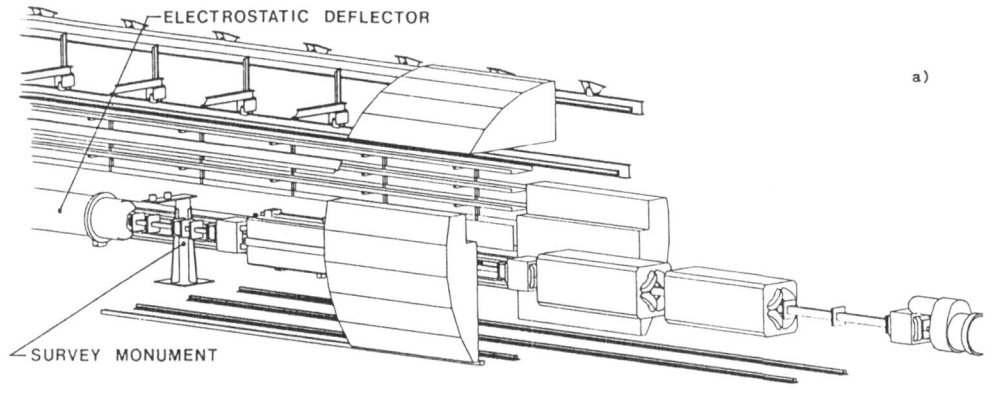

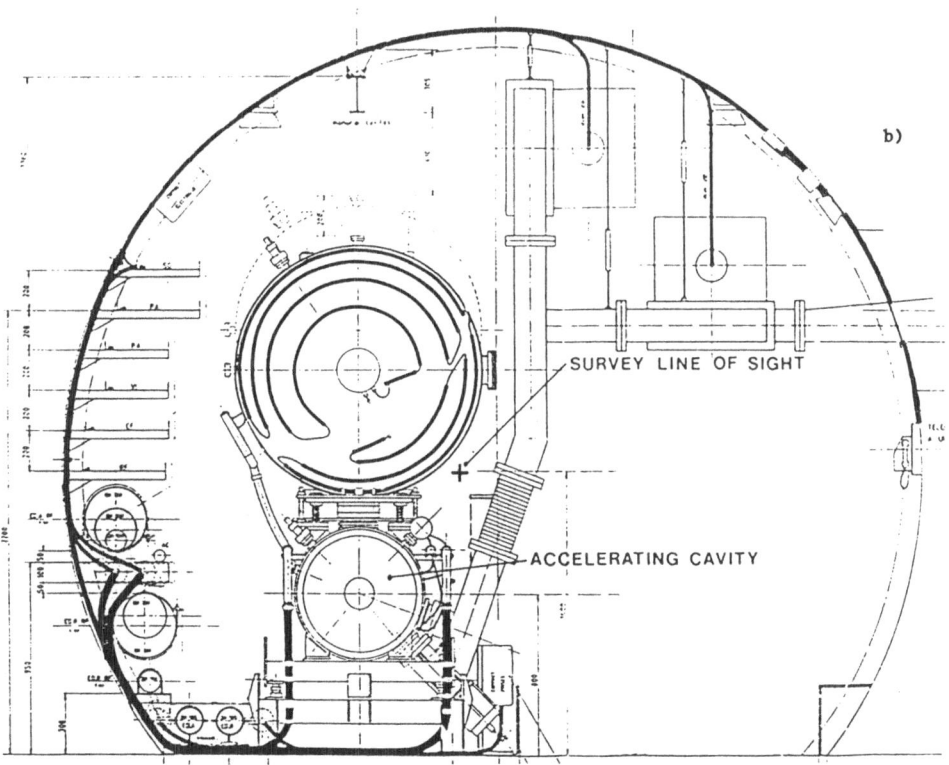

Fig. 10   Typical straight-section elements :
a) Electrostatic deflector
b) RF accelerating cavity

4. CONCLUSION

We have emphasized in the paragraph 2.2 that the new problem with LEP was the significant increase in the number of measurements and the appreciable decrease in the duration of the installation.

Consequently we must improve the rapidity and reliability in taking measurements. For that purpose, we chose to automate some instruments and use on-line computing methods. This will be described in the part II of the lecture.

COMPUTER-AIDED GEODESY FOR LEP INSTALLATION

Part II : INSTRUMENTATION

W. Coosemans
CERN, Geneva, Switzerland

ABSTRACT

The aim of this report is to describe the means by which the metrology work load needed to place the LEP accelerator in position is accommodated.

1. INTRODUCTION

This report follows the complete process of data communication between the measuring instruments and the central computer system, as illustrated in Fig. 1.

The measuring instruments, excluding one, which are used in the LEP surveys are built using the same principles as those developed and implemented over several years by the Applied Geodesy Group.

However, to accommodate the LEP work load and to reduce the measurement time, whilst taking special care in data acquirement and treatment, these instruments have been automated and computerized here at CERN.

Three different types of data acquisition devices are now used :

- the first simply displays the measurements,

- the second is a portable microcomputer,

- the third is a portable computer conceived and constructed at CERN, specially for the survey work in LEP.

For the office computer all the materials come from a well known firm while data treatment is performed on the CERN central computer system.

2. INSTRUMENTS

2.1 Description

2.1.1 Distinvar (Fig. 2)

The distinvar is designed to make accurate differential measurements (± 0.015 mm) of short distances (from 0.40 to 50 m). It consists of three parts: the invar geodetic wire with its clip-in fittings, the remote attachment head and the measuring instrument. The latter is a precision balance mounted on a movable carriage. The principle of the measurements is to find the position of the movable carriage when the balance is in equilibrium. Absolute lengths can be obtained, provided that the complete system (remote attachment head – invar geodetic wire – measuring instrument) is calibrated on a primary base line.

2.1.2 Laser interferometer with self-aligning reflector (Figs. 3 & 4)

At CERN, the primary base line for distance measurements is determined with laser interferometry. Therefore, we make use of a calibration room where its possible to make absolute measurements of distances from 0.40 to 64 m. Obviously, to gain accuracy and time, the primary base line should be used in the field as often as possible.

Since the interferometer only measures change in phase, any interruption in the reception of the laser beam results in a break in the counting system and, hence, a complete loss of the information.

The main difficulty in the use of the movable reflector is that it must be guided within half the diameter of the laser beam. In the calibration room, the movable beam is mounted on a carriage which moves along a rail. Clearly, this cannot be used in the field. Hence, a portable set has been developed, containing the reflector prism which aligns itself on the laser beam. The reflector is mounted on a carriage, which can be moved in the horizontal and vertical directions by two servo-motors controlled by a servo-amplifier receiving signals from a four-quadrant photo-cell. This forms a servo-loop which, by moving the reflector, keeps the reflected laser beam centred on the photocell. In good conditions, measurements can be made up to 50 m with a standard error of less than 0.01 mm.

2.1.3 Horizontal offset measuring device (Fig. 5 & 6)

This device is designed to measure the shortest offset distance from one point to a vertical plane which contains a straight line used as a reference.

The distance range is 0.60 m. and the accuracy is ± 0.03 mm. According to the way the reference line is physically produced, two different types of devices are used.

<u>Nylon wire reference line</u> .- In addition to the 0.25 mm diameter nylon wire, the nylon set consists of three parts each fitted with a locating cylinder to ensure centring in the reference sockets. These parts are : the offset measurement device and two remote devices to attach the wire of which one is equipped with an automatic winder which acts as a wire reserve and applies a constant tension on the wire. The offset measurement device contains a movable carriage driven by a micrometric screw which is driven by a motor and coupled to a displacement counter. An electro-optical system detects the wire and controls the driving motor of the screw in order to centre the movable carriage under it. The distance measured is the displacement of the carriage from a reference point where the shifting counter is reset to the centring position under the wire. The electro-optical system can, to a certain extent, measure the angle between the rule and the reference plane and automatically calculate the reduced distance.

<u>Laser beam reference line</u> .- The laser set consists of two parts : first, the transmitter with its laser optics and power supply, mounted on a locating cylinder to ensure centring in the sockets and to orientate the beam, second, an offset measurement set which has the same function as that of the nylon set except that the wire detecting system is replaced by a photo-cell set for the detection of the laser beam.

2.1.4 <u>Vertical offset measuring device</u> (Figs. 7 & 8)

This device is new in the range of metrology instruments at CERN. It detects on a vertical active rule a horizontally adjusted laser beam. Apart from the position of the laser beam detector, the rule is a copy of the horizontal offset laser measuring device, but in a vertical position. The active rule gives very good results, but the horizontal laser emitter is unsatisfactory and two devices are under development.

In the first, the laser beam is emitted by a GLO laser from WILD and optical fiber directs it to the eyepiece of a WILD ZL zenithal telescope. An automatic device makes the optical axis of the telescope vertical with a precision of one in two hundred thousand. A pentaprism made at CERN horizontally adjusts the laser beam and can rotate, thus making the beam sweep out a horizontal plane (Fig. 9). The use of the zenithal telescope makes two-face observations possible by rotating the telescope through 200 grades without moving the upper prism. This system is easy to handle, but the quality of the beam is inadequate with respect to the required accuracy. The laser spot is not homogenous and we do not obtain a parallel beam with a diameter consistent with the detection cells of the active rule. We are therefore considering the possibility to modify the optics of the ZL telescope.

In the second, the emitter is a classical laser tube centred on a cylindric mechanism. An optical system reduces the divergence by increasing the diameter of the beam. The laser and optics set is mounted on an adjustable support, levelled by an electronic clinometer mounted on two dogs gripping the cylinder of the laser. The average of the measurements obtained by two-face observations of the laser and clinometer eliminates instrumental errors. The quality of the beam is excellent, but on the other hand the handling is arduous and brings back memories of the "NIVEAU LENOIR". However it is presently used for testing the active rule (Fig. 10).

2.1.5 Clinometer (Figs. 11 & 12)

This instrument is designed to accurately measure the tilt in one direction only and it can be used either for absolute or differential measurements. The instrument detects the transverse position of a pendulum suspended by means of five copper wires from a rigid support. The position is read by an electro-optical system using differential photodiodes and amplifiers.

2.2 Automatization and computerisation (Fig. 13)

All the instruments described above are specially designed to be flexible, adaptable to the needs of the user, and easily interfaced. They communicate information via a unique link containing the necessary power cables, as well as permitting standard serial communication. In each instrument a microprocessor circuit board, through adapted incorporated interfaces, controls the electro-optical and electro-mechanical elements and the user dialogue.

Two microprocessor circuits boards have been developed by the Applied Geodesy Group. The first one is truly a microcomputer, programmable in machine code and in basic language, which can be programmed in the field by the user and is specially adapted to the more complicated instrument calculations. The second circuit board is simpler but well adapted to control measuring systems. It is programmable in machine code and the programs cannot be modified by the user in the field. The control programs are interactive and only use the three ASCII characters [CR], [LF], [ESC]. These instruments are not equipped with control keyboard and display, but are operated with the aid of a control box, a terminal or a computer. All the necessary electronics are contained within the instrument thus allowing the user the freedom to choose the terminal or the measurement acquisition system.

3. TERMINALS AND FIELD COMPUTERS (Fig. 14)

In the laboratory, for test measurements, the instrument can be connected to a standard character terminal and a 12 V power supply.

For the unrecorded field measurements, we have developed a control and power box which relays information to the instrument via a unique cable containing power and communication links. A 12 V battery is incorporated in the box, also equipped with a display panel and four control buttons; one being the on/off switch, the other three representing the ASCII characters recognized by the interactive programs incorporated in the instruments.

With a single instrument the collection of measurements is made by connecting a power supply and a portable microcomputer. This type of machine is now available commercially and we have chosen the EPSON PX4 which can also be used as a single terminal.

For the power supply operation control and simultaneous data collection from several instruments, we have developed and built a field computer (Fig. 15), whose main characteristics are the following :

- easily transportable

- can be powered by internal or external battery or main power supply

- possesses sufficient memory allocation to store the data and measurement acquisition programs, the metrological operation control files and the data storage files

- includes an additional stable support memory which can be disconnected from the main system for transportation of data (RAM cartridge of 64 Kbytes)

- offers the option to connect several measurement instruments and a terminal at the same time (five RS232 lines)

- possesses a printer with 40 characters per line

- is compatible with the CERN network of computers, permitting the exchange of data contained in the field computer memory and the central system

- is user programmable with PILS an interacting language developed at CERN

- incorporates transducers for the measurement of temperature, humidity, pressure and the voltage of the battery. This information is accessible from the user program

- the terminal used in the system is the EPSON PX4.

4. THE OFFICE COMPUTER (Fig. 16)

The office machine must complement the data acquisition of the field computer to provide a complete and efficient chain of measurement and calculation. It acts as an intermediary between the field machine and central computer data bases and must allow :

- temporary and local storage of data allowing the user direct and instantaneous access

- local treatment in order to quickly prepare the installation files and to control and validate the measurements made in the field

- easy recuperation of the necessary elements from the central calculation data bases to prepare the installation files

- assistance in the development and maintenance of the field computer operating system

- formulation and transfer of field data to the main computer system for the calculation of the geodetic and metrological networks, or for adding to the data bases.

The system chosen to ensure the above properties is an HP 9540 accommodating several users and tasks. It functions under the UNIX operating system and can be programmed in FORTRAN, PASCAL and C language. It works in conjunction with a hard disc, a rapid input/output tape drive for back-ups, a printer, a double flexible disc drive and six HP 150 microcomputers used as work stations. The computer is linked to the CERN main computer system by four serial lines and an ETHERNET connection. The work stations are equipped with a printer, two floppy disc drives and two RS232 lines for communications with the HP 9540, the field computers and the CERN network.

5. CERN COMPUTER CENTRE

The main computer centre, one of the biggest scientific calculation centre in Europe, houses three large systems : the first using large CDC machines, the second based on large scale IBM and SIEMENS equipment, the third using several VAX machines.

There are many other computers, essentially 32-bit machines shared throughout the CERN site for allocated local tasks. All are connected via CERNET-ETHERNET networks to allow very high speed data transfer between them. For the users, the computers are accessible by a remote batch service named INDEX, which works at 4800 bauds, manages 1300 terminal and also allows connection with the European and American scientific nets, EARN, and BITNET.

We use the facilities of the computer centre for all large calculations generated by the problems of geodesy, metrology and topography, and also to update and consult the data bases.

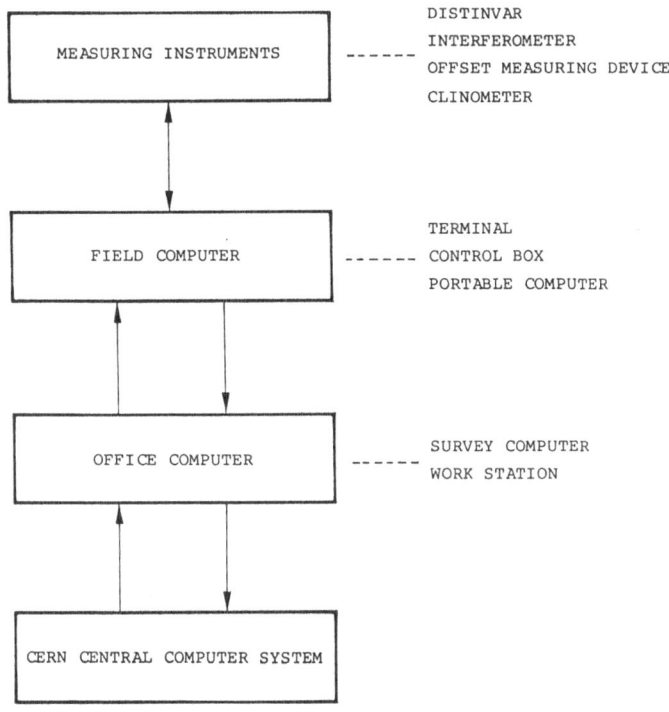

Fig. 1   Process of data communication

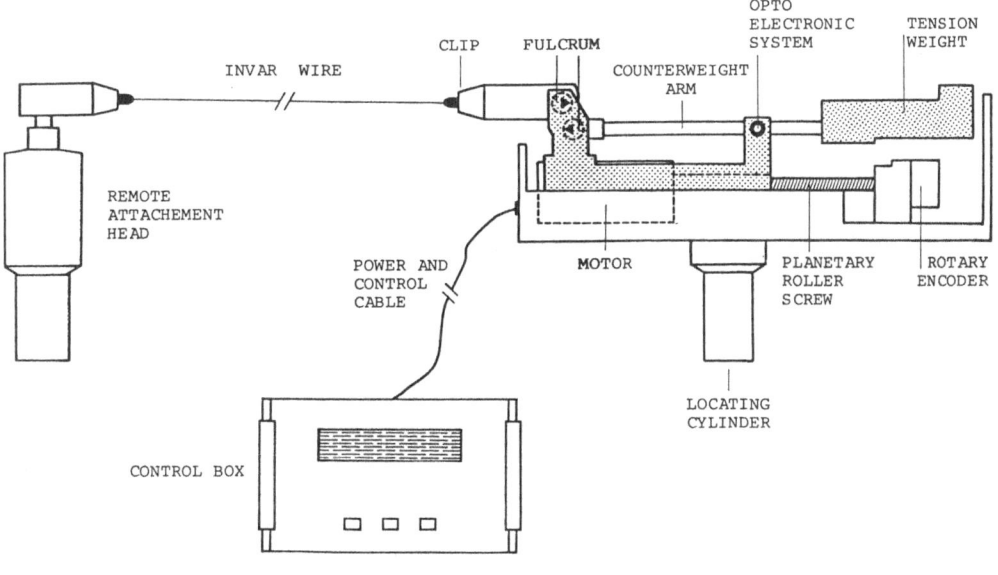

Fig. 2   Distinvar

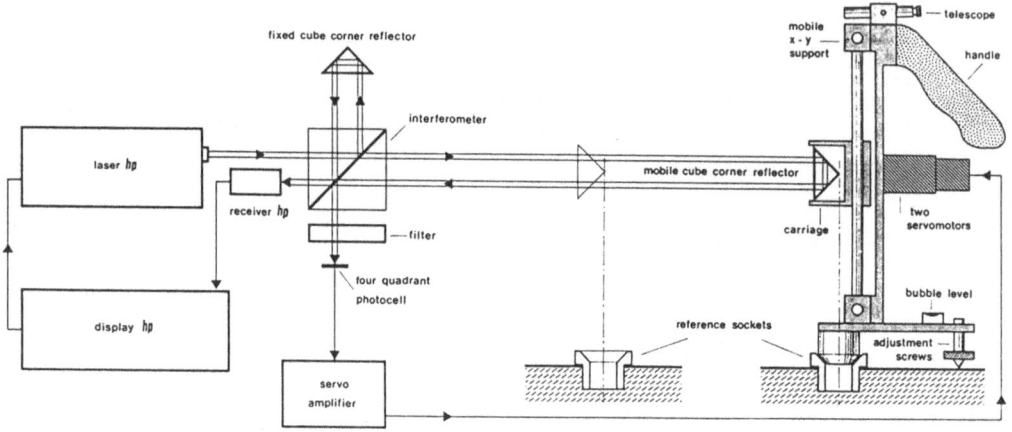

Fig. 3   Scheme of the laser interferometer

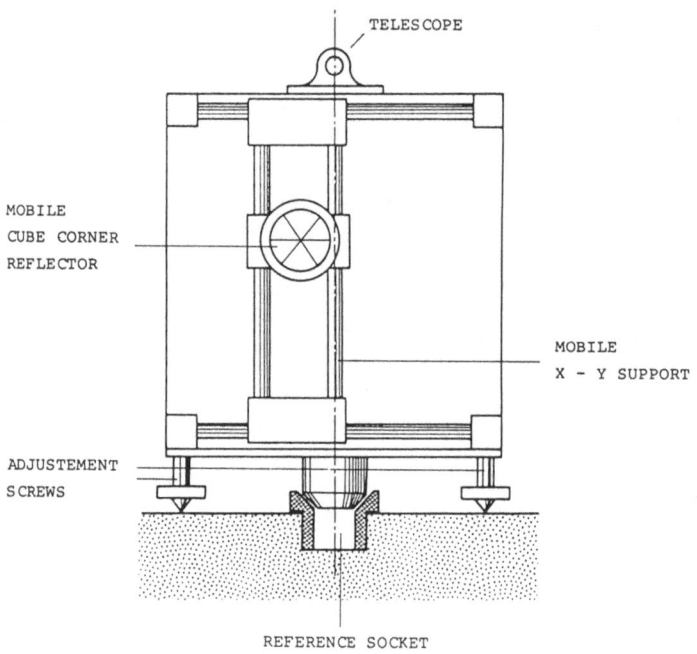

Fig. 4   Self-aligning reflector

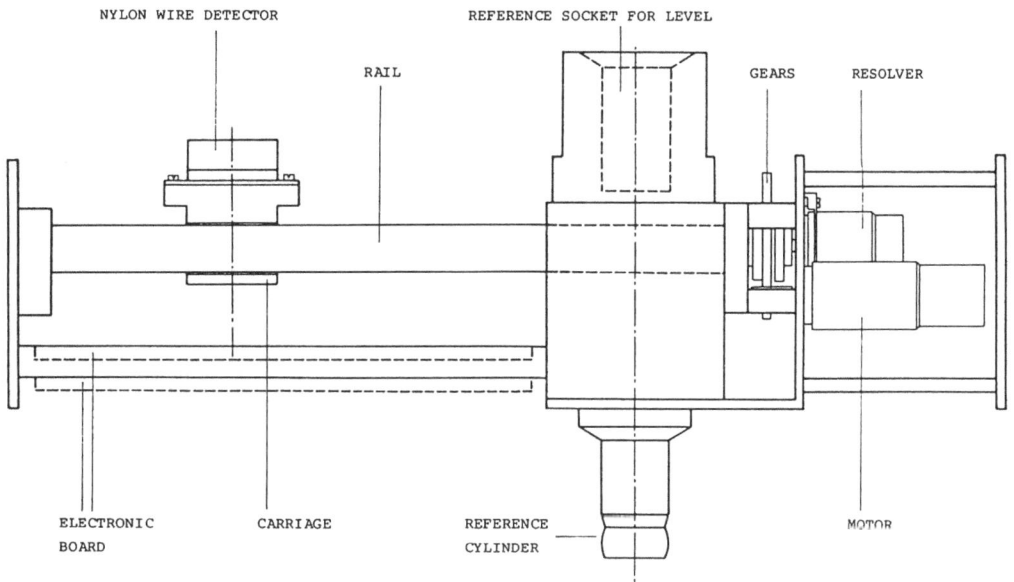

Fig. 5   Profile section of the horizontal offset measuring device

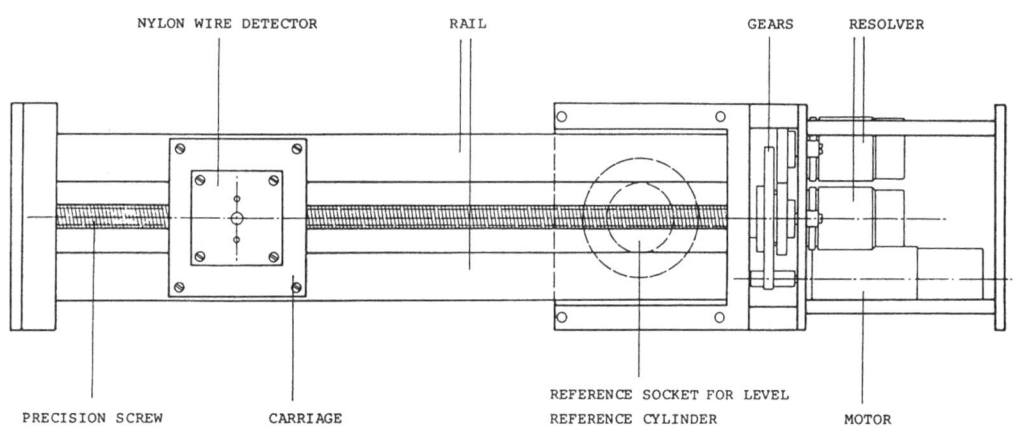

Fig. 6   Plan of the horizontal offset measuring device

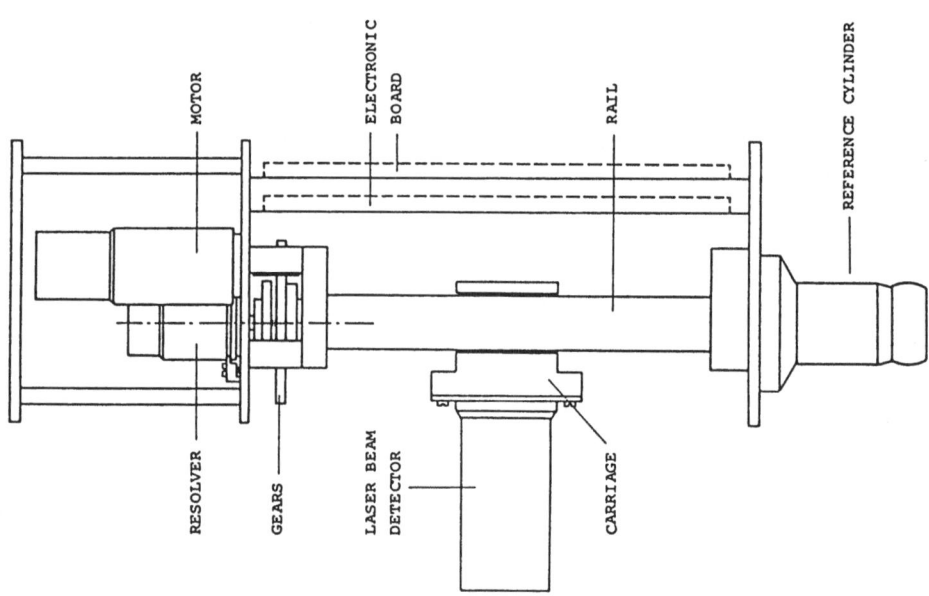

Fig. 8   Profile section of the vertical offset measuring device

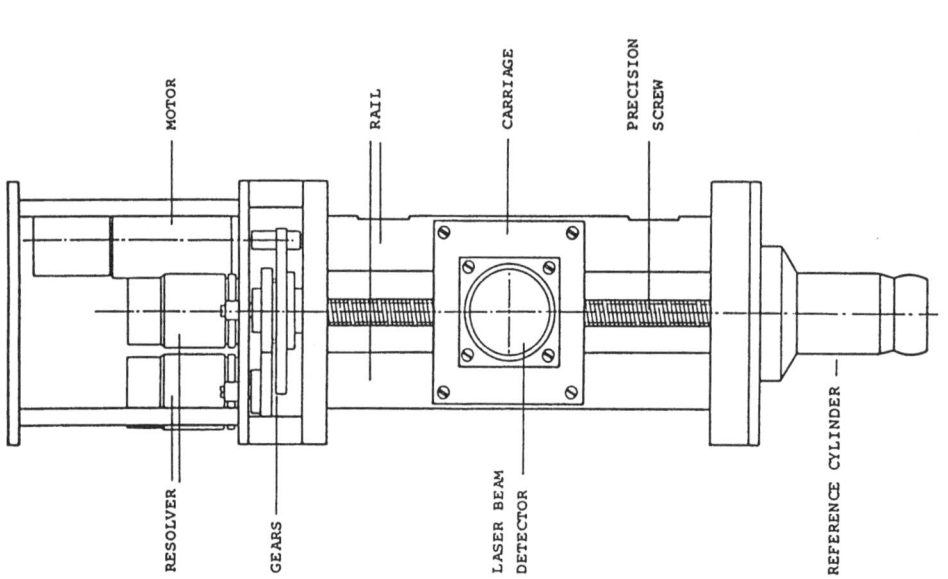

Fig. 7   Front view of the vertical offset measuring device

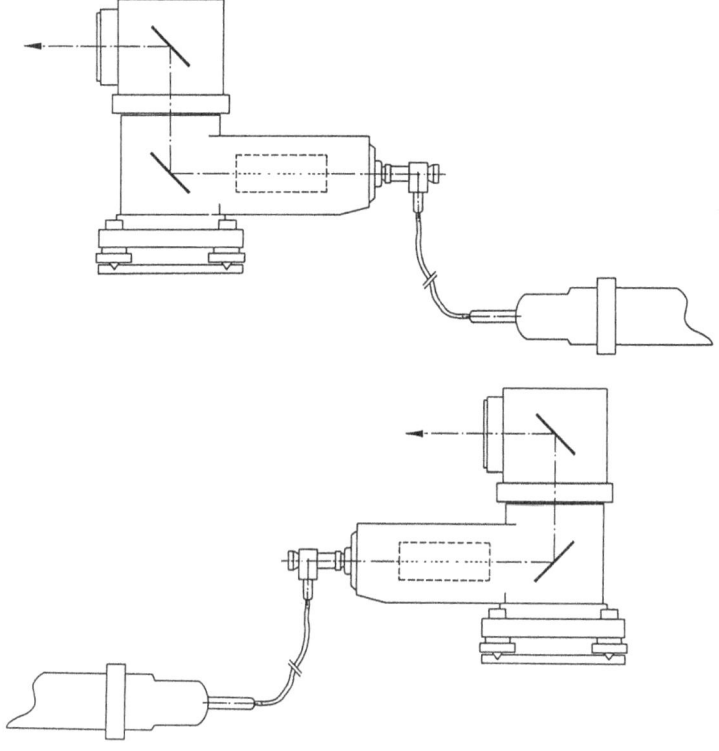

Fig. 9   Laser beam – automatic level

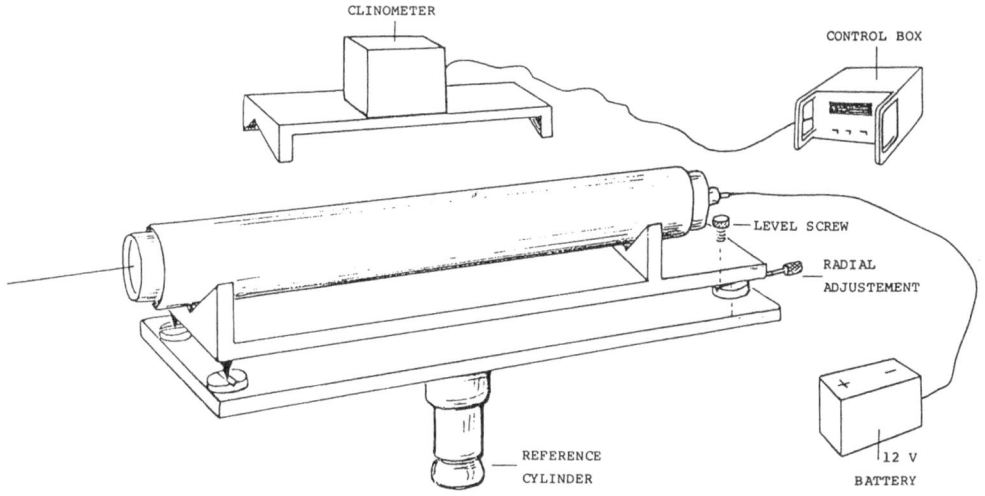

Fig. 10   Laser beam – manual level

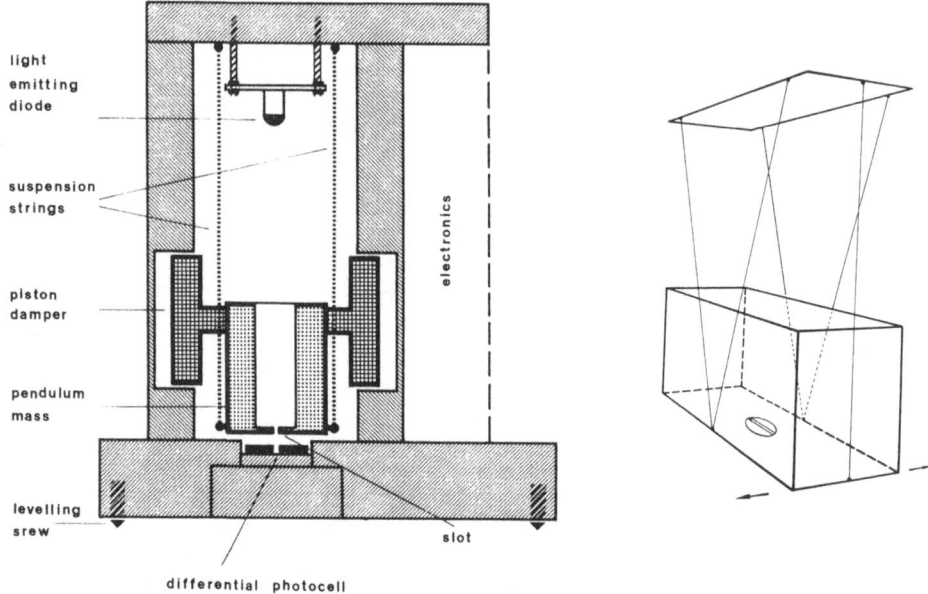

light
emitting
diode

suspension
strings

piston
damper

pendulum
mass

electronics

levelling
srew

slot

differential photocell

Fig. 11   Cross-section of the clinometer   Fig. 12   Diagram of tilts

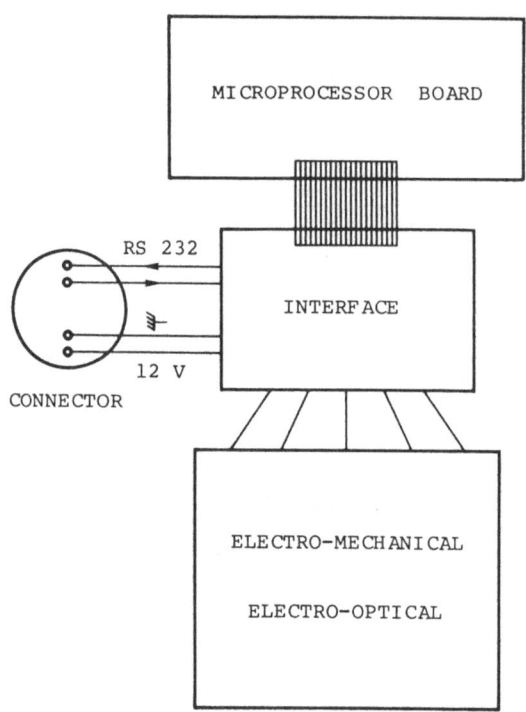

MICROPROCESSOR BOARD

RS 232

INTERFACE

12 V

CONNECTOR

ELECTRO-MECHANICAL

ELECTRO-OPTICAL

Fig. 13   Scheme of an instrument

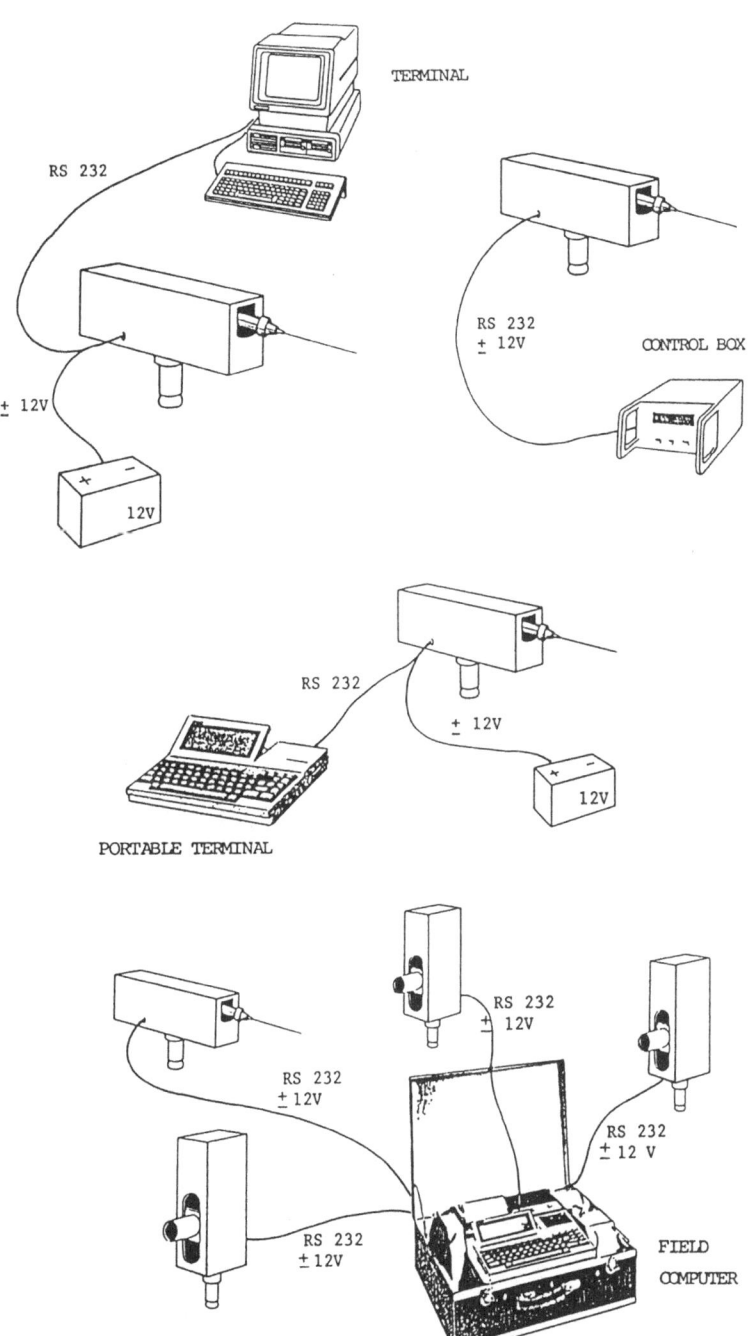

Fig. 14   Terminals, control base and field computers

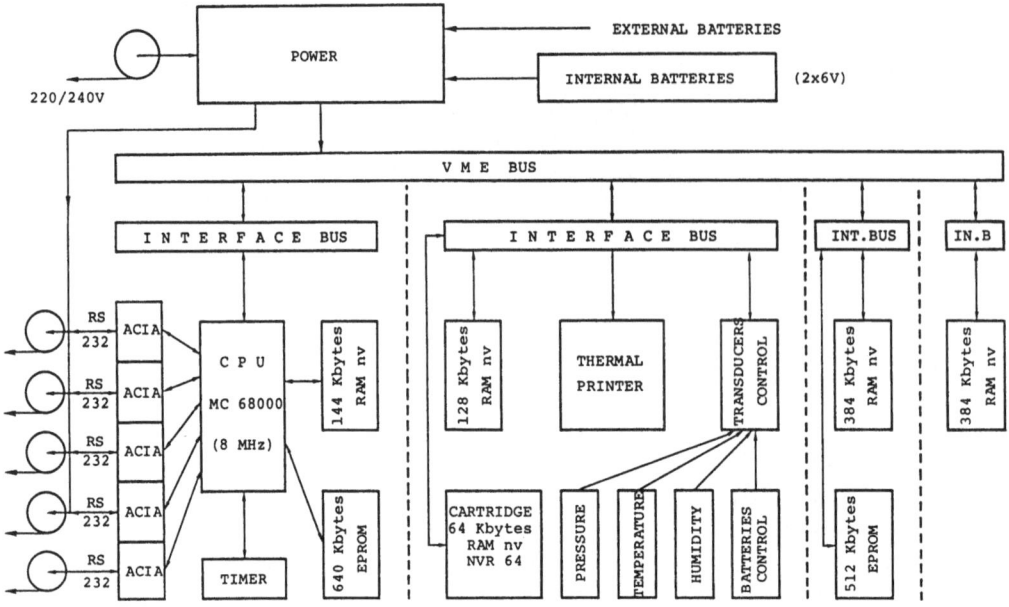

Fig. 15   Scheme of the field computer

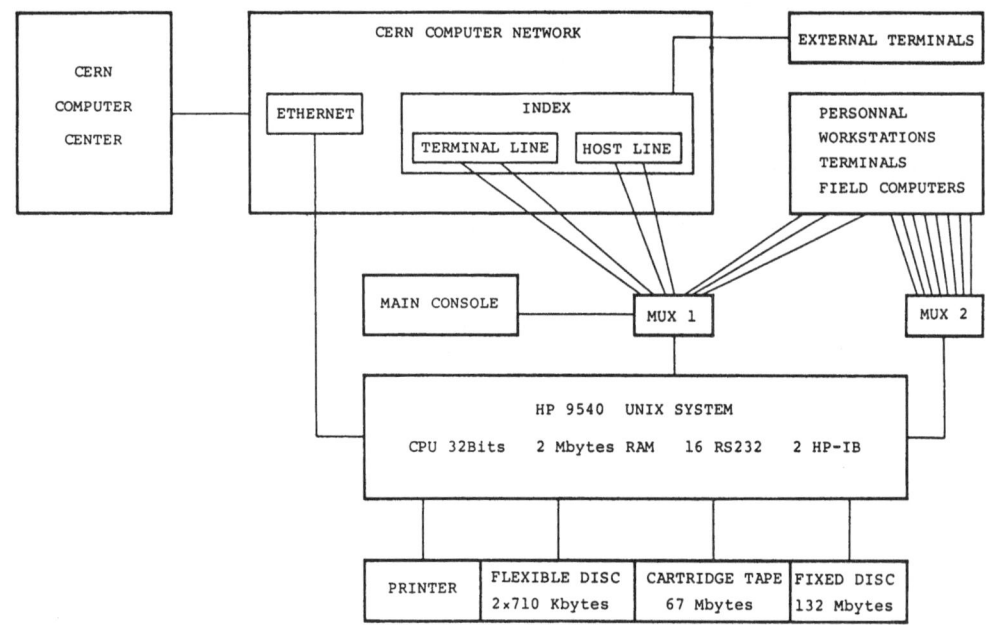

Fig. 16   Scheme of the office computer

APPLIED METROLOGY FOR LEP

Part I : COMPUTING AND ANALYSIS METHODS

M. Mayoud

CERN, Geneva, Switzerland

ABSTRACT

The search for accuracy in geodetic metrology demands that the processing of the data always remains rigorous and takes into account any factor which can affect the precision of the measurements or that of the results. This constant care leads to very precise considerations in defining the absolute positioning of a new accelerator within an area submitted to strong deflections of the verticals, in processing the data of long and flexible linear networks, or in estimating the real accuracy of the alignment of machine elements. The basic principle of the methods developed at CERN for the geodetic positioning of LEP, the processing and smoothing of the alignment data and the stochastic analysis of networks are described in this third part. An estimate of the predicted errors of LEP metrology is given as a conclusion.

1. INTRODUCTION

For each new project, the size of the accelerator to be built has led to reconsideration of several aspects of the methodology. The major changes in concept have been introduced in designing the SPS (2.2 km diameter) and LEP (8.6 km diameter) control networks.

First, in both cases, the computation of the theoretical XYZ coordinates of the machine has involved more and more detailed considerations of the geometry of the earth. For the SPS, a spherical approximation was sufficient to express the effects of the earth's curvature in computing the Z ordinates, correcting the vertical "descent" of geodetic points along the shafts or properly tilting the magnets, in order to obtain a real plane in space. With the LEP project, which partly lies under the Jura mountains, a further step has been to determine the vertical deflections generated by gravity disturbances, and then to express the separation between a reference equipotential surface and a reference (local) ellipsoid. This knowledge provides the necessary corrective factors to convert measured altitude into ellipsoidal heights in 3-D computations, to correct the coordinates of bottom points from the effects of vertical deflections or to reduce the gyro measurements.

One other change in the methodology is that repetitive measurements of the SPS or LEP control networks could no longer be thought of and managed as "absolute" surveys. For such long and flexible ring-shaped figures, the variations of the coordinates issuing from different sets of comparable measurements have no physical meaning for the particles. The trajectory of a beam within an accelerator is mainly sensitive to short-range errors. Survey or alignment are thus "seen" as local imperfections of the guidance magnetic field while long range errors have less effect but are not negligible. Then, the major requirement for the geometry of an accelerator is that relative errors must be as small as possible. In other terms, the figure must be smooth. This smoothing concept is fundamentally involved in a particular refinement process used for the first installation of a large machine and for any new partial or global survey when a re-alignement of components is to be done.

It must also be remembered that the certitude in any accuracy problem cannot be acquired without a thorough knowledge of the stochastic behaviour of the measured networks. Although this statement sounds self-evident, its reality depends on the effective means for estimating the actual errors and deformations that a given network may undergo, due to the effects of both random and systematic errors in the measurements. For that purpose, a simulation method has been developed at CERN on the basis of a statistical analysis with controlled perturbations.

2. THEORETICAL POSITIONING OF LEP

2.1 Transformation of machine coordinates

The theoretical coordinates of the components of the machine are first produced by the MAD program (Methodical Accelerator Design) in an arbitrary plane system. The positioning of this theoretical machine is then made according to local constraints (Fig. 1 & 2) including :

- feasibility of a future link between the SPS and LEP for e-p physics, this fixing the first interaction point
- joining the existing reconnaissance gallery
- confinement within the molasse bank
- optimisation of the depth of pits.

Successive unitary transforms are expressed in relation to these constraints and their final product gives the theoretical geodetic coordinates of the machine (Ref. 1).

2.2 The CERN coordinate system

As the particles practically ignore gravity, the CERN coordinate system is a cartesian system. For historical and logical reasons, its vertex is the centre of the first proton synchrotron built at CERN, the point P_o of the PS machine.

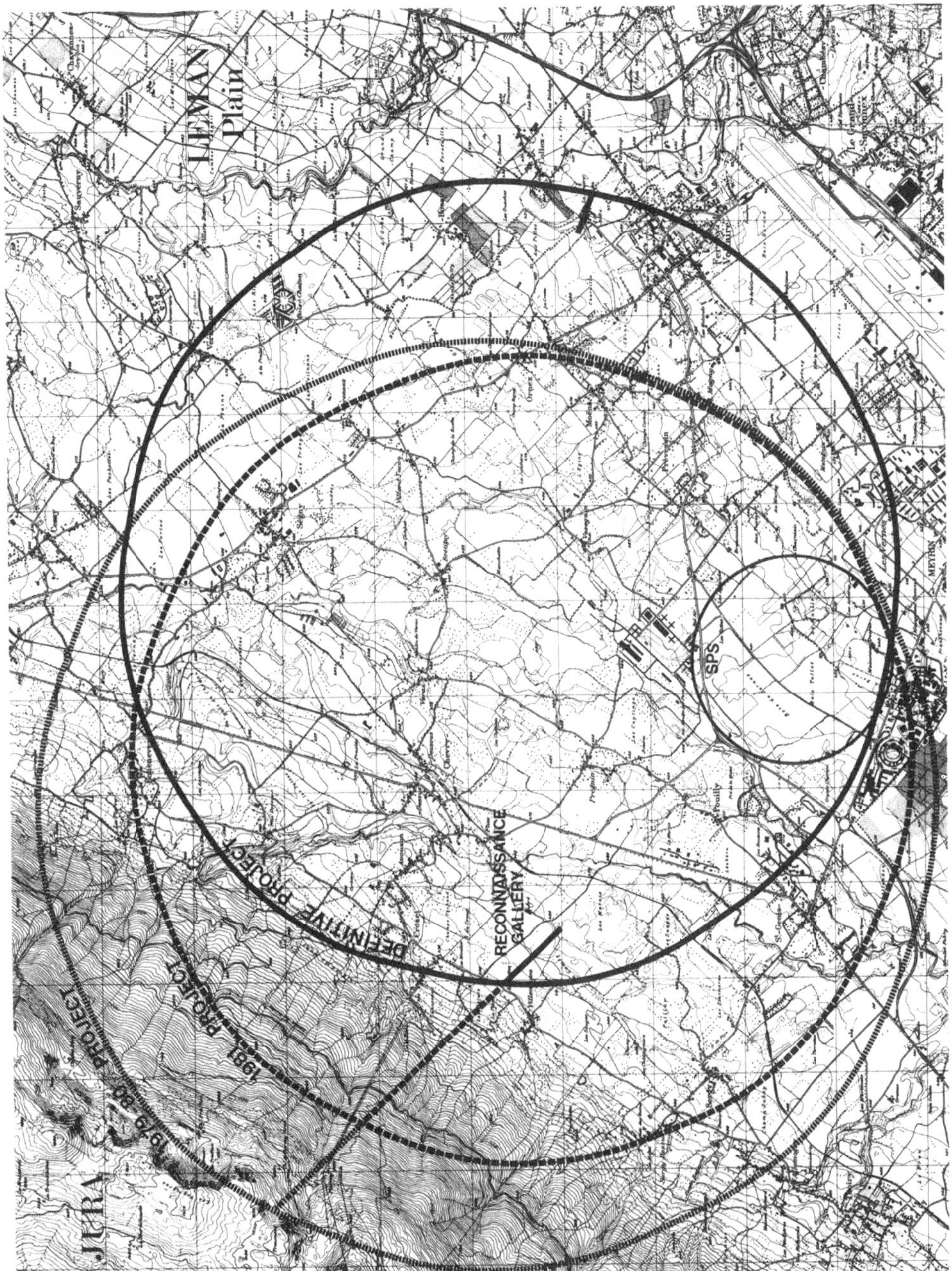

Fig. 1   Successive positions of the LEP project

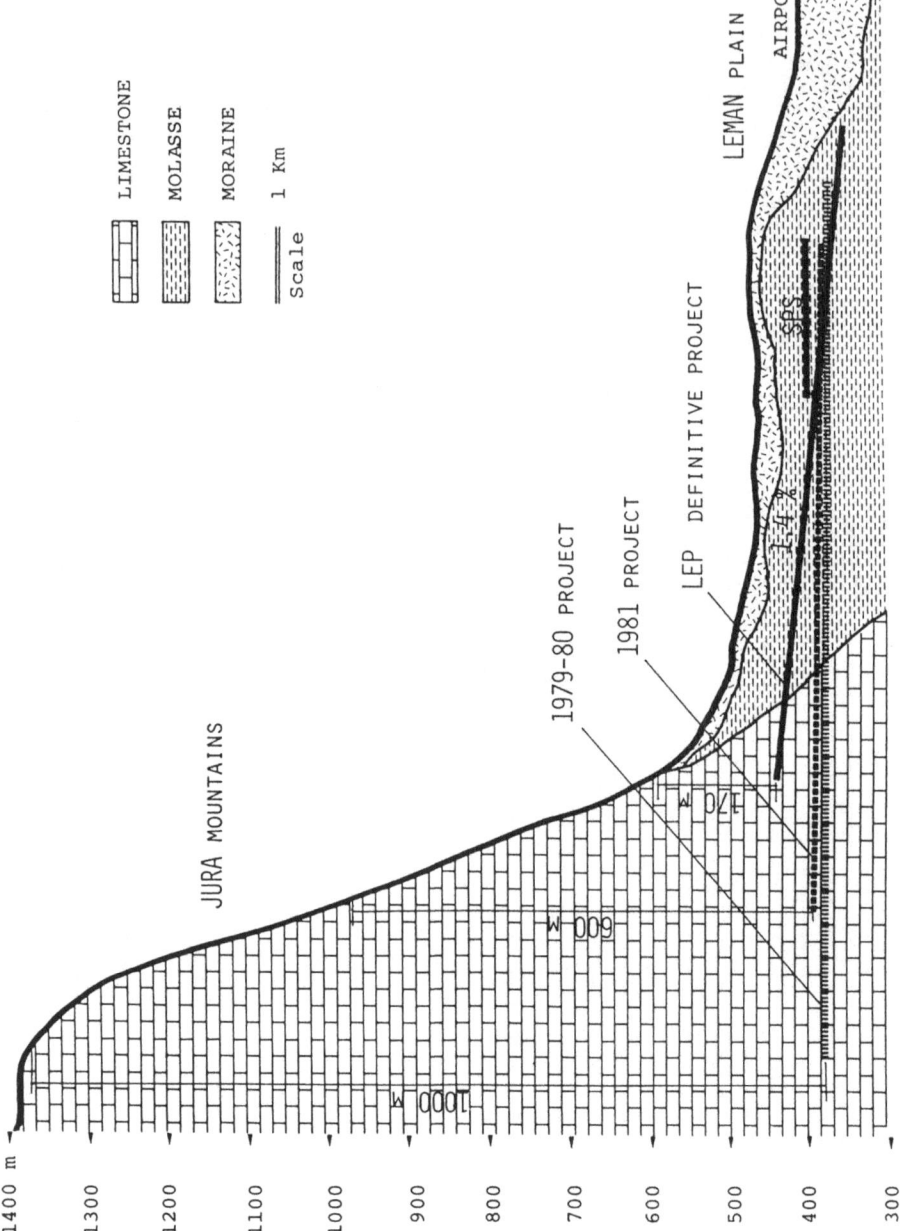

Fig. 2 Geological cross-section of the LEP site

The geographical coordinates of this origin are :

ϕP_0 = 51.36920 gon = 46°13'56".2 N
λP_0 = 6.72124 gon = 6°02'56".8 E

While the azimuth of the Y CERN axis with respect to geographical North is :

G_0 = 37.77864 gon = 34°00'02".8

The Z CERN axis is collinear with the vertical at P_0. To avoid any confusion between the Z_c ordinate and the altitude H, a constant of 2000 m was added to the H values of the PS machine. The local coordinates of the vertex P_0 are :

XP_0 = 2000.00000 m
YP_0 = 2097.79265 m
ZP_0 = 2433.66000 m
HP_0 = 433.65921 m

If, in the past, a spherical model of the earch was sufficient to express the geodetic data of accelerators (Ref. 2), this was no longer the case with respect to the size of LEP (Fig. 3).

The reference system has therefore·been re-defined to accommodate a projection onto an ellipsoid, the radius of curvature of which is dependent upon the azimuth. The figure used at CERN is the IUGG 1980 ellipsoid with the following parameters :

- semi-major axis   a   = 6 378 137 m
- semi-minor axis   b   = 6 356 752 m
- polar flattening   f   = 1/298.26
                       where f = (a-b)/a
- first eccentricity e^2 = 0.006 694 380 022 9
                       where $e^2 = (a^2-b^2)/a^2$
- second eccentricity e'^2 = 0.006 739 496 775 5
                       where $e'^2 = (a^2-b^2)/b^2 = e^2/(1-e^2)$

This ellipsoid is considered tangential to the geoid at P_0, and hence the normal to the ellipsoid is collinear with the vertical at P_0. The radii of curvature of the ellipsoid in the meridian ρ and in the prime vertical ν at latitude ϕ are :

$$\rho = a(1-e^2)/1-e^2\sin^2\phi)^{3/2} \qquad (1)$$
$$\nu = a/(1-e^2\sin^2\phi)^{1/2} \qquad (2)$$

At P_0 the two principle radii are :

ρ_{Po} = 6 368 761.40 m
ν_{Po} = 6 389 299.67 m

The consideration of the osculating circle in P_0 (vertex at zero level) is sufficient, within the CERN area, to express the geometrical relations allowing the conversion between Z ordinates and ellipsoidal heights HE. The radius of this circle, for an azimuth α, is :

$$\rho_\alpha = \nu/(1+e'^2.\cos^2\phi.\cos^2\alpha) \qquad (3)$$

Knowledge of the separation S_0 between the ellipsoid and the geoid is derived from physical considerations : computation of equipotential surfaces in a mass model and astrogeodetic measurements using a zenith camera (Refs 1, 3, 4). This quantity is the necessary link between the observable heights HG above the geoid and the ellipsoidal heights, which are purely geometric (Fig. 3).

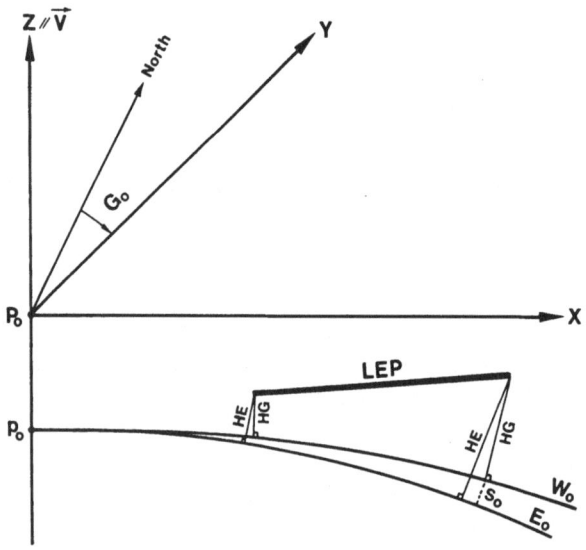

Fig. 3   Ellipsoidal and geoidal heights

2.3 Theoretical altitudes at LEP level

Derivation of ellipsoidal heights from measured altitudes needs two additive terms. The first gives the position of the "zero" equipotential surface – the geoid – with respect to the reference volume V (Fig. 4). If T is the perturbating potential generated by the topographic masses and γ the normal gravity of the reference volume, the separation S_0 at point M is expressed by the Bruns' formula :

$$S_0 = T_M/\gamma_m \qquad (4)$$

The second term is the orthometric correction of the measurements, giving the "normal" altitudes in the Modolensky free air model. For instance, from L to M, this correction would be :

$$C_n = \int_L^B g \cdot dh / \gamma \qquad (5)$$

where g is the actual measured gravity along the traverse.

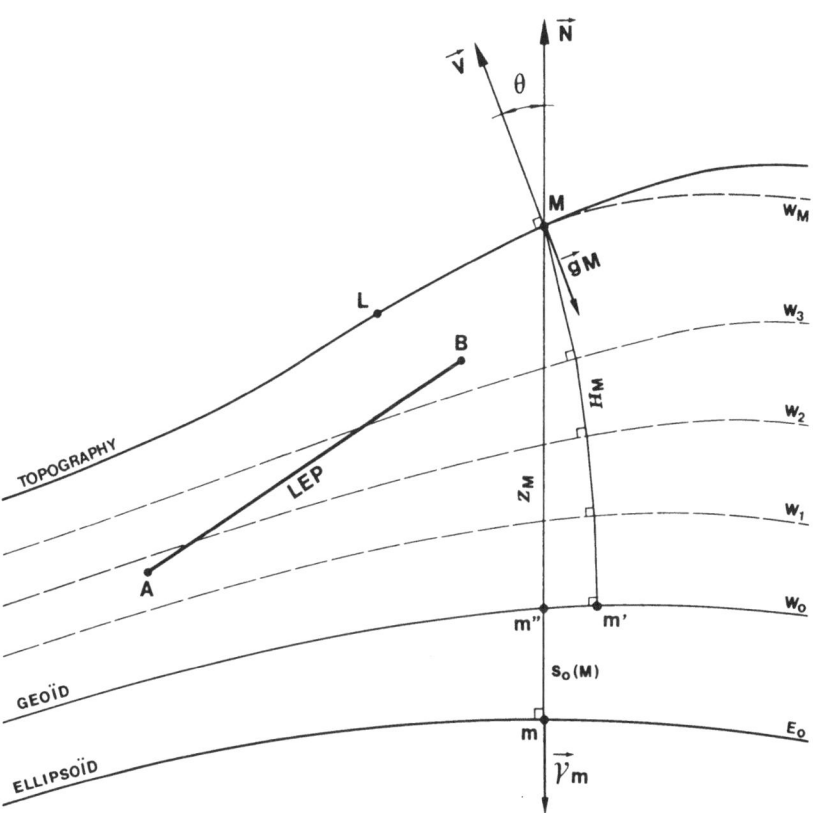

Fig. 4   Equipotential surfaces and vertical deflections

This last term corrects, in fact, the inner divergence of equipotential surfaces which reflects the curvature of the plumblines within the irregularities of the gravity field inside the crustal part of the earth. The values of these curvatures are also derived from the mass model. When no gravity measurements can be made, this being the case before tunnelling of the LEP project started, these values give the possibility to anticipate the evaluation of the normal correction and therefore to compute the predicted altitudes of the project in order to get a true plane in space.

For that purpose, the deflections θ of the verticals have been expressed along the machine, onto its tilted plane, using the computed gradients of curvature. A similar reduction has been applied to astro-geodetic measurements. Integrating these values with a least-squares fitting, the corrective paraboloid so obtained expresses the sum of the two additive terms $S_o + C_n$ and allows the computation of the theoretical altitudes of the project elements.

3. **ADJUSTMENT OF OBSERVATIONS**

3.1 **Adjustment programs**

Many adjustment programs have been successively written to satisfy the geodetic needs for CERN accelerators : planimetric (XY) or altimetric (H) programs, tridimensional adjustment strictly limited to micro-geodesy, processing of large matrices (Ref. 5, 6), Helmert transforms, etc.

The size of the LEP project called for a new and rigorous computational tool, fully adapted to the processing of all kinds of geodetic data in a local system and whose main features are the following :

- local 3-D adjustment on the ellipsoid GRS 80. Altitudes are referred to the local geoid, which has been investigated and parametrized

- generalised least-squares processing of all types of available data, some being very peculiar to the metrology of large objects

- statistical and variance analysis of the results - generation of random and/or systematic perturbations for simulations

- flexibility in data handling : free formatting and intensive use of keywords.

A detailed description of the program can be found in the contribution from J.C. Iliffe (Ref. 7). The principles of the CERN simulation method will be explained in Section 4.

3.2 **Radial and vertical smoothing**

When installing the machine components, the first determination of the control network gives the displacement vectors between their actual "rough" position and their theoretical one. In fact, magnets are positioned around an unknown mean trend curve (one among an infinity) contained within the envelope of maximum errors. The polynomial degree of the curve depends on redundancy, overlap of measurements, and the bridge distance between control (fixed) points. The final relative errors are a quadratic combination of those of the network itself and those of the positioning, i.e. installation errors. Their statistical nature is essentially gaussian : the aligned elements are randomly and normally distributed around this mean trend curve (Fig. 5).

As the major requirement for the geometry of an accelerator is that the relative errors must be very small ($\sigma \leq 0.1$ mm), an obligatory step for surveyors is to check the installation by measuring and - if needed - improving the smoothness of the machine alignment.

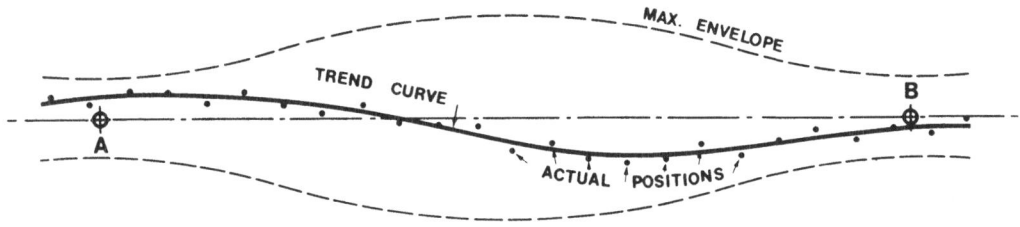

Fig. 5   Position of magnets with respect to theoretical orbit

Another important consideration is that, when making successive surveys of long and flexible figures, absolute comparisons would be a nonsense. The difference between trend curves, corresponding to each survey, must be analytically eliminated without inducing systematic or harmonic errors, which are critical for an accelerator. Therefore, a special smoothing algorithm has been developed in order to process the local data in a purely relative way, i.e. without any absolute involvement in coordinates (Refs.8, 9).

The smoothing process consists of a set of radial - or vertical - measurements (the last measured values after installation) which are treated in the following way:

Considering for instance the radial observation scheme (Fig. 6), the relationship between three points H, I and J with an off-set measurement on I is :

$$\frac{1}{1+k} \, dRh - dRi + \frac{k}{1+k} \, dRj + St - Sm = vi$$

where :

   $k = \dfrac{proj \ (HI)}{proj \ (HJ)}$ along (HJ)

   dR = unknown radial discrepancy with respect to mean curve

   St = theoretical sagitta of these three points

   Sm = measured sagitta

   vi = residual of the measurement

Each point receives three overlapping measurements resulting in a good redundancy. Nevertheless, these measurements only cover six points (i.e. three quadrupoles) and the global system would be rather poorly conditioned if a determination of coordinates was required. But as the purpose is to get local and purely relative information, this fact is not critical if the normal system is solved under the double condition ||dR|| and ||v|| minimum.

The condition ||dR|| minimum constrains the reference line of the ordinates dR to be the mean curve. This condition cannot be directly expressed in the adjustment. Neither can it be linearized, since the differential increments, the unknowns dR, and the residual v are, nearly, of the same order of magnitude. The convergence would consequently be slow and the results uncertain.

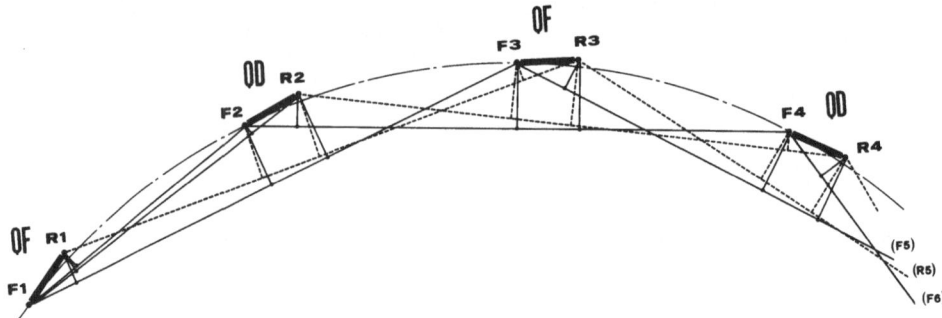

Fig. 6   Scheme of smoothing measurements

The only way which has been found for solving such a system is to move dR_I onto the right-hand side of the equation, introducing for each central measurement on I a new observation equation :

$$1/\sqrt{3} \ (1/1+k \ dRh + k/1+k \ dRj + St - Sm) = dRi \ (+vi) \qquad (6)$$

The weight $1/\sqrt{3}$ comes from the fact that each point receives three measurements.

The normalisation of these additional equations is equivalent to $\Sigma(dR + v)^2$ minimum, i.e. :

$$(\Sigma dR^2 + \Sigma vi^2 + 2\Sigma dR \ vi) \ \text{minimum} \qquad (7)$$

Σvi^2 is set to its minimum by the normalisation of the first equation, for each measurement. These residuals vi constitute a Gaussian distribution [0, σ_m] and the radial ordinates dR are also expected to be a set of Gaussian variates with, of course, a zero mean-value. ΣdR^2 will consequently be a minimum if the quantity ($\Sigma dR \ vi$) tends to zero.

For two Gaussian samples $U(0,\sigma_1)$ and $V(0,\sigma_2)$ of the same dimension N in their initial (random) order, it can be easily proved that if, in general, X and Y are two random variables with density functions a(X) and b(Y), the expectation of the product is :

$$E(XY) = \sum_{ij} P_{ij} X_i Y_j = \sum_{ij} P_i P_j X_i X_j = \sum_i P_i X_i \sum_j P_j Y_j$$

$$E(XY) = \int\int f(X,y).X.Y dX dY = \int\int a(X).b(Y) XY dX dY = \int a(X) X dX . \int b(Y).dY$$

whence E(XY) = E(X).E(Y) = X.Y

for   X = Y = 0,

then   E(XY) = 0.

If the sample V has dimension n = N/3, this is more difficult to prove. But it can be checked easily with a program using the computer's Gaussian generator.

This global procedure is a kind of compromise, which is not quite rigorous but nevertheless satisfactory. Simulations and real computations have given acceptable results both for checking the first installation and for successive and comparative surveys of the machines.

For vertical positions, smoothing has been first made "visually", by drawing a mean curve on a plot of differences H measured - H theoretical. A similar procedure has been tried by generating pseudo-observations of vertical off-set values, derived from adjusted altitudes, with the same overlap. Results are, here also, satisfactory :

- unfavourable sign sequences are located,

- outstanding dR values can be pointed out.

4. EXPECTED RESULTS FOR LEP METROLOGY

4.1 The CERN simulation method

Starting from the a priori standard deviations on measurements, derived from experience, the well-known tools of stochastic analysis are :

- unit weight variance : $\sigma^2 = V'PV / n_0-n_u$,

- covariance matrix : $C_X = \sigma^2 N^{-1}$, which gives expression of absolute and relative error ellipses,

- histogram of residuals, estimated accuracy of (groups of) observations, elimination or refinement methods,

- confidence intervals on estimates.

Nevertheless, this classical way of analysis may leave some interpretation problems on mixed networks. Significant distortions can occur on a posteriori estimates and it is difficult to appreciate the relative "strength" of each group of observations. Furthermore, no signal is given to detect the systematic errors, which are not modelled in this process. This kind of simulation does not give a clear view of the behaviour of a perturbated network and one must remain rather wary when interpreting the resulting statistical figures.

A pragmatic way to obtain a picture of the true situation is to simulate on a computer all the perturbations which can affect the geometry of a figure : Gaussian errors in measurement, artificial generation and addition of systematic errors, controlled constraints, etc. (Ref. 9).

For this purpose, the provisional coordinates (or theoretical ones for accelerators) are taken as ideal. Measurements are supplied in the input file as for a normal computation of the network. The program computes the ideal measurements and a gaussian generator adds random errors scaled on a-priori variances. The data is then processed as usual. Repeating these operations gives a "Monte-Carlo" generation of 'n' sets of hypothetical measurements of the same network.

Empirical statistics carried out on the results give very interesting estimates to be compared with the known (and controlled) a priori values of the variance of each group of observation or, even, to allow a direct analysis of the effect of errors on the coordinates. Tests can also be made to appreciate the agreement of actual results with predicted values. Such a method gives a clear idea of the "response" of a complex network to random errors and provides some corrective factors to apply to the various estimates in order to make a correct scaling of the predicted errors.

When adding systematic errors and/or controlled constrains, the simulations also give a true image of the distortions suffered by the network. The effect of each constraint can then be evaluated. The resulting shifts of the mean values of the residuals, with respect to any selected systematic error in each group of measurements, gives an idea of the "warning lights" to watch for when actual measurements are to be processed.

This pragmatic method has been used for years at CERN and it is a very helpful tool for the engineer who needs a real knowledge of the network he has to design and optimise, and then observe and compute.

4.2 Preliminary study of the LEP network

The LEP underground network is a 27 km long chain (Fig. 7) determined with invar lengths and off-set measurements, which is constrained by eight control points coming from the surface geodetic network. Each octant, between two control points, is therefore 3.3 km long.

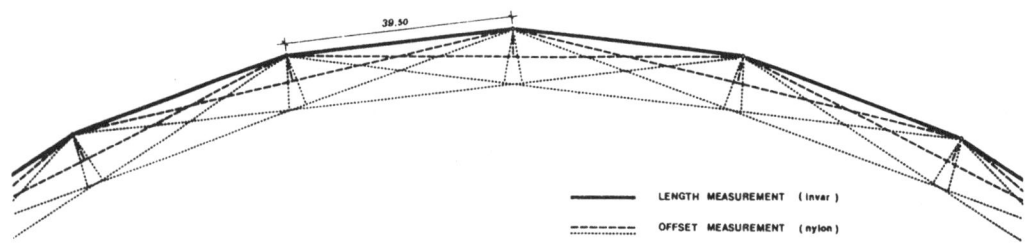

Fig. 7   Scheme of measurement

The preliminary study of the error envelope of this network has been made with the following assumptions and means :

a) Radial and longitudinal relative errors at control points are taken as $\sigma_r = \sigma_p = 2$ mm. These figures come from the simulations of the surface network and experience in "descending" points.

b) Along the figure, the radial errors ϵ_n and ϵ_{n+1} of the ends generate radial discrepancies which can be estimated at positions 1/4, 1/2 and 3/4 of the octant (Fig. 8).

$$\epsilon_{1/2} = (1/2) \ (\epsilon_n + \epsilon_{n+1}) \Rightarrow \sigma_{1/2} = [(1/4) \ (2^2 + 2^2)]^{1/2} = 1.4 \text{ mm}$$

$$\epsilon_{1/4} = \epsilon_{1/2} + (1/4) \ (\epsilon_{n+1} - \epsilon) \Rightarrow \sigma_{1/4} = [(9/6) \times 2^2 + (1/16) \times 2^2]^{1/2} = 1.6 \text{ mm}$$

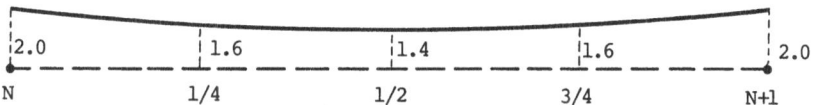

Fig. 8   Effects of radial errors

c) A longitudinal error dL in control points has a distorting effect resulting in radial errors dR, because of the flexibility of the figure. For LEP networks, this effect has been simulated and, at the middle of one octant, it produces an error dR = 1.8 dL (Fig. 9a).

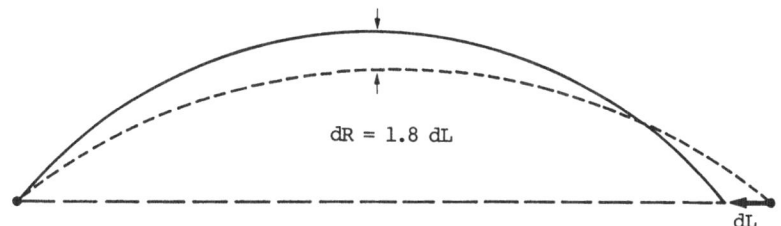

Fig. 9a   Longitudinal flexibility

The estimate of dL comes from longitudinal errors in ends n and n+1 :

$$dL = \epsilon_n + \epsilon_{n+1} \Rightarrow \sigma_{dL} = (2_2 + 2_2)^{1/2} = 2.8 \text{ mm}$$

The resulting dR along the figure are expressed in Fig. 9b.

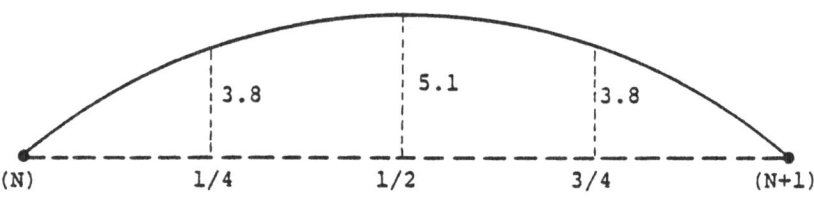

Figure 9b   Effects of longitudinal errors

d) The radial errors of the metrological chain itself are evaluated through repeated simulations of the underground network. Initial assumptions are $\sigma = 0.05$ mm for invar measurements and $\sigma = 0.08$ mm (short) or $\sigma = 0.15$ mm (long) for off-set measurements. The values issued from confidence ellipses and confirmed by statistics on coordinates are summarized in Fig. 10.

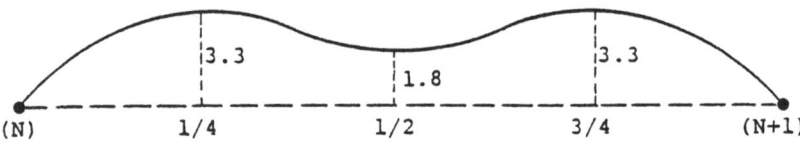

Fig. 10

It should be noted that the bumps are caused by a higher degree of flexibility near the ends, even with the overlaping measurements which provide the continuity and homogeneity from one octant to the next. The final r.m.s. envelope of the radial errors is the quadratic composition of these independent errors (Fig. 11).

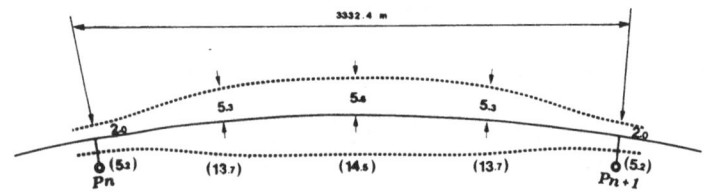

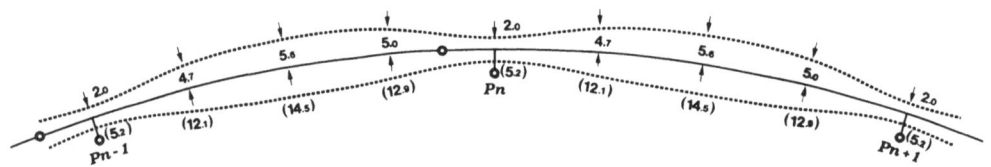

Fig. 11   Final errors before and after link

4.3 Full simulation of ten hypothetical surveys of LEP

In order to have a general check on these estimates, ten simulations of the whole machine network have been carried out with all random perturbations (errors in control points and in measurements). The values found at positions 0, 1/4, 1/2, 3/4 and 1/1 were respectively 2.1, 3.9, 5.2, 4.7 and 2.1 mm. A Fischer test of the comparison between these empirical estimates and the predicted ones gives a full agreement at 95% confidence level (Fig. 12).

Further studies have been made on the polynomial degree and the harmonic Fourrier content of these ten simulated surveys of the machine. The aim was to extract the trend of the mean curve of the positions in order to discern the contributions from long-range (absolute) and short-range (relative) errors. The resulting data has been introduced into the program simulating the orbits of the particles. Analysis of these simulated orbits has shown that the perturbations generated by the alignment errors can be damped by the planned correcting magnets. It has also established the fact that absolute long-range errors have more effect than previously thought in accelerator theory.

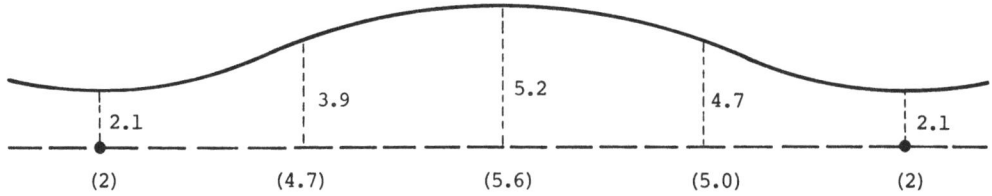

Fig. 12   Empirical estimates derived from simulations

*   *   *

REFERENCES

1) M. Mayoud, Géométrie théorique du LEP, LEP Note N° 456, (1983).

2) J. Gervaise, M. Mayoud, E. Menant, Système tridimensionnel de Coordonnées utilisé au CERN, CERN 76-03, (1976).

3) W. Gurtner, Das Geoid in der Schweiz, Institut für Geodäsie und Photogrammetrie, Mitteilungen Nr 20, Zurich, (1978).

4) V. Achilli, P. Baldi, Computation of Local Anomalies of the Vertical Deflections in Geodetic Networks, Survey Review No 205, Vo. 26, (1982).

5) W. Coosemans, M. Mayoud, E. Menant, D. Roux, Processing of the CERN Ring-Shaped Geodetic Networks, Workshops on Special Large Matrices for Geodesy, CERN-CAST, 20 av. A. Einstein, F 69 Villeurbanne, (1977).

6) H. M. Dufour, Résolution des Systèmes linéaires par la Méthode des Résidus conjugés, Bulletin géodésique, No 71, (1974).

7) J. C. Iliffe, Three-dimensional Adjustments in a Local Reference System, these proceedings.

8) M. Mayoud, Programme de Lissage radial, Private communication, (1974).

9) M. Mayoud, Applied Geodesy for CERN Accelerators, Part II - Seminar on High Precision Geodetic Measurements, University of Bologna, (1984).

APPLIED METROLOGY FOR LEP

Part II : DATA LOGGING AND MANAGEMENT OF GEODETIC MEASUREMENTS WITH A DATABASE

J.-P. Quesnel

CERN, Geneva, Switzerland

ABSTRACT

This paper describes the data base which has been created, with the ORACLE system [1], to manage the data from the positioning and surveying of the CERN particle accelerators in order to check the true shapes of the machines and the alignment of their components. A comparison between data handling with the database and with a classical file architecture is given.

1. INTRODUCTION

Today, some 8000 elements, laid end to end, make up CERN's particle accelerators. This represents a total beam line length of more than 20 km. These quantities will be doubled when the LEP machine starts. To manage all the data connected with the geometry of the accelerators, the Applied Geodesy Group has now established a relational database. The data architecture will be first described and then its efficient use examined.

2. DATA ARCHITECTURE

In the following sub-paragraphs, we shall describe the data which is stored and then how it is structured.

2.1 Stored data

All the data related to the geometry of the particle accelerators and the survey measurements is stored in one of the following categories:

Geometric definition of the accelerators.- Different algorithms allow the accelerators to be defined, in accordance with the required characteristics and the chosen strategies. Programs describe the beam lines as a succession of elements laid end to end along the reference orbit. Each point of entrance and exit of the beam in or out of the element is defined by coordinates, and the three angular parameters of the orbit at these points are given in the reference coordinate system used at CERN.

Geometric definition of the elements.- Each element to be aligned is equipped with two sockets which guarantee the centring of all the metrological instruments and whose positions are known exactly in relation to the theoretical beam path in the element. Thus, with a three-dimensional rotational matrix, we can compute the coordinates of these targets in the CERN system.

Deliberate misalignments.- The accelerators are designed for specific characteristics of the particle beams. If these characteristics are modified, or if the mode of use is changed (for instance, accelerator mode changed to collider mode for the SPS), it is sometimes necessary to modify the alignment of some quadrupoles, in order to adjust the orbit. These deliberate misalignments are stored in the database.

Geodetic network coordinates.- To ensure the correct absolute and relative positioning of several accelerators and each of their beam lines, CERN is covered by a geodetic network which extends on the surface and also underground. All the points are known either in ellipsoidal coordinates or in the CERN coordinate system. The altitudes are defined with reference to the ellipsoid or the geoid.

Measurements.- It is important to distinguish between two sorts of measurements. On one side, there are those usually made for the determination of a geodetic network, while on the other, are the measurements taken during the component's alignment. The former have to be used together, the latter are independent and connected only to the element itself.

The measurements of a geodetic network or a machine are of several types :

- distance, made with an electro-optical distancemeter, a distinvar or a calibrated tape

- vertical or horizontal angles

- offset

- levelling.

Each type of measurement has a code and is stored with the date, the number of the instrument and the place where it has been made.

When an element has just been aligned, we immediately make a complete survey of its position in relation to the reference points used. Each set of measurements is dated, stored and archived.

True position of the elements.- All the measurements described above can be computed to give the true position of each component. In fact, we store the comparison between the true position and the theoretical one, along and radially to the beam.

2.2 Structure of the Tables

Numerology.- A strict and precise numerology had to be defined for all the CERN machines. All the machines or transfer lines, all the components to be aligned, and all the targets have names which satisfy the rules adopted.

Tables.- All the information contained in the database is stored in the form of tables, each made up of a row for each entry and as many columns as required. The rows are not ordered in any way and when deciding the structure of the tables, some columns can be chosen as indices in order to help select the data. A unique index is also needed to be sure that a row is not recorded twice in the table.

In addition, the tables are independent in the database, but their architecture has to satisfy the relational language which allows access to several tables at the same time, and also the unique storage of the data. All around this main architecture are other tables which contain data devised to aid the connections between tables and the selection of data. Thus, we have tables which contain the name of the machines, the list of different sorts of targets, the calibration values of the invar wire, etc. (Fig. 1).

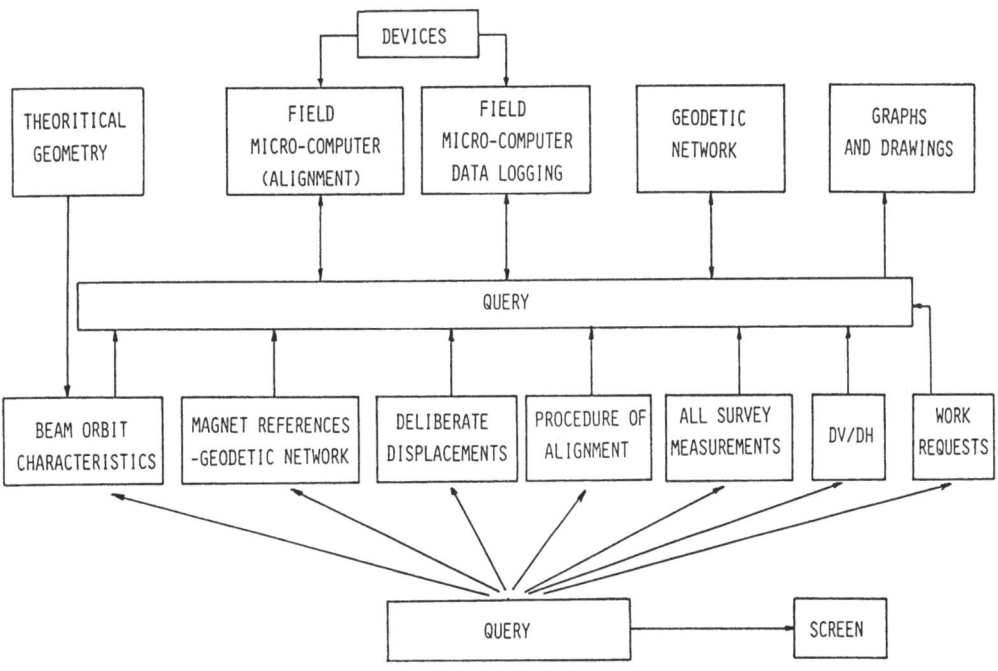

Fig. 1   Organigram

3. USING THE DATABASE

A database is useful only if it is active. It must allow anyone access at any time. Further, it must not be necessary to copy the information, in order to use it, and updating must be safe and easy. The database of the Applied Geodesy Group has been established with the ORACLE system [1]. It is closely connected with a specific software tool kit developed according to the needs of the Group.

3.1 Updating

To ensure that the procedure of updating remains simple, only the "primary" data are stored. These are theoretical data on accelerators and non-derived measurements. All the data resulting from a computation of data stored in the database is obtained by activating the corresponding programs. For theoretical data, the tables are updated by completely clearing the rows related to the modified machine, and recording the new values; even if only one row is to be changed. This procedure is the safest and the fastest for this purpose. It is driven by programs which can handle errors and do not allow an update when somebody else is using the table.

The measurements are added directly to the tables, and it is possible to check if the row is not already there. They are transferred to the database either directly from the field microcomputer or by hand from the keyboard of the terminal. A "menu" enables access to several updating programs.

3.2 Data Access

ORACLE has three tool kits with which forms can be defined on the terminal screen. These forms are used to present the information usually needed for consultation.

Moreover, the user has a direct access to the system and, at any time, can give the commands he wants in SQL language.

3.3 The Transition from database to data processing

However complete a database may be, it is of little benefit if consulted only as a dictionary. It is very important not only to get the data on the screen, but also to be able to use it directly in existing FORTRAN or PASCAL programs by writing selecting commands in the programs. This possibility exists with ORACLE system (Fig. 2) which also permits the storage format of the measurements of a geodetic network to be changed before computation.

Figures 3 and 4 give an example of the storage of the measurements.

```
1  SELECT * FROM CALIBRATION
2* WHERE NO_FIL='EPA5'


   NO_FIL          REPORT     APPOINT JOUR       TEMPERATURE
   --------------- ---------- ---------- ---------- ------------
   EPA5             16757      2108 07-JAN-86        21


LIEU    JOUR      NO_APPAREIL          CODE POINT1          POINT2            VALEUR ECART_TYPE NO_FIL
------- --------- ---------------- ---------- -------------- -------------- ---------- ---------- --------
EPA    04-FEB-86 INVO2             5 EPA    0000006 EPA0000001       4531    .00005 EPA5
```

```
UFI> SELECT POINT1,POINT2,ROUND(((REPORT/1000)+(VALEUR/100000)-(APPOINT/100000)),5)
   2   FROM MESURES M,CALIBRATION C
   3   WHERE C.NO_FIL='EPA5'
   4   AND M.NO_FIL='EPA5'
   5   /


   POINT1          POINT2          DISTANCE
   -------------- -------------- --------------
   EPA    0000006 EPA0000001       16.78123
```

Fig. 2   Use Example

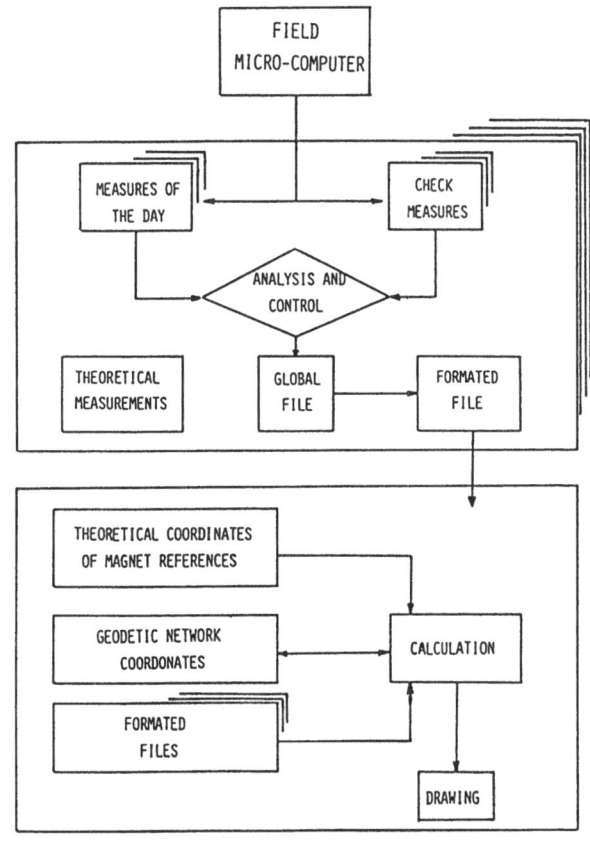

Fig. 3   Classic handling

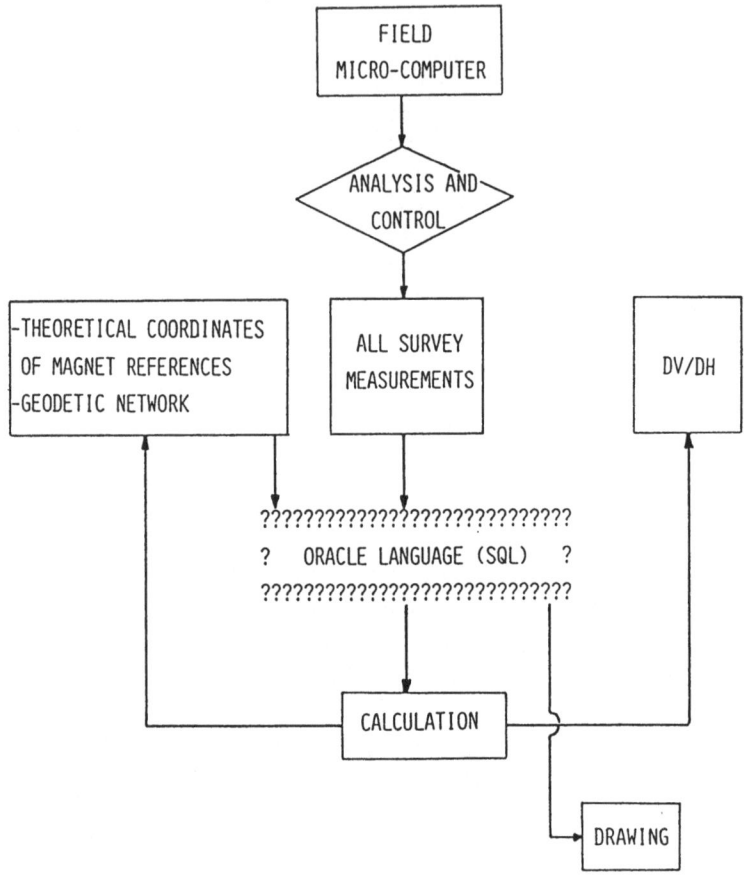

Fig. 4   Handling with a data base

When a database is not used, a multitude of files is created each day, containing the data for each sort of measurement. All these files have to be handled and linked together and the space needed is usually double that necessary because of the copies. Furthermore, after checking the measurements – which usually follows the transfer of the data from the field microcomputer into the main computer – it is often necessary to repeat some measurements, for instance if they are wrong or not precise enough. These new data are not usually presented in the same form as the others since the procedure of the second measurement is not so systematic. Thus it is difficult to manage with an automatic procedure. When a database is used, this management is much easier. All the measures are stored as they arrive in only one table. An interactive program of selection gives access to all the data at the time of the computation.

The selection program can be included in the computing program. Thus, the data are used directly; it is not necessary to copy them. For the compensation programs, which are important, it is preferable that this search of data takes place before the computing, and separately. The data is then grouped into files to be used normally by the computing program, which saves a lot of time for the machine. In fact, it is quite expensive to run large programs interactively.

4. CONCLUSION

Today's instruments, with the ease of data transmission and automatic collection are greatly increasing the quantity of data. However, a relational database simplifies enormously all the problems linked to the management of the ever more numerous measurements. In the case of the information relating to the geometry of the CERN accelerators, it is the guarantee of safe storage and an easy and rational manipulation of the information.

A database must be evolutive and able to adapt itself to new methods of measurement or computing, which can be introduced at any time. It should no longer be regarded as a background technical issue, but should be approached as a pragmatic strategy for surveying and implementing integrated data-oriented systems.

The survey database fulfils exactly these requirements.

*   *   *

REFERENCE

1) ORACLE Reference Manuel, Relational Software Incorporated, 1983.

METROLOGY FOR EXPERIMENTS

C. Lasseur
CERN, Geneva, Switzerland

ABSTRACT

This paper outlines the approach to metrology of the Experiments
Section of the Applied Geodesy Group at CERN. This procedure has
been developed to deal with the continuous evolution in size and
complexity of modern particle physics experiments. The precision
in construction and the resolution of these experiments
necessitates the adoption of geodetic methods to achieve the
required sub-millimetric positioning accuracies. The philosophy
behind the adopted approach and the methodologies applied are
discussed with respect to collider physics in general and the
future LEP experiments in particular. The instruments employed
are briefly outlined stressing the usage of a complete and
versatile chain of data capture and calculation.

1. INTRODUCTION

In October 1984, the Nobel Prize for physics went to Carlo Rubbia and Simon van der
Meer for the discovery of the W and Z bosons, which produced experimental proof of the
electroweak theory. These discoveries rank among the greatest achievements in the history
of science and were the climax of technological excellence and teamwork on a scale never
before seen in the field of pure science. Among the many elements required to facilitate
this discovery was the design and construction of a new machine, the Antiproton
Accumulator, the adaptation of existing machines, namely the SPS and PS, to fulfil roles
they were not designed for, and also the construction of general purpose detectors of
hitherto unseen size and complexity.

The Applied Geodesy Group was fully involved at all stages of this challenging and
exacting project; in the extension of the geodetic network required by the new works, in
the civil engineering work both above and below ground, in the metrology of the
accelerators new and old, and finally in the construction of the two new experiments UA1
and UA2 in underground caverns on the main SPS tunnel. These experiments had to be
assembled with great accuracy (often sub-millimetric) in a confined space and this
challenging precision was demanded for experiments of an unparalleled size and complexity;
in the case of UA1, a large box 10 x 5 x 7 m which weighs over 2600 tons (Fig. 1), and for
UA2 a mere 600 tons but more complex and innovatory in structure.

Fig. 1   UA1 view in opened position

The purpose of CERN is to provide opportunities for research in particle physics. Accelerators give the particles the required kinetic energy (hence the name high energy physics) before being brought into interaction with other particles. The properties of particles are deduced by looking at the decays of collision by means of equipment called detectors. The experiments can be divided into two groups, following the way the accelerator is used (Fig. 2) :

- fixed target experiments : beams are ejected tangentially from the accelerator and strike a target. The particles so produced continue (for the most part) in the same direction as the beam, hence the detectors are grouped downstream of the target;

- collider experiments : in this case two beams travelling in opposite directions are brought into collision. By this means, the highest collision energy ever achieved in controlled conditions is realised. The secondary particles produced in such a collision fly out in all directions, hence the detectors surround the interaction point.

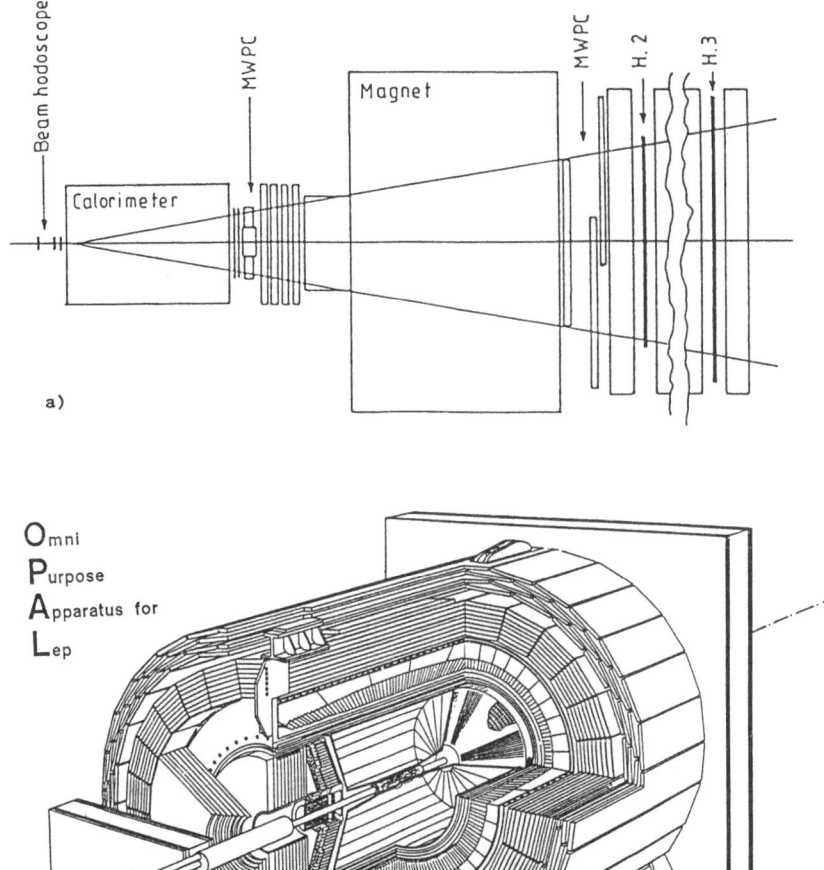

Fig. 2   a) Fixed target position
          b) Collider experiment

The LEP project will operate in collider mode. The experiments under construction will surpass those of the SPS in size, accuracy and complexity. For example L3, a five storey structure weighting over 12000 tons and permanently placed on the LEP ring is predicted to produce a resolution of several hundreds of microns. The three other experiments (DELPHI, ALEPH and OPAL) will be mobile, less massive (around 10000 tons) but produce a similar resolution.

2. HOW EXPERIMENTS ARE CARRIED OUT AT CERN

To analyse what happens, the experimental physicist must be able to measure, with great precision, the properties of the particles emerging from the collisions, such as their direction, time of passage (speed), energy, electric charge and mass. A variety of techniques can be applied.

A selection of these is summarized in the following table :

PURPOSE	TYPE OF DETECTOR	PROCESS
Time of passage measured to the accuracy of the best electronics (mm survey precision)	Scintillator	Charged particles produce tiny flashes of light when they cross blocks of plastic and this is detected by photomultipliers.
Direction/location accuracies of a few hundreds of microns (sub-mm survey precision)	Multiwire proportional chamber Drift chambers Time projection chamber	Ionisation caused by charged particles passing through a gas is picked up by an electric field between closely spaced planes of wires.
Mass/momentum sign of charge (mm survey precision)	Magnetic field (cryogenics) Magnet (a few tesla)	Magnetic fields curve the path of charged particles in opposite directions depending upon the sign of their electric charge. The amount of bending measures the particle's momentum.
Energy (sub-mm survey precision)	Calorimeters (hadron and electro-magnetic)	The energy deposited in a gas by a charged particle can be deduced from the amount of ionisation it produces. Iron plates can be interspaced with apparatus for measuring energy deposited by absorbed particles.

Every one of the CERN experiments involves complex detection systems of the types mentioned above; they surround the region where the collisions or the hits occur and record properties of the emerging decay particles. One can describe proton-antiproton detectors (UA1 and UA2) and the future LEP experiments as sets of Russian dolls, each being designed to detect a certain phenomena such as those described in the above table. Experiments of this type are designed to cope with large numbers of particles, collecting unbiased information from collision products collected over a solid angle as large as possible. UA1 contains 6000 sense wires, 30 km of extruded aluminium required by special positioning detectors; UA2 includes 240 cells pointing towards the centre of the interaction region and 5000 sense wires. To sift and analyse the massive amounts of data produced requires the speed and power of digital computers. The wealth of information recorded makes for prodigious data handling problems bearing in mind that each collision produces enough information to fill a telephone directory.

3. THE SURVEYOR'S ROLE

At the outset, the support given to the experimental teams in setting up physics experiments was purely empirical.

In order to obtain good results in the analysis of events (reconstruction of particle paths), a precise knowledge of detector positions, both with respect to the beam and respect to each other, is mandatory. Experience has shown that doing this using off-line alignment programs based on cosmic events or track fitting is tedious.

The large and complex structure of a modern physics experiment, made up of many individual modules, requires the use of a careful and systematic surveying and alignment procedure (often sub-millimetric) to derive the full benefit from the quality of the detector components and off-line software. The use of coordinates provided by the Experiment Section of the Applied Geodesy Group as approximate coordinates (raw data) in the off-line alignment programs allows a considerable saving in computer time.

The surveyor has two main responsibilities :

- towards the physics collaboration to ensure that the dimensional parameters of the experiment are fulfilled during fabrication and assembly and to give the final position once the whole detector is set in the particle beam (data-taking position).

- towards the machine group to ensure that common equipment (accelerator / experiment) such as experimental targets, magnets and virtual interaction points are positioned inside the required tolerances with respect to the theoretical beam position.

3.1 Geometrical approach of building a physics experiment

The use of increasingly complex detectors precludes a "Do it yourself" approach. For the physicist, the definition of the accuracy of the results is of primary interest. The global precision for an experiment is a combination of, firstly, the internal precision of each of the detectors and, secondly, the relative positioning accuracy. The surveyor's service is of importance in the second category.

The individual components of an experiment at CERN are rarely manufactured in the same place and are first fitted together on site during the final assembly of the experiment. Thus good accuracy of fit is required to avoid last minute machining and modification which is costly both in time and money.

There are a large number of measuring devices available for small items. When large objects have to be measured the erection or creation of large (often specialised) measuring devices has to be envisaged. These devices often require a considerable associated infrastructure, can be the cause of undesirable side effects (interference, work interruption, deformation) and do not give a global view of the object in its correct position.

The 3-D metrology method gives a picture of an object in a certain position at a given instant. This picture is defined in a referential frame whose main directions are known either in a general reference system or one related to the object itself and the coordinates so produced can be compared to the theoretical values, so ensuring that tolerances are respected. The method is both direct and of universal application; spatial coordinates of specified points on an object providing a suitable basis for the computation of spatial distances, body shapes and dimensions.

The methods and principles that will be explained here are, for the most part, based on our experience with UA1 and UA2 and our preparation for the LEP collider experiments. If we continue with the analogy of detectors as Russian dolls, the scheme of survey operations is that as each is inserted, one inside the other, their relationship is established geometrically. Once assembled, only the largest one is visible; the position of the smaller ones is reconstructed, assuming the geometrical relationship established during assembly is maintained.

The definitive precision of the experiment must include the various precisions in the geometrical relationship whose inaccuracies must be masked by the "noise" in the definitive precision in the working position. Reproducable accuracy (when remeasuring) must be equal to stability (of the object and the reference frame) accuracy, both of which have to be less than or equal to the relative precision between the components. This requires :

- the use of high precision instruments and methods at all stages (manufacturing, assembling, definitive survey).

- comprehensive surveying computation methods to determine an object by spatial coordinates.

We will particularly focus on these features of our work.

3.2 Necessity of good understanding

An agreement between the physicist and surveyor has to be built up from an early stage so that they understand the service we offer and, of course, the limits of that service. Basic working premises have to be clearly expressed, such as the global positioning precision required by the experiment, the precision and position of certain critical elements and, of course, working conditions. Every step should be defined in advance, in writing, and be included in the planning, from the initial to final stages. A comprehensive agreement has to be established covering all steps of the assembly and subsequent measurements. Any weakness in the overall project coordination can create problems, such as invisible references, means of adjustment unattainable, lack of preparatory measurements; the result being detectors either unmeasurable or only with reduced precision.

Prior to the installation of any element, be it a detector module or even a major support structure, it is mandatory that a computer file is prepared on the internal survey data. This gives a complete description of the unit (position, shape, dimensions) with respect to a set of fiducial marks. The latter are physical and exterior points attached mechanically to the object itself, whose position (rectangular or polar coordinates) are known with respect to a reference of use to the physicist (for a wire chamber this could be the principle sense wires). There must be enough points to fully describe the detector.

The external survey data file is a set of geometrical parameters describing the spatial position of a detector with respect to the theoretical particle beam position when in the data-taking position. It is an adaptation of the internal survey data file following measurements when the experiment is on-line. The definitive document is called the external survey data-base and it is the surveyor's role to determine i.e. to provide the spatial coordinates of the whole set of fiducial marks. Therefore, for the surveyor, an experiment is a list of XYZ coordinates. He has to ensure the real significance of this data, keep them up to date and be able to define the quality of the results.

4. SURVEYING PROCEDURE

The most frequently used procedure in surveying for the determination of spatial distances, body shapes and dimensions is the so-called micro triangulation-trilateration method. Theodolites or length measuring devices are set up at the end points of a baseline whose length and height difference are known to the required degree of accuracy. Precise measurements of the horizontal and vertical angles and/or distances permit determination of the X, Y and Z coordinates of the reference points on the object. If the measurements have to be repeated or if several sides of the object have to be surveyed (i.e. additional bases are necessary), it is worthwhile establishing a network of fixed points whose coordinates are known from a preliminary survey. Fixed points can be installed either as instrument stations, pillars, brackets on walls, or as sighting marks. The establishment of a network has the advantage that the location of the instrument can be freely selected, depending on the shape, size and position of the object to be measured.

To achieve sub-millimetric precision, the Survey Group has made certain choices with regard to instrumentation and methods. These include the intensive uses of distance measurements, calibrated instruments, force centring, well structured reference networks, redundancy, rigorous computer programs and a geodetic approach to the problem.

We will describe certain instruments and methods used by the CERN Survey Group : not all of them are universally applicable, indeed often they were specially designed for high-precision measurements in the experiments.

4.1 Precision requirements

The accuracy of the position of an intersected point depends on the accuracy of the calibration scale, from which the baseline data are derived, and from the observation distances (when the coordinate computation is based on the measurement of angles). Further accuracy determining factors include the magnitude of the intersection angle, the quality of the object point marking, the stability of the instruments, illumination conditions and of course the shape of the object itself.

4.1.1 Difficulties in measuring angles

The prerequisite for measuring angles is the absence of temperature gradients along the optical path; curvature of a light ray varies with the angle between its trajectory and the normal to the surfaces of equal temperatures : over a distance of 100 m and for a gradient of 1°C, it is deviated by 1 mm. Gradient of temperature, grazing sights and unequal lengths of sight all make optical measurements unreliable.

An ordinary theodolite telescope is unsuitable because of its unstable collimation which can change by as much as 30 seconds of arc throughout its range of focus, usually 1.8 m to infinity. When one seeks very high accuracies (0.1 mm or better), the position of the line of sight must be known very accurately with respect to some physical point. Moreover, the three axes, nominally orthogonal, do not, except by chance, pass through a common point. The optical tooling telescope seems to be essential for all serious accurate work (Taylor-Hobson : line of sight central to the cylindrical tube to an accuracy of 0.005mm and collimation maintained within 2 arc seconds). However this equipment needs special features (top plates, trivet stands) which are not flexible enough for our purposes.

For all these reasons precise surveying networks have to be formed as much as possible of linear measurements.

Angle measurements are still used for the "details" in experiment metrology because the sighting distances are very short - a few metres and most of the halls at CERN have a temperature regulation system. Also in some cases, pure triangulation is the only suitable method because the object to be measured is not accessible directly or cannot be subjected to undesirable loads (deformations); the angle measurements made externally exert no influence on the structure and can be carried out on an object in use. Nevertheless, it is prudent to employ a combination of angle and distance measurements whenever possible.

4.1.2 Forced centring systems

To obtain high accuracy, and because of the short distances involved, it is obvious that the entire geodetic system must use forced centring systems. The standard socket used is machined to better than a few microns and the top of the socket can receive either a positioning cylinder (with which every instrument is equipped) or Taylor Hobson spheres. The bottom part of the system can be adjusted precisely in a given theoretical position, the reference being the centre of the 3.5 inch Taylor Hobson sphere set in the cup. The top part can be adjusted to the vertical so that all measurements are processed in the horizontal plane.

4.1.3 Object point marking

The type of reference is crucial for the accuracy of the determination. If there are no well defined points on the object, such as an edge, a corner or a punch mark, artificial marks must be added prior to measurement. In many cases the Taylor Hobson sphere is too big and heavy, and requires special additional equipment which renders it unsuitable for object point marking.

Precise reference holes of different diameters designed to receive plug-in target marks adapted to the measuring device (angles, distances) and easily visible from many directions are of more universal use. This system offers the most flexibility in the choice of measuring pattern because the forced centring disposition allows us to change the targets without affecting the repeatability of the measurement. Other possibilities are self-adhesive sighting marks (crosses, concentric rings, symbols taken from transfer films), while various target patterns are available for different conditions of sight distances and illumination. They do not require any special machining but only angle measurements can be made with them.

Sometimes it is impossible to make marks or attach references to the object. A possible solution is to use a laser whose beam is projected via the optical axis of the theodolite so that the target point thus created can be intersected by other theodolites.

Forced centring systems as developed at CERN appear to be the essential condition in high-precision measurement because of their reproducibility and flexibility of usage. They have helped us to perform measurements whose quality is similar to those achieved in laboratory conditions.

4.2 Technical surveying milestones of an experiment

4.2.1 Reference network

Setting up a micro reference network seems to be one of the more convenient ways to determine three-dimensional patterns. From the design stage up to normal running of the experiment, it ensures that the required accuracies will be attained. It is successively used for provisional installation, alignment of detectors within the tolerances, measurement of detectors in the data taking position. It also allows survey of the stability of these components; successive measurements, based on the network, attempt to maintain a correlation between the evolution of the components and of the geometry itself.

The network is really the fundamental frame of all survey activities. It acts as a stable and precise large calibration bench which surrounds the volumes to be measured so that the critical points are visible and accessible physically. This bench is defined by pillars, brackets and/or rivets, whose positions must comply with a certain number of criteria :

- reference points must be easily accessible. The density of points has to be great enough to cover the objects to be surveyed.

- structure must allow simple, precise, easy and quick measurements of whatever kind on the points to be determined and between the reference points themselves.

- network structure must from the very beginning take account of the different steps of installation and be evolutionary enough for it to grow relevant to any future needs.

- key positions must be occupied in respect to the layout and be adapted to the point's projected usage (angles, distances, direct levelling).

- creation must be made as soon as possible and in one operation for the homogeneity to be maintained; completion in hindsight must be avoided.

- installation avoids unstable areas while allowing rapid measurements to be performed to detect and evaluate casual movements or accidents.

The assembly and positioning tolerances guide the homogeneity and the precision to be obtained when measuring and controlling the network, the threshold being given by the most sensitive elements of the experiment (which are generally the most difficult to reach practically). It is useful to attain the best possible accuracy under current conditions since even if this is not strictly necessary, it can be of great help in the future as the experiments evolve. Homogeneity means that the precision of the bench must be identical at any location of the network. Redundant measurements must be performed to ensure quality and uniformity of accuracy. These points must be considered when setting up the network since good homogeneity and density provide us a greater flexibility of methodology and control.

The form of the network depends on the type of the experiment. For fixed target experiments, the network consists of two lines parallel to each other and to the beam. The method of defining these lines depends on the precision required, for example, the line can be defined by survey sockets inset into the floor or, in a less demanding case, be simply drawn on the floor with several rivets placed along its length. For collider physics the network consists of several levels of references commanding both the maintenance and the on-line position of the experiment.

In the case of the UA1 and UA2 experiments, references are situated on three distinct levels (+ 6.0 m, 0.0 m, - 3.0 m with respect to the beam). The geometrical links between the floors were achieved using high precision verticality measurements. These were carried out between brackets on the same vertical (within several centimetres) for the case of UA1, whereas in UA2 sloping invar wires were used. The following results were obtained :

- UA1 : 38 brackets - 160 measurements - range of accuracy .08 -.17 mm
- UA2 : 56 brackets - 270 measurements - range of accuracy .08 -.18 mm

These results were only possible because all the requirements quoted above (length measurements preferable to angle measurements, forced centring) were met. Unfortunately, the network cannot always be established in its ideal form; it must be compatible with the other services such as ventilation conduits, cabling and cooling systems which greatly reduce the options available.

4.2.2 Detector assembly

Preparation phases.-

A detector is rarely a single unit. It is often divided into component parts which we can term sub-detectors which geometrically are considered as single, rigid and non-deformable units. Each assembly of sub-detectors is considered to produce another rigid object.

To plan the measurement procedures, it is of overriding importance that the steps of assembly are clearly defined, this normally being determined by the structure of the detector. Specific discussions can then start on any remaining ambiguities, and the following questions answered :

- is the detector a single unit or composed of several independent modules ?

- is it rigidly mounted on another detector? How is the relationship between the two units to be made (mechanically or by means of surveying) ?

- is its determination to be made sequentially (in a laboratory, in the experimental area) ?

- have external reference marks to be created ? If so how many and where ?

- is there a supporting structure or a mechanical adjustment ?

- has the detector (as a unit or modules) to be placed at a theoretical position, in respect to what (beam, surrounding detectors) and with what precision ?

- what precision is needed for the different detectors relative to each other ?

The surveying procedure can then be decided upon; the best place for reference marks, the redundancy required, the most suitable moment and place for the work during the assembly of the detectors, etc.

Steps of the geometrical determination.-

To obtain the final version of the survey data base, a cascade of operations is needed, from the first availability of individual objects up to the experiment's data-taking stage. The steps of geometry result from principles whose sequence follows closely the following assembly phases.

Internal geometry.-

Each individual detector element (chamber or counter module) is equipped with fiducial marks defined in its own coordinate system (for example : the main axis of symmetry). They are deemed to be internal because they are part of the object. These marks can be the accessible sensing part of the detector, for example the cells of a calorimeter or wires directly visible from the exterior of the chamber.

Another possibility is reference marks directly connected to the sensing part of the detector; for example the pin holes used to centre planes of wires. Each plane is held within a frame and at each corner there is a reference hole whose diameter is precisely machined and whose position is known with respect to a wire (measurements carried out in the laboratory). If the chamber contains several planes of wires the relationship between each plane is assured by calibrated cylinders which fit precisely in the reference holes. From the measurement of the position of these cylinders, the position of each plane of wires and the position of each wire can be deduced.

Features of internal geometry are that it is carried out as an integral part of the manufacturing process and is used for geometry when easily, directly and permanently accessible. It is the preferable way of defining the geometry of the detector as a direct method of determination.

Link geometry.-

Measurements in this category are carried out for one of two reasons. Firstly, if the internal marks are not convenient for surveying, special survey references must be installed. These have then to be situated with respect to the internal geometry by geometrical, optical or mechanical methods, carried out in the laboratory, workshops or assembly halls before the definitive installation of the detector. Secondly, when a detector consists of several individual modules, then for reasons of economy (many sub-detectors), accessibility of the sub-detectors or to derive parameters for the physics data banks, link measurements between adjacent modules become necessary. These are related to marks (at least three) which will be accessible when the detector is in the data taking position and can be either visible internal reference marks or specially installed auxiliary references. The position of every element of the detector can then be deduced from the measurements of several references.

Features of link geometry are :

- the intermediate step required for the geometrical knowledge of a detector

- the difficulty to foresee whether it is really necessary since it is dependant on the environment (encumbrance of zone etc.)

- that auxiliary marks are not an integral part of the detector (link geometry has to be repeated or plug-in system foreseen)

- the use of mixed step link geometry, for example a mechanical plug-in system to extend internal reference marks

- that, if possible, it should be carried out with a precision greater than, or equal to, that of the original references; this is often not possible hence loss of precision.

Transformation geometry.-

Before being moved onto the beam line, the detectors are installed one by one in their definitive relative position. Due to the planning requirements, detectors can be assembled at different places in the mounting area before they are fitted together. This allows us to verify that the theoretical parameters of the experiment are respected, and to determine the relationship between different detectors which will later be hidden. An element unmeasured at this stage means that the position of one of the detectors will not be known when the experiment is on the beamline. Equally annoying is a relationship known with insufficient precision since all detectors internal to the one measured cannot then be located with sufficient accuracy.

Features of transformation geometry are :

- the establishment of the relationship between successive layers of detectors

- that, by this stage, the internal and link geometry of the individual detectors must be known (internal data base)

- the measurements are related to an assembly network during system assembly

- the relationships must be recorded at each step.

Definitive geometry.-

Once on the beam line, the visible references are measured from a network whose zero is the theoretical beam interaction point of the accelerator. Consequently, the position of all the detectors and sub-detectors can be calculated using transformation programs, whose truth depends on the care with which the measurement procedures were planned and the rigidity of the detectors. These results constitute the external survey data base required by the physicists.

The main feature of definitive geometry is that the position of all detectors is deduced from the measurement of the visible references of the experiment in the system of the machine.

The geometrical phases described above require much work and it is important that the number of cascaded operations should be reduced as much as possible. To this end non-essential movements of the experiment should be avoided as this requires new measurements and spatial transformation programs to determine the new shape and dimensions of the ensemble.

The final result, if it does not come from direct measurement, will be affected by the successive adjustments introducing slight differences due to the computations. To reduce this loss of accuracy, a high precision (homogeneity of network) in measurement is needed so that intermediate survey steps are not biased by successive determinations.

Possible geometrical method.-

Methods must be comprehensive enough to give a true 3-D picture of the object. The techniques used in most cases will be a mixture of the following : angles, distances (horizontal, vertical and spatial), offset measurements (optical or physical, in the horizontal or vertical plane). To the measurements made in the field can be added internal distances. That is, high precision values known from the manufacturing of the object or from calibration measurements performed in the laboratory. These values can be used in several ways :

- verification : to check that there has been no distortion of the object after movement (transportation, handling) by a comparison of the deduced internal distances and the known internal values. They can serve equally to verify the scale factor of the network (with an appropriate weighting). It could be envisaged to measure an object using auxiliary theodolite stations where the precise scale of the base is derived from the internal distances of the object.

- determination : these redundant distances are often known to a higher degree of precision than we can achieve in the field. They are also direct measurements made on the object itself. Thus, through their precision and their redundancy, they optimise the 3-D determination. In the case where points are unattainable or too numerous, internal distances between unobserved and observed points are a mean of deducing the unmeasured points.

Methods depend on the following considerations :

Characteristics of the object.- Its shape, dimension and behaviour (a priori instability in movement or a light structure).

Immediacy.- Because of fragility or limited availability of the object, rapid results may be required for verification.

Reference system.- When lab or preparatory works are necessary a local reference system is required. The choice depends on the spatial situation of the object and whether a direct knowledge of the parameters of the object are needed. Three cases can be distinguished for an object in any position :

- direct knowledge : the referential is directly linked to the object itself. The parameters are measured by trilateration or by triangulation but in the latter case the vertical axis of the theodolite must be parallel to the axis of the object;

- object can be adjusted and direct knowledge : it is obviously much more convenient to adjust the object, for instance with jacks, to produce a truly vertical axis;

- indirect knowledge : the object is measured from a reference system unrelated to the object itself. To obtain dimensions and relations in the object reference system, a transformation is required.

Environment.- For example not enough space around object for comprehensive coverage; the use of artificial lights with the risk of phase error; temperature gradient rendering triangulation unreliable.

Number, location, accessibility and type of survey marks used.- Visibility and required accuracy determine the most suitable method. Standard multi-purpose targets appropriate to any kind of measuring are very useful in this type of work, especially when access is difficult or installation of the point has to be permanent. Pattern and dimensions must be chosen in respect of the length of sight and thus have important considerations for the network. Insufficient room around the detector means too many stations (time-consuming) or the need to use special lenses to allow focusing on the objects with consequent loss of precision.

Non rigidity of the object.- Unfortunately sub-detectors are not always rigid. This can be due to the environment of the object (local vibration) or just a weakness in the structure of the detector (the more sensitive detectors in particular are built with a minimum of material). The non-rigidity can be controlled either by using instruments like gauges, clinometers or by using geometrical methods. The former give punctual information of a high precision, can be placed at critical points, and are independent of the geometrical results. However, this information cannot be directly included in a 3-D determination; Moreover, that needs special equipment support and many instruments to give a complete picture. The geometrical method can provide a relationship between adjacent modules but requires a certain preparation and is relatively time consuming. It requires marks on the object which are not necessarily linked to the internal references. Adding marks in hindsight should obviously be avoided but because of the complexity of the detectors is sometimes the only solution.

The question of whether the detector or the network has moved is obviously a problem, in particular in underground experiment halls where movements of the order of mm's continue for many years after the completion of the civil engineering work. This requires periodic verification of the reference network, often in conditions less favourable than when the network was established (experiment and services in place, lines of sight blocked). Redundancy becomes more and more essential but less easy to achieve. In any case with the geometrical method it is impossible to determine the precise moment of the movement, the amount and direction at any given time to be allowed for in the calculations and the rotation/translation centre. To avoid some of these problems it is a good idea to use instruments giving punctual information, especially if used to furnish on-line information to monitor the situation. Once the deformation exceeds a certain limit a general geometrical survey can be made. All in all, it emphasises the need for a good preparation of the object, to detect at an early stage any possible movements and allow the possibility to install special equipment if necessary.

Special tools.- Precise extension plugged into reference socket, various shaped "jigs" or pins, plug-in brackets, adjustable supports, targets with scales, self-illuminating marks, ball bearing fitting system to measure a distance in any direction are all of considerable help. Precision of the machining, a rigidity of structure, adaptability, adjustable if necessary and simplicity of use are generally required.

4.2.3 Survey and assembly sequences

During installation, it is essential that coordination is established between surveyor and the person in charge of the installation to ensure timely interventions and that allowance will be made for the need for space and time for survey measurements. A communication system has to be arranged to allow a fast access to the survey data at all steps of the geometrical process. Hence possible inconsistencies between different measurements can be detected during installation so that measurements can be repeated if necessary. This survey file should be used to test off-line alignment programs between the modules already in place (e.g. by Monte Carlo simulations).

4.2.4 Data-taking period : diffusion of results

Survey measurements have to be repeated often because of position instabilities and movements on replacement of modules. The survey results must be entered as soon as possible in the experiment's data-base. It is updated following the progress of the installation and must reflect any last minute changes or modifications. It can be considered as the definitive document between the surveyor and his customer. At the request of the latter, further information such as the precision of the results must be provided. Publication of the results must be meticulous, in particular when designations of the measured points are concerned. Graphics plotting in 3-D can help in the understanding of the results.

In fact we also need intermediate data banks - specially constituted by the link measurements - and we have to create an adapted computing chain that will be described later.

5. MEANS AVAILABLE IN THE FIELD FOR THE SURVEYING OF LARGE PHYSICS EXPERIMENTS

5.1 Angle measurement

Theodolites (Wild T2 and T3, Kern E2) are used over short distances (a few metres) in conjunction with forced centring systems for detector assembly of medium accuracy, in particular when references are non-accessible or unreliable. Extensive use of vertical angles in conjunction with direct levelling improves the accuracy perpendicular to the base. Angles are also used in network observation when required to improve the strength of the geodetic figure.

5.2 Distance measurement

5.2.1 High precision (accuracy better than .1 mm)

The three main instruments for providing high precision distance measurements are listed below together with their relative merits and drawbacks.

Distinvar.-

This instrument can measure from .02 m up to 50 m with a rms error of .05 mm; due to the instability of the invar alloy, the elongation is directly proportional to usage, hence for work of the highest order the wire must be calibrated before and after measurement on a precise bench. The invar exerts a force of 15 kgf on an object being measured so the structure has to be rigid and requires a standard forced centring socket. It can be utilized on slopes up to 11%. It is rarely used on detectors unless prior provision has been made to accommodate it. Its main use in metrology for experiments is in the measuring of reference networks.

Self aligning interferometer.-

It is used for distances of .1 m to 60 m with an accuracy of .01 mm and does not apply traction to the detector but requires a free path and accessibility between reference sockets. It cannot be used for distances on slopes More time is   necessary but the instrument has the advantage of requiring straight line access rather than the catenary curve which may prohibit use of the distinvar.

The self aligning interferometer system is used for in-field auxiliary bases or for measuring high-precision networks when the reference points are all at the same height and freely accessible (linear network).

High precision bar.-

This new device (A digital micrometer based on differential condensers) will be used for distances up to 2 m and the accuracy expected is .05mm. Any mechanical adaptation can be fitted to make measurements between any kind of reference holes and in any plane, so that is is possible to increase length using precise extensions. Because of its light structure and easy handling, it does not deform the object.

This device is still under development and is expected to be of great use for precise micro-trilateration, in particular for link geometry in the metrology of experiments.

5.2.2 Medium precision (several tenths of a mm)

A stadimetric method has been developed which, although not of the accuracy of the devices mentioned above can cheaply produce good accuracy (.04 mm/m for distances up to

15 m and slopes of up to 20%). It consists of a 3 m invar tape   mounted on an H bar. Four accurately aligned marks act as references on the tape, the distance between these points being known from a laboratory calibration. A staff is held vertically in a reference socket, vertical angles being observed to the marks. From these observations (vertical angle resection) can be calculated the distance between the sockets, difference in the height (bottom of staff and theodolite) and the non-verticality of the staff. Corrections can be calculated from the latter. The main attractions of this method are the versatility (especially when working in bad conditions) and the possibility of calculating the distance in the field using a pocket calculator.

5.2.3 Millimetric precision

This is usually achieved by electronic distance measurement (EDM) using, for example, the Kern DM 502/4 instrument. Its main features are an accuracy of 1 mm rms for distances up to 15 m and that it can be used to determine approximate coordinates or 1-2 mm metrology, while Kern mirrors can be adapted to fit any reference holes.

To benefit fully from the EDM, frequent calibration is required (bench or an invar network) in the range of the measured lengths. The instruments used are adjusted at the factory to reduce phase errors over short distances, analysis of calibration measurements having confirmed this for distances up to 30 m. Thus, using linear regression, the zero constant and the scale error can be calculated. Calculation of the standard errors of the zero constant and the scale error allow estimation of degree of confidence. Important points are :

- the calibration constants can be calculated using a pocket calculator (though attention must be paid to rounding errors)

- additional information about the zero error can be gathered from residuals computed and plotted with respect to distance

- it is important to calibrate often since the zero error changes with time and rough handling

- the possibility to use self-adhesive "scotchlite" tape (up to 15 m) is very advantageous

- the absolute rotation angles of a large object (verticality, perpendicularity) can be obtained by means of differences in distances.

5.3 Levelling

Here we use an automatic high-precision levelling instrument such as Zeiss Ni2 with the limitations that long traverses cannot be made and intensive use is required of redundant measurements (nodal points) while frequent calibration is necessary since sights rarely of equal length. For convenience a vertical translation stage is used to measure directly on the point without using a staff though for experiments the large height differences often encountered require the use of 3/4 m staffs or vertical invar tapes.

5.4 Special instrumentation

5.4.1 Offset measurement

The offset is designed to replace angular observations by the precise measurement of the shortest distance (offset) from one point to a straight line defined by a nylon wire. Offsets of up to 500 mm can be measured in this way with an accuracy better than .1 mm. The method can be applied over large distances (up to 100 m) but requires stable air conditions, i.e. no draughts.

This instrument is used in conjunction with distinvar and/or interferometer measurements for the determination of linear networks and we intend to use it over short range and in any plane, in particular for redundancy in link measurements when they are combined with spatial distances done on the object itself. However, this special application will demand adaptation of the forced centring device.

For optical offsets a vertical plane is described by a theodolite telescope and an offset reading is made on a ruler which is approximately perpendicular to the plane. These measurements are used to give the verticality of an object and/or redundant measurements. The precision achieved is of the order of .3 mm over short distances.

An active system using a laser has also been used when turbulent air conditions prevent the use of a nylon wire. A rms error of .1 mm over 100 m is possible but this precision decreases rapidly over shorter distances.

5.4.2 Clinometers and hydrostatic levels

Clinometers give an accurate measurement of tilt angles in one direction and the value can be read by remote control. Accuracy is .01 mm/m and tilts of up to 2 mm/m can be measured.

The hydrostatic level is designed to measure the difference in height between several stations. The instrument installed on each station can be operated and read either directly or remotely and can be controlled either manually or by computer. The accuracy is of the order of .05mm over distances of several tens of metres.

These instruments give periodic or permanent survey of altimetry, deformations or micro-movements. However, they require auxiliary equipment for their installation while generally their size and weight limit their application.

5.4.3 Vertical lines and plumbing methods

When a network is composed of several distinct levels it is necessary to find a way to transfer coordinates between floors. This can be achieved using the distinvar if the slopes do not exceed 11%. In other cases the transfer is done using vertical lines between stations which are on approximately the same vertical using for example :

- a plumb-bob damped in an oil bath

- a specialized optical instrument : precision nadir plummet with mercury horizon (Wild GLQ)

- a nadiro-zenithal telescope

- optical methods using special adapters such as :

  - the diagonal eye piece for measuring zenithal distances in several planes between the vertical line of the theodolite and the point but only possible if measuring upwards.

  - the pentaprism which when fitted to the telescope of a theodolite describes a plane perpendicular to the optical axis. Observations (horizontal and vertical angles) describe planes whose intersection give the observed point coordinates. For good accuracy, the height difference between the points must be known precisely and the planes described must be well distributed.

Both the pentaprism and diagonal eye-piece methods use redundancy in the determination which can be calculated with a pocket calculator. Accuracies of better than .2 mm have been achieved over distances of up to 20 m.

6. COMPUTER PROCESSING

6.1 Importance to metrology

The variety of measurements and the redundancy, the number of points to be handled mean that least-squares adjustment programs are an essential instrument for calculating definitive coordinates. All surveying operations in metrology depend on this since the calculation gives the observed object as a truly geometrical figure. The definitive form is the one which fits most closely to the observations. Some differences will remain because of the inaccuracy of the methods used but the resultant form is the most probable (statistically speaking).

The following features are essential for a least-squares program adapted to the metrological method :

- capable of handling the wide variety of observations such as horizontal and vertical angles, horizontal, vertical and spatial distances, offsets in any plane, punctual information given by non-geometric instruments (autocollimation, clinometer etc.)

- of sufficient size to adjust the largest data set in one go

- do not influence the method in the field

- give statistical analysis of the results including redundancy factor, standard error a posteriori of the measurements, error ellipses, residuals after compensation, criterions of homogeneity (relative error ellipses)

- independent weighting of variables

- good, clear presentation of the results (histograms, plotting etc.).

6.2. In-field programs for metrology

A chain of programs has been developed to aid the work in the field; their simplicity and versatility of usage being a considerable asset. These programs were first developed on HP41 calculators but the increasing availability of portable computers such as he Epson HX20 has allowed the development of more comprehensive and versatile complementary programs. For example, the determination of one point in X and Y by least squares (resection, intersection, distances and angles) can be programmed easily. This possibility of field computing offers a flexible setting out method which can be adapted to the working conditions.

6.3 Data capture and computation programs

When a large number of points have to be measured in a limited time, automatic recording programs become almost obligatory. They reduce operator strain and the possibility of error both in data taking and in subsequent manipulation of the data, and ensure data is taken in a consistent way. The connection of a portable computer to an electronic theodolite means that angles (and distances) pass directly to the computer, the data is then checked (verification, between two pointings, collimation of the instrument) automatically and directly. In the case of error the operator has the possibility to verify and correct the observation.

An extension of this idea is to calculate automatically (intelligent recording program) the results in the field (distance/bearing, intersection, resection, levelling), which are then stocked in the memory of the computer. This is used when there is little or no redundancy, and means that no computing is required in the office. If the measurements are periodic the previous values can be stored in the memory thus allowing a direct check before replacement by the new value. Observations or coordinates are then transferred to a larger computer for distribution and eventual further calculation.

6.4. Main core compensation programs, specific concepts

6.4.1 XYZ determination

Comprehensive planimetric, altimetric and spatial compensation programs have been developed over many years and are being continuously updated to take account of future needs. All are based on the variation of coordinates algorithm and specially cater for :

- offset measurements in any plane defined by the three connected points and offsets taken from a theodolite set on a given bearing

- gyroscopic bearings

- plumb-bob observations

- computation of any network as a free network (one fixed point and one specified direction), thus a minimum of constraints

- correction of distances by a scale factor or by a systematic constant after calibration

- the a posteriori transformation of angle residuals into radial vectors which are much more appropriate to short distance determination

- tracing of remaining mistakes by entering approximate coordinates as fixed points, big discrepancies between observed and calculated values highlight problems.

6.4.2. Simulations

The possibility of simulating future survey operations is intensively used (especially for the preparation of the LEP experiments). The methodology is established (set of measurements, choice of points), the program then calculates the theoretical measurement and adds an appropriate error calculated from a random number table and the a priori weight of the observation. The "observations" are then adjusted and the results calculated. The program repeats this several times (normally 10 is sufficient) using different random numbers. The precision of the definitive network can be seen plus the range of values for the definitive coordinates. If the program is used on several possible data sets, the most favourable pattern can be found and it can be seen if the required accuracies can be achieved. This allows planning and equipment requirements to be established in a rigorous way and well in advance.

6.4.3 Spatial-adjustment program

Spatial-adjustment programs are imperative in metrology since they give the complete set of coordinates even if only some points are remeasured from a previously established relationship and the possibility to change the reference system following needs (providing the relationship between the main axes is known). To define a coordinate system without ambiguity seven coordinates distributed among three points are needed (i.e. X1, Y1, Z1; X2, Y2, Z2 and Z3). Alternatively, six coordinates and a distance constraint have the same effect. In all of these cases, the relative geometry is not distorted and observation residuals are always the same. In an over abundant situation the network will be adjusted after the principle of least squares; this will produce inevitable inconsistencies in the residuals of the observations though these will be small or negligible. A set of three points known in X, Y and Z constitutes a redundant control set.

The decision to use minimal or redundant control should be based upon confidence in the control coordinates versus that in the measuring system and has important implications for the definition of methodology. If an item (network, detector) is periodically remeasured, then two cases are possible. First, all the points are used to fix the parameters of adjustment and the residuals dx, dy and dz prove the item to be stable or not. Second, due to changes in the environment the quality of the measuring method can not be maintained so that the relationships established from a previous measurement (in better conditions) have to be imposed to improve the quality and reliability of the results. However, in an over abundant situation the distortion of the results by nonconform points is always a danger (micro-movements, instability). An a priori limit has to be defined to allow the rejection of suspect values which would otherwise bias the adjustment parameters (translation, rotation in the three planes and scale). Thus stable points can be selected and new coordinates computed for passive points.

6.5. Possible computable steps in metrology

6.5.1 Applications to periodic survey

In frequently repeated operations a methodology has to be developed to explain discrepancies in position. A comparison with the previous data set allows rapid detection of gross errors and gives an idea of possible discrepancies in position. Then computing a network as a free system (one fixed point one fixed direction) and adjusting the whole set of coordinates of the new network in relation to the old network is a possible way to detect stable points. The values of the residuals give a guide for introducing rejection criteria and give information on the accuracy; so a new adjustment (using so-called fix points) provides a direct comparison (absolute and relative positions).

Often the items to be measured are more stable than the network itself. This emphasises the need to use non-geometrical devices to control stability and the fact that a reference system related to the object itself (directly or through the use of internal distances) is often the most suitable way of obtaining true positions. When internal distances on the observed object impose unacceptable constraints the stability of the item or the so-called fixed points must be open to doubt (particularly if the network was not remeasured). Then, solutions have to be found to resolve the dilemma, such as computing network and object together (free calculation, one point fixed and one direction fixed) and processing successive spatial adjustment of the ensemble, taking different values of rejection criteria into account. The object may thus be used to relocate the network. After final adjustment the magnitude of the residuals dx, dy and dz gives an idea of the "exactness" of the final solution but, to envisage a solution along these lines, there must be redundant information otherwise casual deformations can not be brought into evidence. In any case these can only be detected if they are greater than the rejection value.

Example of link-transformation geometry adjustment.-

Let us consider the detector D3 which is part of the experiment UA2. The detector is composed of four identical modules each composed of four boxes containing three detector chambers : the physicists require the position of each of these 48 units. To achieve this, four distinct operations were required :

- laboratory measurements (carried out on a calibration bench) to an accuracy of a few hundredths of a mm (internal geometry),

- two separate operations of triangulation to establish link and transformation geometry,

- final triangulation to find position of the detector D3 in the experimental network (definitive geometry).

This entailed three successive adjustments, using the intermediate sets of coordinates, to provide the position of the detector chambers in the experimental network, the whole operation requiring more than 2700 angle measurements. However, in the following table, we can see that despite the number of operations, a sub-millimetric accuracy was still achieved

<u>Residuals after each of the adjustments</u>

(values in mm)

	Measure 1	Measure 2	Measure 3
X	0.19	0.16	0.16
Y	0.05	0.17	0.10
Z	0.06	0.10	0.10

7. SPECIFIC FEATURES FOR THE LEP EXPERIMENTS

7.1 Computerisation

The size and complexity of the LEP detectors and the constraints imposed by the assembly and installation schedules require an extension of the geometrical methods. All relevant theoretical data and results files have to be available in the field, allowing a direct comparison of theoretical and measured parameters. Thus, the problem is one of data management rather than field technique and requires the use of a comprehensive system of data capture and calculation.

7.1.2 Features of the system

An internal disc drive allows permanent storage of theoretical parameters of an experiment (detectors, sub-detectors), coordinates of network, measurements and results. Use of the Unix system of exploitation enables logical subdivision into as many parts as there are units, while least-squares 3-D compensation and adjustment programs can be introduced into the computer.

The system can be interfaced with different configuration such as main frame computers, (IBM, VAX), electronic theodolites (master and/or receiver), and micro-computers (HP-IPC, Epson HX20 and PX4) used as data loggers. Multi-task functions allow the system to be used in a versatile way (acquisition, control of instruments, calculations, communication lines) to reduce data handling, while special routines and facilities help in producing comprehensive description and analysis of the results.

7.1.3 Expected benefits

Compensation and adjustment programs facilitate on-line comparisons either with internal fiducial-mark coordinates (theoretical values) or with previous results if the current measurement reveals displacement. Since verification is made in the field, any discrepancy (residuals given by the on-line capture and computations) can be checked, thus giving greater confidence in the final results. This produces considerable time saving since assembly, measurement, verification and correction are carried out simultaneously.

7.2 Instrumentation

The complete system consists essentially of two parts, namely electronic theodolites and computer. The former is the Kern E2 (with DM502/504). It is used singly or in combination (several E2s can be independently controlled through one interface RS232 and via ASCII SINGLE BUS System). Readings are taken and the circle can be set remotely. Computer equipment includes the data loggers and the central unit of management. The HX20 and PX4 will continue in their role as individual data loggers either remotely or through keyboards. Their portability, robustness and practicality plus the battery of existing programs make them ideal for data capture and simple calculations, both in the field and in the office. The management unit will use the HP-IPC as its master unit. Its compact design (internal disc drive, screen, printer), its capacity (1 Mbyte), its versatility (programmable in BASIC, FORTRAN and C, system of exploitation UNIX, easily interfaced with other apparatus) make it well suited for this role.

8. CONCLUSION

Progress at CERN is not limited to the domain of physics and the different methods and instruments presented in these proceedings underline this. The rapid evolution of high energy physics constantly poses new problems to the surveyor and in meeting this challenge proress is made. Metrology as described in this paper has benefited greatly from recent advances in electronics, data handling and computing. The new automatic instruments (theodolites and short distance devices) have greatly aided us. This however, has not blinded us to less "glamorous" but sometimes more effective solutions. Organisation of field and office work becomes more and more important in order to meet the needs of our customers in a more efficient way. Full benefit from the possibilities of computing requires considerable forethought, organisation and self discipline.

It is indispensable in our case to have a realistic idea of the precision (standard errors, confidence limits). This requires redundancy of measurement and often the only possibility of achieving this is by a mixture of "unusual" observations requiring comprehensive and flexible compensation programs. To handle any wide ranging alignment, portable computers are essential in order to manage the flow of data, calculate the results and to stock the results in a data bank.

ACKNOWLEDGMENTS

The author would like to thank D. VEAL for his help and his large contribution to this paper.

IV. Mathematical Geodesy

NEW TRENDS IN MATHEMATICAL GEODESY

H. Moritz

Technical University, Graz, Austria

ABSTRACT

Modern mathematical geodesy is faced with increasingly large data
sets. This requires the use of large computers as well as an
adequate theoretical understanding of the underlying problems. For
this purpose, linear functional analysis, a generalization of linear
algebra to infinite-dimensional spaces, furnishes powerful mathematical tools, especially Hilbert space methods and Fourier transform
techniques. A general review of such methods and their applications
is given.

1. HANDLING LARGE DATA SETS

Geodetic analysis is called upon to handle data sets which rapidly increase in size.
Sets of the order of a million data are no longer unrealistic. There are several reasons for
this:

− Modern computers are able to perform rigorously extensive calculations which formerly
had to be broken up, approximately, into a number of smaller tasks. An example is the contemporary readjustment of the North American triangulation, which involves several hundred
thousand stations.

−Another reason lies in the nature of modern geodetic measuring techniques which often
furnish a data stream which flows almost continuously in time. Satellite techniques such as
Doppler, GPS or satellite altimetry serve as examples for this phenomenon.

Thus n , the number of data or the number of unknowns, may be very large. To get rid
of a definite number n , it may well be useful to let $n \to \infty$. Surprisingly enough, this
may frequently simplify matters and facilitate understanding the problem, with positive
effects on the actual numerical treatment of the data. It is like considering a definite
forest, rather than a set of 713 692 trees.

To give a more appropriate illustration: difference equations are known to be mathematically very difficult. For this reason, one tries to approximate them by differential
equations for which both a highly developed theory and efficient numerical techniques exist.
The truth of this is strengthened, rather than impaired, by the dialectical fact that these
numerical techniques again use finite-difference methods: their development has become
possible only on the basis of the theory of differential equations.

Going to the infinitely large is conceptually almost the same as going to the infinitely small as in the case of differential equations. Now we have, for instance, a data set which forms a vector of n elements, n being a very large number. If we formally let n → ∞ , we get a vector in an infinite-dimensional space which is called <u>Hilbert space</u> (provided certain mathematical conditions are satisfied). Hilbert and similar infinite-dimensional spaces (e.g. Banach spaces) are very fashionable in contemporary mathematics. They are the subject of <u>functional analysis</u> which is almost as well developed nowadays as classical analysis (differential and integral calculus). In fact, some abstract modern treatments ("Bourbaki", Dieudonné) introduce differential and integral calculus right away on such general spaces ...

If the data are, in some way, regularly distributed, for instance with equal sampling intervals, then a branch of linear functional analysis, namely spectral analysis or Fourier transform techniques, provides very powerful theoretical and computational tools. Fourier transforms are an analog of diagonal matrices and can therefore be handled very well. You have certainly heard of <u>Fast Fourier Transforms</u> which are increasingly used in geodetic applications as well as in numerical mathematics.

Letting n → ∞ is not only a mathematical artifice but comes natural in many instances. Consider the earth's gravity field. Its complete mathematical description requires an infinite number of parameters, e.g., the coefficients in a series of spherical harmonics. Other examples will be given later.

2. THE POWER OF SYMBOLISM

Consider the matrix equation

$$Ax = y \tag{1}$$

and its solution

$$x = A^{-1}y \; . \tag{2}$$

If A is a regular square matrix (of nonvanishing determinant) of dimension n × n , and if x and y are n-vectors (more precisely, column vectors n × 1), the (2) is the standard solution of a regular system of n linear equations for n unknowns, A^{-1} being the ordinary inverse matrix of A .

But the formalism (1) and (2) also works if A is a rectangular matrix m × n , whether m is smaller or larger than n . Then the corresponding system of linear equations has the form

$$a_{11}x_1 + a_{12}x_2 + a_{13}x_3 + \cdots + a_{1n}x_n = y_1 \; ,$$

$$a_{21}x_1 + a_{22}x_2 + a_{23}x_3 + \cdots + a_{2n}x_n = y_2 \; , \qquad (3)$$

$$\vdots$$

$$a_{m1}x_1 + a_{m2}x_2 + a_{m3}x_3 + \cdots + a_{mn}x_n = y_m \; .$$

In the sequel, we shall always consider the case $m < n$. Then the system (3) admits an infinite number of possible solution vectors $x = [\, x_1, x_2, x_3, \ldots, x_n \,]^T$, a column vector being considered as the transpose T of a row vector.

Equation (2) can still be interpreted to hold, but in the sense of a <u>generalized inverse</u> matrix; (cf. Bjerhammer, 1973)[1] From the theory of generalized inverses it is known that A^{-1} has the form

$$A^{-1} = KA^T(AKA^T)^{-1} \; , \qquad (4)$$

where K is an arbitrary $n \times n$ square matrix such that the $m \times m$ square matrix AKA^T has a regular inverse (verify by substitution!). Various matrices K give various admissible solutions.

An "underdetermined" system of type (3) is furnished by the linearized condition equations in adjustment by conditions: there are less conditions than observations. The least-squares solution in classical:

$$x = KA^T(AKA^T)^{-1}y \; , \qquad (5)$$

K being the inverse weight matrix and x being the residuals (usually denoted by v). It has the form (2) with (4).

Let us go one step further and let $n \to \infty$. Then the matrix A, originally

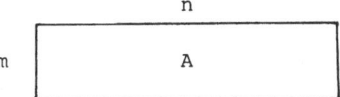

becomes

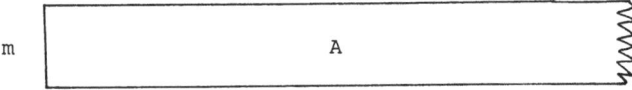

extending to infinity towards the right. Then obviously x must be a Hilbert space vector, whereas y remains a m-vector.

Then (3), on $n \to \infty$, becomes a system of m equations for infinitely many unknowns, and now comes a curious fact. The solution can still be written in the form (2), with the generalized inverse again given by (4)! Of course, now K will be an infinite square matrix, but it can be arbitrary as long as all sums, which are now infinite series, will converge.

Infinite systems of such type do occur in geodetic practice, namely in the determination of spherical-harmonic coefficients from satellite observations; (cf. Moritz, 1980, sec. 21)[2]. In fact, there are infinitely many harmonic coefficients, to be determined from a large but finite number of equations. This is a special case of least-squares collocation, to be mentioned in the next section.

So the simple and innocent equations (1) and (2) are of much wider validity and applicability than intended by the inventors of regular square matrix inverses. A good symbolism has a peculiar power which goes far beyond the original intention, almost autonomously tending towards generalization. Sometimes children outgrow their parents beyond their boldest expections.

3. HILBERT SPACE TECHNIQUES

3.1 Error propagation.

Consider a linear function F of a vector f :

$$F = h^T f = \sum_{i=1}^{n} h_i f_i \quad ; \tag{6}$$

the real number F may be regarded as the inner product of f with an auxiliary vector h or as the matrix product of the row vector h^T (transpose of the column vector h) with the column vector f .

If the variance-covariance matrix of f is denoted by Σ , of elements σ_{ij} , then the standard error m_F of F is well known to be

$$m_F^2 = h^T \Sigma h = \sum_{i=1}^{n} \sum_{j=1}^{n} h_i h_j \sigma_{ij} \quad . \tag{7}$$

Consider next a linear functional F of a function $f(u)$:

$$F = \int_{u=a}^{b} h(u) f(u) du \quad . \tag{8}$$

Obviously there is a close analogy between (6) and (8): there corresponds:

$$
\begin{array}{lll}
\text{function } f(u) & \text{to} & \text{vector } f_i \quad , \\
\text{variable } u & \text{to} & \text{subscript } i \quad , \\
\text{integral } \displaystyle\int_{u=a}^{b} & \text{to} & \text{sum } \displaystyle\sum_{i=1}^{n} \quad .
\end{array}
\tag{9}
$$

By analogy with (7) we expect the standard error of the functional F to be given by

$$m_F^2 = \int_{u=a}^{b} \int_{v=a}^{b} h(u) h(v) \sigma(u,v) du dv \quad , \tag{10}$$

$\sigma(u,v)$ denoting the _error covariance function_ of $f(u)$, which is the continuous analogue of the matrix σ_{ij} .

So far we have exploited the analogy between the _continuous_ function $f(u)$ and the _discrete_ vector f_i . An even closer analogy can be found in the following way.

Expand $f(u)$ into a series of _orthonormal functions_ $\phi_i(u)$:

$$f(u) = \sum_{i=1}^{\infty} f_i \phi_i(u) \quad . \tag{11}$$

For instance, let $a = 0$ and $b = 2\pi$ and

$$\phi_1(u) = \frac{1}{\sqrt{2\pi}} \quad ,$$

$$\phi_2(u) = \frac{1}{\sqrt{\pi}} \cos u \quad , \qquad \phi_3(u) = \frac{1}{\sqrt{\pi}} \sin u \quad ,$$

$$\phi_4(u) = \frac{1}{\sqrt{\pi}} \cos 2u \quad , \qquad \phi_5(u) = \frac{1}{\sqrt{\pi}} \sin 2u \quad , \tag{12}$$

$$\phi_6(u) = \frac{1}{\sqrt{\pi}} \cos 3u \quad , \qquad \phi_7(u) = \frac{1}{\sqrt{\pi}} \sin 2u \quad ,$$

$$\cdot \ \cdot \ \cdot$$

Then (11) is nothing else than a Fourier series, and f_i are the Fourier coefficients of the function $f(u)$.

If we similarly expand

$$h(u) = \sum_{i=1}^{\infty} h_i \phi_i(u)$$

$$\tag{13}$$

$$\sigma(u,v) = \sum_{i=1}^{\infty} \sum_{j=1}^{\infty} \sigma_{ij} \phi_i(u) \phi_j(v) \quad ,$$

then (8) and (10) may be readily seen to become

$$F = \sum_{i=1}^{\infty} h_i f_i \quad , \tag{14}$$

$$m_F^2 = \sum_{i=1}^{\infty} \sum_{j=1}^{\infty} h_i h_j \sigma_{ij} \quad . \tag{15}$$

Now the analogy to (6) and (7) is perfect; the only difference is $n \to \infty$!

Thus we see that we have two isomorphic kinds of Hilbert spaces: the space of (square integrable) functions $f(u)$ and the space of infinite vectors f_i , the isomorphy (one-to-one equivalence) being established by (11). The theoretical physicist, of course, will smile: this is the equivalence of Schrödinger's wave formulation and Heisenberg's matrix formulation of quantum mechanics!

So far, so good, but what is the use of all this in geodesy? When I introduced these ideas in my doctoral thesis in 1958, I was working in the cadastral service where areas of pieces of land were still determined by an oldfashioned apparatus called a polar planimeter, which essentially is an analog integrator. So I started from the error theory of integration (Moritz, 1961)[3].

The most natural application of error theory in function spaces is, of course, in physical geodesy, since the gravitational potential, gravity anomalies, etc., are such continuous functions, the corresponding infinite vectors being formed by the coefficients of spherical-harmonic series, as we have already mentioned in the preceding section. Since these applications are now standard, I shall not dwell on them here too much.

However, other applications present themselves readily. Integrations are involved, e.g., in integrated Doppler measurements or in inertial surveying where coordinate differences are obtained by twice integrating measured accelerations.

In the terminology of mathematical statistics, the transition to infinite-dimensional space corresponds, of course, to the transition from random vectors to stochastic processes.

3.2 Adjustment Computations. Least-squares collocation, now very popular in physical geodesy, may be regarded as an adjustment in Hilbert space, quite similar to an adjustment by conditions (Krarup, 1969[4], Moritz, 1980, sec. 25 [2]). This is understandable since it involves the gravitational potential, which is a continuous function; see also the simple example given in sec. 2.

In fact, least-squares adjustment in Hilbert spaces can be developed in an abstract and rigorous way, both adjustment by parameters and adjustment by conditions, cf. (Eeg, 1983)[5]. I need not give you the formulas: they look exactly the same as in usual adjustment computations, only that vectors may now be continuous functions or infinite vectors, and that matrices may be infinite or, more generally, will be linear operators. Don't be afraid: linear operators are in Hilbert space what matrices are in finite-dimensional spaces.

The geodesist who may have to apply these formulas will be gratified with their familiar look: this is another example of the power of symbolism. The mathematician, however, will pay attention to the fact that transition to the infinite involves nontrivial questions of convergence, existence, etc.

Let me finally mention an example where continuous adjustment by parameters comes natural. Consider an "integrated navigation system" for a land vehicle, a ship, or an airplane, consisting of an inertial system combined with a Global Positioning System (GPS) receiver. In principle, both systems can provide continuous recordings of position; in both cases, the observations could be considered continuous functions. Combining them by a suitable adjustment can provide elimination of systematic effects of either system and even information on the gravity field.

3.3 <u>Nonlinear Functional Analysis.</u> Linear functional analysis is the mathematical domain that treats Hilbert and Banach spaces, linear operators and functionals, etc. Nonlinear functional analysis, of course, is much more difficult mathematically, but formally sometimes surprisingly simple, because nonlinear operators are formally quite analogous to nonlinear functions: note again the power of symbolism ... So I cannot refrain from mentioning that a very simple first conceptual understanding of Molodensky's boundary value problem, the dread and delight of physical geodesists, can be obtained in this way, (cf. Moritz, 1980, sec. 40)[2].

4. SPECTRAL METHODS

4.1 <u>Diagonal matrices.</u> Consider again the linear equation system (1), $Ax = y$, or if A is a square 3×3 matrix:

$$
\begin{aligned}
a_{11}x_1 + a_{12}x_2 + a_{13}x_3 &= y_1 \ , \\
a_{21}x_1 + a_{22}x_2 + a_{23}x_3 &= y_2 \ , \\
a_{31}x_1 + a_{32}x_2 + a_{33}x_3 &= y_3 \ ,
\end{aligned}
\tag{16}
$$

or briefly

$$
y_i = \sum_{j=1}^{3} a_{ij}x_j \quad ;
\tag{17}
$$

y_i is obtained from x_j by a <u>"discrete convolution"</u>, that is, multiplication followed by summation.

Matters are much simpler if A is a diagonal matrix: (16) reduces to

$$
\begin{aligned}
a_{11}x_1 &= y_1 \ , \\
a_{22}x_2 &= y_2 \ , \\
a_{33}x_3 &= y_3 \ ,
\end{aligned}
\tag{18}
$$

or if we write $x_i = x(i)$, $y_i = y(i)$ —this may be motivated by the analogy (9)—and also $a_{ii} = a(i)$, then instead of (17) we simply have

$$
y(i) = a(i)x(i) \ ,
\tag{19}
$$

that is, <u>convolution is replaced by a simple multiplication.</u>

The simplification is particularly striking if we consider the inverse of a diagonal matrix:

$$
x = A^{-1}y
\tag{20}
$$

reduces to

$$x(i) = \frac{y(i)}{a(i)} \quad , \tag{21}$$

that is, inversion reduces to a simple division.

We could, of course, bring every square matrix to diagonal form by an appropriate linear transformation, so that the simple relations (18) through (21) hold--clearly, $n = 3$ was chosen only for simplicity and the procedure is valid for any n --, but finding such a linear transformation is in general certainly no less difficult than solving the original system $Ax = y$.

In certain cases, however, such a diagonalization is rather simple, and this is the basic idea of <u>spectral analysis</u>. Not only matrices, but also linear operators, their equivalent in Hilbert space, can be diagonalized in this way.

4.2 <u>Fourier Series and Integrals</u>. We have met with Fourier series already in eqs. (11) and (12). Using the well-known Euler-Moivre identity

$$\cos\alpha + i\sin\alpha = e^{i\alpha} \quad (i = \sqrt{-1}) \quad , \tag{22}$$

we can instead write

$$f(t) = \sum_{k=-\infty}^{\infty} c_k e^{ikt} \quad , \tag{23}$$

where the Fourier coefficients c_k are given by

$$c_k = \frac{1}{2\pi} \int_0^{2\pi} f(t) e^{-itk} dt \quad . \tag{24}$$

Here we have denoted the independent variable by t instead of u ; in fact, t is often, but by no means always, time: it may also be distance along a spatial profile. Accordingly we say that the function f is defined in a "time domain" or in a onedimensional "space domain".

We note a considerable analogy between (23) and (24), similar to (9): both formulas have a similar structure, an integral corresponding to a sum. This analogy may even be emphasized by putting

$$c_k = \bar{f}(k) \quad ; \tag{25}$$

the set of these Fourier coefficients forms the <u>spectrum</u> of the function $f(t)$.

It is well known that the expansion (23) holds for periodic functions with period 2π . Similar Fourier series can be obtained for functions with an arbitrary period T. By letting $T \to \infty$ we get functions that are not periodic at all. For them, Fourier series are replaced by <u>Fourier integrals</u>: instead of (23) and (24) we then have

$$f(t) = \int_{-\infty}^{\infty} \overline{f}(\lambda) e^{i\lambda t} d\lambda \quad , \tag{26}$$

$$\overline{f}(\lambda) = \frac{1}{2\pi} \int_{-\infty}^{\infty} f(t) e^{-i\lambda t} dt \quad . \tag{27}$$

Here $\overline{f}(\lambda)$ denotes the spectrum, or Fourier transform, of the function $f(t)$; note that we now have a continuous spectrum instead of the discrete spectrum (25).

Now the symmetry between (26) and (27) is almost perfect; it is disturbed only by the inessential factor 2π ; both i and $-i$ have essentially the same nature, both being square roots of -1 .

The continuous analogue of (17) is the formula

$$g(t) = \int_{-\infty}^{\infty} h(t,\tau) f(\tau) d\tau \quad ; \tag{28}$$

especially important is the case

$$h(t,\tau) = h(t-\tau) \quad , \tag{29}$$

which is a function of the difference $t-\tau$ only.

Then

$$g(t) = \int_{-\infty}^{\infty} h(t-\tau) f(\tau) d\tau \tag{30}$$

denotes a <u>convolution</u> in the proper sense: multiplication followed by integration.

We are purposely disregarding all questions of convergence, etc., which justly are a major concern to mathematicians but can be omitted in our heuristic introduction.

Now we have a beautiful and fundamental fact, which reminds us of the diagonal matrices mentioned at the beginning of this section. Consider the Fourier transforms, or spectra, (27) and

$$\overline{g}(\lambda) = \frac{1}{2\pi} \int_{-\infty}^{\infty} f(t) e^{-i\lambda t} dt \quad , \tag{31}$$

$$\overline{h}(\lambda) = \frac{1}{2\pi} \int_{-\infty}^{\infty} h(t) e^{-i\lambda t} dt \quad .$$

Then the convolution in the <u>"time domain"</u> is equivalent to a simple multiplication

$$\overline{g}(\lambda) = \overline{h}(\lambda) \overline{f}(\lambda) \tag{32}$$

in the <u>"frequency domain"</u> (λ has the character of a frequency).

The equivalence of (30) and (32) constitutes the basic theorem of spectral analysis. Spectral analysis, or harmonic analysis, not only holds for functions on the line, i.e., of a variable t --this leads to Fourier series and Fourier integrals--but also for functions defined on the plane or on the sphere. This leads to two-dimensional Fourier series and Fourier integrals for the plane and to spherical harmonics for the sphere. They represent transformations from a "space domain" (plane or sphere) to a frequency domain.

The basic integral formulas of physical geodesy, such as Stokes' formula, are convolutions on the sphere or in the plane. This makes the usefulness of spectral methods in physical geodesy immediately obvious; the evaluation of integral formulas by the "Fast Fourier Transform" (cf. Brigham, 1974)[6] is now well established (cf. Sideris, 1985)[7].

Similarly, matrix inversions are replaced by division in the frequency domain, which makes "frequency domain collocation" very attractive, although it is restricted to rather regular data distributions (cf. Colombo, 1979)[8].

However, in line with my endeavor to conceal (not very successfully) my bias towards physical geodesy, I shall finally consider an example from the theory of

4.3 Geodetic Networks. Sufficiently regular networks can be successfully studied by spectral methods (cf. Meissl, 1976)[9]. This presupposes a regular network structure, but gives extraordinary mathematical insight which can be used for understanding networks of a more irregular structure.

I shall try to illustrate this by means of a very simple example (taken from Sünkel, 1985)[10]. Consider a regular levelling line consisting of equidistant points:

It is very long, so that it can be regarded as extending to infinity in both directions.

The observation equations are

$$h_{k+1} - h_k = \ell_{k,k+1} \quad , \tag{33}$$

h denoting the heights and ℓ the observed differences. Let us form the normal equations in the way familiar from adjustment computations:

$$-h_{k-1} + 2h_k - h_{k+1} = -\ell_{k,k+1} + \ell_{k-1,k} = r_k \quad . \tag{34}$$

For an infinite levelling line, k may assume all positive or negative integer values.

The normal equations (34) may be written in the matrix form

$$Ah = L \tag{35}$$

or, more explicitly

$$
\begin{bmatrix}
\ddots & \ddots & \ddots & \ddots & \ddots & \\
\dots & 0 & -1 & 2 & -1 & 0 & \dots \\
& \dots & 0 & -1 & 2 & -1 & 0 & \dots \\
& & \dots & 0 & -1 & 2 & -1 & 0 & \dots \\
& & \ddots & \ddots & \ddots & \ddots & \ddots
\end{bmatrix}
\begin{bmatrix}
\vdots \\
h_{-1} \\
h_0 \\
h_1 \\
\vdots
\end{bmatrix}
=
\begin{bmatrix}
\vdots \\
r_{-1} \\
r_0 \\
r_1 \\
\vdots
\end{bmatrix}
\tag{36}
$$

The matrix A is infinite but has a very regular band structure; it is a Toeplitz matrix. In fact, we have

$$a_{ij} = a_{j-i} \tag{37}$$

(e.g. $a_{oo} = a_{11} = a_{22} = \dots = 2$, $a_{10} = a_{21} = a_{32} = \dots = -1$, etc.), reminiscent of (29). Because of this regularity, we may introduce the function

$$\bar{a}(\lambda) = -e^{i\lambda} + 2 - e^{-i\lambda} = 2(1-\cos\lambda) \tag{38}$$

to represent the Fourier transform of the matrix A, the coefficients being the non-zero elements in any row (or column) of A, and similarly

$$\bar{h}(\lambda) = \sum_{-\infty}^{\infty} h_k e^{-ik\lambda} \quad, \tag{39}$$

$$\bar{r}(\lambda) = \sum_{-\infty}^{\infty} r_k e^{-ik\lambda} \quad. \tag{40}$$

Since (36) is a "discrete convolution" (cf. (17)), the Fourier transform of the infinite system (36) is the simple equation

$$\bar{a}(\lambda)\bar{h}(\lambda) = \bar{r}(\lambda) \quad, \tag{41}$$

corresponding to (32). Its solution is, of course,

$$\bar{h}(\lambda) = \frac{\bar{r}(\lambda)}{\bar{a}(\lambda)} \quad,$$

so that the unknowns h_k are simply given as the Fourier coefficients of $\bar{h}(\lambda)$

$$h_k = \frac{1}{2\pi} \int_0^{2\pi} \bar{h}(\lambda)e^{ik\lambda}d\lambda = \frac{1}{2\pi} \int_0^{2\pi} \frac{\bar{r}(\lambda)}{\bar{a}(\lambda)} e^{ik\lambda}d\lambda \tag{42}$$

by (24); note that the roles of "time domain" and "frequency domain" are now interchanged.

The substitution of (40), written with j as summation index:

$$\overline{r}(\lambda) = \sum_{-\infty}^{\infty} r_j e^{-ij\lambda}$$

into (42) finally gives h_k as a linear combination of the observations r_j defined by (34):

$$h_k = \sum_{j=-\infty}^{\infty} b_{k-j} r_j \tag{43}$$

with

$$b_k = \frac{1}{2\pi} \int_0^{2\pi} \frac{e^{ik\lambda}}{\overline{a}(\lambda)} d\lambda = \frac{1}{4\pi} \int_0^{2\pi} \frac{\cos k\lambda}{1-\cos\lambda} d\lambda \quad , \tag{44}$$

by (38) and by the symmetry of $\overline{a}(\lambda)$. Using a trick to eliminate a singularity, the integral can be evaluated to give

$$b_k = -\frac{1}{2} |k| \quad , \tag{45}$$

so that (43) becomes simply

$$h_k = -\frac{1}{2} \sum_{j=-\infty}^{\infty} |k-j| \; r_j \quad . \tag{46}$$

Finally we note that the difference forming the left-hand side of (33), and the "second differences" appearing on the left-hand side of (34) are discrete analogues of first and second derivatives. Such analogies are useful if we make the formal transition from regular nets of high point densitiy to "continuous nets", which can be treated by differential equations as pointed out by Krarup and others; (cf. Grafarend and Krumm, 1985)[11].

5. CONCLUDING REMARKS

Contemporary mathematical geodesy is characterized by an interplay between the finite and the infinite, between the discrete and the continuous, between time (or space) domain and frequency domain, and between deterministic and stochastic aspects. Continuous measurements are discretized for digital processing; on the other hand, discrete, regularly distributed quantities may be approximately considered continuous to get a deeper mathematical insight.

This interplay furnishes the basis of a fruitful cooperation between theoreticians and numerical analysts in geodesy and in other applied sciences as well.

*   *   *

REFERENCES

1) Bjerhammar, A., Theory of Errors and Generalized Matrix Inverses, Elsevier, Amsterdam (1973).

2) Moritz, H., Advanced Physical Geodesy, Herbert Wichmann, Karlsruhe, and Abacus Press, Tunbridge Wells, Kent (1980).

3) Moritz, H., Fehlertheorie der graphisch-mechanischen Integration: Grundzüge einer allgemeinen Fehlertheorie im Funktionenraum, Sonderheft 22, Österr. Zeitschrift für Vermessungswesen, Wien (1961).

4) Krarup, T., A contribution to the mathematical foundation of physical geodesy, Publ. 44, Danish Geodetic Inst., Copenhagen (1969).

5) Eeg, J., Continuous methods in least squares theory, Proceedings of the VIII Symposium on Mathematical Geodesy (Como, Sept. 1981), Firenze (1983).

6) Brigham, E.O., The Fast Fourier Transform, Prentice-Hall, Englewood Cliffs, N.J. (1974).

7) Sideris, M.G., A Fast Fourier Transform method for computing terrain corrections, Manuscripta Geodaetica, $\underline{10}$, 66 (1985).

8) Colombo, O.L., Optimal estimation from data regularly sampled on a sphere with applications in goedesy, Report 291, Dept. of Geodetic Science, Ohio State Univ., Columbus (1979).

9) Meissl, P., Strength analysis of two-dimensional angular Anblock networks, Manuscripta Geodaetica, $\underline{1}$, 293 (1976).

10) Sünkel, H., Fourier analysis of geodetic networks, in: Optimization and Design of Geodetic Networks (E.W. Grafarend and F. Sanso, eds.), Springer, Berlin (1985).

11) Grafarend, E.W., and Krumm, F.W., Continuous networks I, in: Optimization and Design of Geodetic Networks (E.W. Grafarend and F. Sanso, eds.), Springer, Berlin (1985).

INVERSE REGIONAL REFERENCE SYSTEMS:

A POSSIBLE SYNTHESIS BETWEEN THREE- AND TWO-DIMENSIONAL GEODESY

H.M. Dufour

Institut Geographique National, Saint-Mandé, France

ABSTRACT

For reasons perhaps more practical than theoretical, the early geodesists
tended to consider horizontal and vertical coordinates separately. This
introduced complications when it was decided to transform a straight line,
in space, to a geodesic on an arbitrary ellipsoid.   The appearance of
three-dimensional geodesy in the last 30 years has permitted a simplification of the formulae - but the interest in the two groups of parameters
remains, at times, fundamental, notably for the comparison of two networks
(ancient or modern). The introduction of regional three-dimensional reference systems, defined, in principle, in the neighbourhood of a point O, by
an inversion, permits the conservation of the simple formulae of three-
dimensional geodesy. It also allows the division of the parameters into two
categories: stereographic coordinates (X, Y) on a regional projection, in
the plane tangent to O, and Z coordinate close to an altitude. Some practical applications are presented (for example the representaton of the local
gravity field).

1. THE HISTORICAL PERSPECTIVE OF THE PROBLEM

For many reasons, both theoretical and practical, geodesy has for a long time remained divided
into distinct domains: horizontal positions defined by two parameters (two dimensional geodesy), and
the third dimension characterised by levelling, giving an altitude HN, the evaluation of which has
developed almost independently to that of the horizontal parameters (longitude λ, latitude ϕ). This
separation was clearly not total, since a connection was made in using a "best-fit" ellipsoid for a
given nation or region, these surfaces being estimated to minimize the deviations of the vertical.

$$\xi = \phi(\text{Astro}) - \phi(\text{Geod.})$$
$$\eta = (\lambda(\text{Astro}) - \lambda(\text{Geod.})) \cos \phi$$

which implies the existence of a geometrical height H. The difference DH = H - HN was considered
negligible.

One may wonder why the geodesists of the past, including among them the greatest mathematicians
of their time, did not directly set the problem in three dimensions, but obstinately replaced the
straight line AB in Euclidian space, by the geodesic linking the projections of "a" and "b" - the
extremities of the segment AB on an ellipsoid. This geodetic line was a particularly difficult tool,
even though its study permitted the development of remarkable theories on curved surfaces, such as
Legendre's theorem concerning the spherical excess of a geodetic triangle, or the famous Gaussian

integral on a closed contour. I believe that the reasoning of our ancestors was more pragmatic: they had practically no means of handling matrix calculations, even of such small dimensions as 3 × 3, however the two dimensional field permitted the use of formulae of plane and eventually spherical trigonometrical tables.

The advent of spatial satellite surveying and the need for an improved precision in geodetic networks has induced the precise definition of a third dimension, distinctly different from the levelling height which is in fact, only a transformed potential. The geometric height H associated with geographical coordinates (λ, ϕ) must, when integrated with the latter, supply the cartesian coordinates (U, V, W):

$$
\begin{aligned}
U &= (N + H) \cos \phi \cos \lambda \\
V &= (N + H) \cos \phi \sin \lambda \\
W &= (N(1 - e^2) + H) \sin \phi
\end{aligned}
\tag{1}
$$

(N being the maximum radius of curvature at latitude ϕ, and e the eccentricity of the ellipsoid).

The transformation from a network (1) to a network (2) supposes that both are correctly measured in order to provide the principle translation parameters T (TU, TV, TW), or, for more complex problems, the translation, scale change, and rotation. For example:

$$
\begin{aligned}
U2 &= U1 + TU + a \cdot DU - r\ DV + q\ DW \\
V2 &= V1 + TV + r\ DU + a\ DV - p\ dW \\
W2 &= W1 + TW - q\ DU + p\ DV + a\ DW
\end{aligned}
\tag{2}
$$

with DU = (U1 - U0), DV = (V1 - V0), DW = (W1 - W0), (U0, V0, W0) being the coordinates in network (1) of a central point (barycentric for example) of the working region. The scale change is represented by the parameter (a) and the rotation by the coefficients (p, q, r). [This type of formulae has been introduced by a number of authors such as: WOLF, BURSA, MOLODENSKY, BADEKAS, VEIS, etc ... The combination of Cartesian coordinates with those of a local system has been especially brought to the forefront[1-3]).

Cartesian coordinates are irreplacable in their use for worldwide representation, or for large expanses. On the other hand they can also lead to difficulties in the cartographic representation of results, which remain two dimensional. Until now, the cartographic representation has not evolved logically, treating the horizontal (λ, ϕ) coordinates separately from the vertical (H) coordinates. Most representations are simple with respect to (λ, ϕ) for example the Mercator (X, Y) projection

$$X = \lambda\ ; \quad Y = \text{function of } (\phi)\ .$$

But few of them really have the necessary spatial characteristics, that is to say they do not allow an acceptable transformation to a three-dimensional reference system.

In order to make use of these cartographic representations at the time of calculation, and for the presentation of results, it is necessary that they display the following quality: the uniqueness of classical coordinates (λ, ϕ, H) must be replaced by systems of multiple coordinates, whilst retaining interchangeability for spatial transformations. One can therefore talk of the "dynamics" of reference systems.

There are really only two systems which are capable of allowing such transformations: the gnomonic and stereographic representations.

There exists a set of three-dimensional space representations which defines this space using three positional axes and an arbitrary orientation. Together these representations form a group having for their conjugate relationship a linear transformation (translation, rotation).

$$
\left|
\begin{array}{l}
X' = aX + bY + cZ + d \\
Y' = a'X + b'Y + c'Z + d' \\
Z' = a''X + b''Y + c''Z + d'' .
\end{array}
\right.
\tag{3}
$$

In two dimensions one can consider all the projections, from the same centre I, to be from space to different planes π. Hence one obtains the gnomonic representations, the link between them being the homographic relationships of the real variables (a, b, c, a', b', c', P, Q):

$$
\begin{aligned}
D \cdot X' &= a + bX + cY \\
D \cdot Y' &= a' + b'X + c'Y \\
D &= 1 + Px + QY .
\end{aligned}
\tag{4}
$$

The gnomonic projection conserves straight lines in space but introduces important distortions proportional to the distance from the central point (orthogonal projection of point I onto the plane of projection π).

The fundamental characteristics of the three-dimensional stereographic representation will be considered as:

- Choosing a reference plane π and centre O (the origin of the image space situated in the plane π).
- Performing from point P (the inversion pole), situated on the normal to π at O, an inversion (P, $4R^2$) of pole P and of power $4R^2$ (with, in principle 2R = OP, but this is not obligatory).
- The coordinates (X, Y, Z) in image space are defined from the two axes (OX, OY) situated in π; the third axis OZ being determined by OP (in principle OZ is of the same sign as OP).

The advantage of the method lies essentially with the efficiency of the representation of the space to the proximity of an approximately spherical surface around O: with the following conditions:

- Plane π is tangent (semi-rigorously) to the level surface at the origin O.
- The pole P is (semi-rigorously) at the antipode of O on the surface.
- 2R = OP.
- The coordinates (X, Y) then have the character of planimetric stereographic map coordinates.
- The coordinate Z is approximately an altitude.

In their general form; the different inverse three-dimensional representations (X, Y, Z) of the same object space form a group obtained by the following type of relationships:

$$
\left|
\begin{array}{l}
D \cdot X' = a + bX + cY + dZ + e(X^2 + Y^2 + Z^2) \\
D \cdot Y' = a' + b'X + c'Y + d'Z + e'(X^2 + Y^2 + Z^2) \\
D \cdot Z' = a'' + b''X + c''Y + d''Z + e''(X^2 + Y^2 + Z^2) \\
D = P + QX + RY + SZ + T(X^2 + Y^2 + Z^2) .
\end{array}
\right.
\tag{5}
$$

In considering that relationships of the same type permit a direct passage from general cartesian coordinates (U, V, W) to any X, Y, Z system, one realises that the geographical coordinates (λ, ϕ, H) can never be used in the calculations. There remains a few possible uses, but their main use in calculations with trigonometrical tables is no longer necessary.

It is important to note that all the envisaged transformations [Eq. (5)] are three dimensionally conformal, and that the properties of the figure in the image space (X, Y, Z) are simply deduced from the only knowledge of the scale factor K (at a point Q in the object space):

$$
K^2 = (dX^2 + dY^2 + dZ^2)/(dU^2 + dV^2 + dW^2)
$$

$$
K = \frac{4R^2}{(PQ)^2} .
\tag{6}
$$

An important subset of transformations by inversion are those where the planes π are tangent to the same sphere (Σ) at the points 0, the corresponding poles P being found at the antipoles of 0 on the sphere Σ. In Eq. (5), the important simplifications are characterized by the separation of the coordinate Z from the rest of the equation [see Eq. (23)] and the possibility to uniquely define the subgroup of variables (X, Y) under the following form:

$$
\left|
\begin{array}{l}
D \cdot X' = a + bX + cY + e(X^2 + Y^2) \\
D \cdot Y' = a' + b'X + c'Y + e'(X^2 + Y^2) \\
D = P + QX + RY + T(X^2 + Y^2)
\end{array}
\right.
\tag{7}
$$

which takes a remarkably different form if rewritten as:

$$
X + iY = z ; \quad X' + iY' = z'
$$

$$
z' = \frac{\alpha + \beta z}{1 + Pz}
\tag{8}
$$

[z, z', α, β, P ε C (complex numbers)].

An extension of these formulae for the ellipsoid (E) requires a definition (of a two-dimensional nature) of the stereographic representation of the ellipsoid.

This general appraisal of the stereographic type transformations, has indicated the principle properties involved, in going from the simple to the complex. The appraisal of two-dimensional problems, thus appears worthwhile before considering the most general cases.

2. TWO-DIMENSIONAL PROBLEMS ON THE SPHERE

At an origin O (λ, ϕ) of the sphere, one considers the local system (XS, YS, ZS) which is deduced by rotation and translation from the terrestrial cartesian axes (U, V, W):

$$\begin{matrix} XS \\ YS \\ ZS \end{matrix} = RT \begin{matrix} U \\ V \\ W \end{matrix} - \begin{matrix} 0 \\ 0 \\ R \end{matrix} \qquad (9)$$

with

$$RT = \begin{vmatrix} -\sin \lambda & \cos \lambda & 0 \\ -\sin \phi \cos \lambda \ ; & -\sin \phi \sin \lambda \ ; & \cos \phi \\ \cos \phi \cos \lambda & \cos \phi \sin \lambda & \sin \phi \end{vmatrix} . \qquad (10)$$

The following formulae give the STEREO (X, Y) coordinates

$$X = XS/E \ ; \quad Y = YS/E \ ;$$

$$\frac{I}{K} = E = 1 + \frac{ZS}{2R} , \quad K = 1 + \frac{X^2 + Y^2}{4R^2} . \qquad (11)$$

We will let z = X + iY.

The whole of the STEREO representations of the sphere form a group for which the complex z co-ordinates are interrelated by the homographic transformations in the set of complex numbers C.

A simple formula relates the two elements of the group (z1, z2):

$$\frac{z1}{z1(2)} + \frac{z2}{z2(1)} + t^2 \frac{z1z2}{z1(2)z2(1)} = 1 \qquad (12)$$

z1(2) = coordinate of the origin of (2) in system (1)
z2(1) = coordinate of the origin of (1) in system (2)

$t = tg \dfrac{\theta}{2}$; θ = the angular distance between the normals to the sphere from origins (1) and (2)

$$t^2 = \frac{z1(2)\overline{z1}(2)}{4R^2} = \frac{z2(1)\overline{z2}(1)}{4R^2} \quad (\overline{z} = \text{conjugate of } z) .$$

3. EXPLOITATION OF THE STEREO REPRESENTATION OF THE SPHERE

All the corrections necessary to transform measurements made in the image space (on the pro-jected plane), to the object space (on the sphere) have been obtained from formulae involving the expressions of the scale factor K.

Let $\widehat{AB}$ be an arc of the sphere, the image of $\widehat{ab}$ (of chord $\overline{ab}$) on the plane.

We will assume the general form $X/2R = x$, $Y/2R = y$. Then one has:

$$K = 1 + x^2 + y^2 , \quad E = \frac{1}{K} \quad \text{(Scale factor)} . \tag{13}$$

Length of Great Circle

$$\widehat{AB} = 2R \text{ Arc sin } \left(\frac{\overline{ab}}{2R\sqrt{K(a)K(b)}}\right) . \tag{14}$$

Arc to chord correction $\quad [\rho 1 = \rho 2 = dV/2]$

$$D \cdot tg\left(\frac{dV}{2}\right) = x(a)y(b)-y(a)x(b)$$
$$D = 1 + x(a)x(b) + y(a)y(b) . \tag{15}$$

Correction of a direction

This apparently unknown formula is easily deduced from the preceding correction.

At point a, the tangent to the image $\widehat{ab}$ of the great circle is the straight line $\overline{ab}'$, b' given by the vectorial relationship:

$$\frac{\vec{ob}'}{\vec{ob}} = \frac{1}{1 - \alpha} , \quad \text{with } \alpha = (ab)^2/(4R^2K(a)) . \tag{16}$$

Bearing (C) from a meridian

$$-D \cdot tgC = 2x(y + tg\phi)$$
$$D = 1 + x^2 - y^2 - 2y \, tg\phi . \tag{17}$$

Division of the arc of a great circle

The division of the arc into two elements (p, q) according to the length relationship (p + q = 1) is possible, but only (p = q = 0.5) is simple.

The mid point G of the arc has an image g:

$$D \cdot X(g) = E(a)X(a) + E(b)X(b)$$
$$D \cdot Y(g) = E(a)Y(a) + E(b)Y(b) \tag{18}$$
$$D = E(a) + E(b) - 2 \sin^2 \frac{\alpha}{4}$$

with $\alpha = \widehat{AB}/R$.

Equations (14) to (18) are valid and rigorous for $\alpha < 180°$.

4. TWO-DIMENSIONAL PROBLEMS ON THE ELLIPSOID

One can define a number of representations of the ellipsoid which can be qualified as "stereographic". However, we will only discuss one possible group in terms of functions of three parameters: $F(\phi_0; \lambda, \phi)$:

- First, one considers the conformal representation of the ellipsoid on the sphere bitangential at parallel ϕ_0, preserving the longitudes and the latitude of the origin ϕ_0.

- One then considers the different STEREO representations of the image sphere possible at the points 0 on this sphere, which are images of the coordinates (λ, ϕ) of the ellipsoid.

From the results one can deduce the following:

- All the F representations form a set linked by a homographic relationship of the complex variables.
- One passes from the polar form of subgroup (ϕ_0) to those of subgroup (ϕ_0') by a scale change.

- For each element $(F(\phi_0, \lambda, \phi)$ there is a corresponding element $(F(\phi_0', \lambda, \phi')$ such that the relationship between the two reduces to a linear transformation (translation and scale change).

- There is an important point here as far as cartography is concerned. It seems useful to define a cartographic system

$$F(0; \lambda, \phi)$$

allowing local maps of different (λ, ϕ) i.e. local maps;

and a regional cartography: $F(\phi_0 \neq 0; \lambda, \phi_0)$

This consideration is interesting in applications close to latitude ϕ_0 but the practical realization of regional maps is not necessary, since any local map of the general cartographic system is also (to a translation and scale change approximation) a regional map.

- The exploitation of stereographic representations of the ellipsoid necessitates the addition to the spherical corrections of residual terms. The use of the third dimension gives much more elegant solutions.

5. THREE-DIMENSIONAL PROBLEMS ON THE SPHERE

Figure 1 illustrates the three-dimensional stereographic representation on the sphere, in general: the origin of the projection is the point 0 of the sphere and the plane of the figure is defined by the great circle containing 0 and the point Q of the space represented.

XS, YS, ZS are the local cartesian coordinates derived from the general cartesian coordinates (U, V, W) by the relationship:

$$
\begin{array}{ccc}
XS & U & 0 \\
YS & = RT \ V & - \ 0 \\
ZS & W & R
\end{array}
\qquad (19)
$$

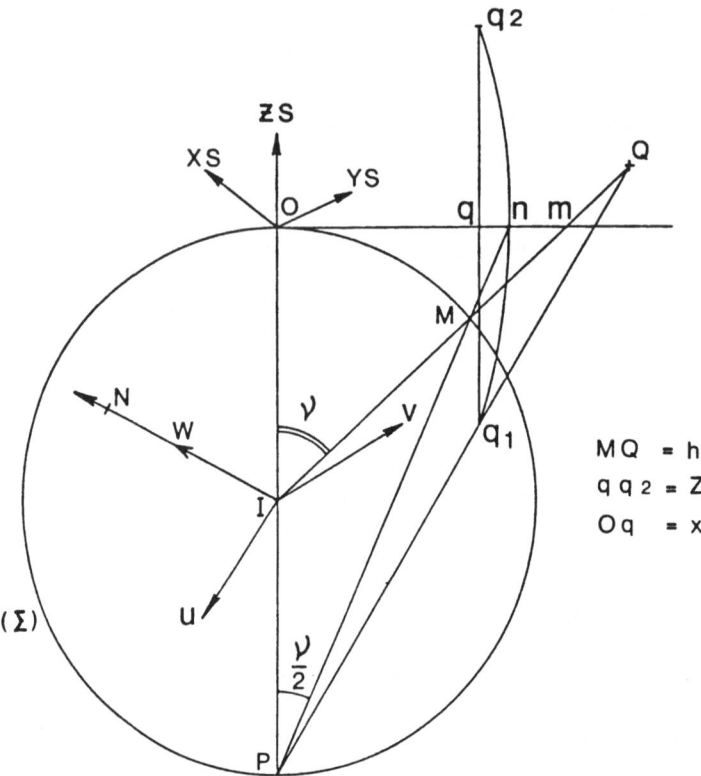

Fig. 1 Three-dimensional stereographic representation

The normal to the sphere passing through a general pont Q cuts the surface of the sphere at M, and must become in the inversion $(P, 4R^2)$ the point q_1 and by symmetry with reference to the tangential plane π the point q_2 (the 3D - STEREO image of the general point Q).

If (XS, YS, ZS) are the coordinates of Q in the regional system, one has the relationships (in the same reference system):

$$D \cdot X = 4R^2(XS)$$
$$D \cdot Y = 4R^2(YS)$$
$$(2R - Z)D = 4R^2(ZS + 2R)$$
$$D = XS^2 + YS^2 + (ZS + 2R)^2 . \qquad (20)$$

With these formulae one can associate the following expressions of the radial distance $x = X^2 + Y^2$ and of the variable Z as a function of the angular distance v and of the real altitude h:

$$D \cdot x = (R + h) \sin v$$

$$D \cdot Z = h(1 + \frac{h}{2R}) \qquad (21)$$

$$D = \cos^2 \frac{v}{2} (1 + \frac{h}{R}) + \frac{h^2}{4R^2} .$$

The transformation from a reference origin $O(\lambda, \phi)$ (coordinates X, Y, Z) to a reference origin $O'(0, \phi')$ (coordinates X', Y', Z') corresponds to the following formulae, in which O represents the angular distance of the arc $\overparen{O'O}$, A' being the direct azimuth and A the reverse azimuth:

$$D \cdot X' = a + bX + cY + e(X^2 + Y^2 + Z^2)$$
$$D \cdot Y' = a' + b'X + c'Y + e'(X^2 + Y^2 + Z^2)$$
$$D \cdot Z' = Z$$
$$D = P + QX + RY + T(X^2 + Y^2 + Z^2)$$

(22)

with

$a = R \cos \phi \sin \lambda$ $= R \sin \theta \sin A'$

$b = \cos \lambda$

$c = - \sin \phi \sin \lambda$

$e = - \cos \phi \sin \lambda / 4R$ $= - \sin \theta \sin A'/4R$

$a' = $ $= R \sin \theta \cos A'$

$b' = \sin \phi' \sin \lambda$

$c' = \cos \phi \cos \phi' + \sin \phi \sin \phi' \cos \lambda$ (23)

$e' = $ $= - \sin \theta \cos A'/4R$

$P = $ $= \cos^2 \dfrac{\theta}{2}$

$Q = - \cos \phi' \sin \lambda / 2R$ $= \sin \theta \sin A/2R$

$R = $ $= \sin \theta \cos A/2R$

$T = $ $= \sin^2 \dfrac{\theta}{2} /4R^2$

Note the behaviour of the variables (Z, Z') in these formulae, which on the sphere (Z = Z' = 0) correspond to the complex variable formula, linking the parameters (z_1, z_2) [Eq. (12)].

6. GENERAL SPATIAL PROBLEMS

One recalls that it is possible to create, at a point anywhere in space, a regional inverse, where the points will be defined by the planimetric elements (X, Y) and a third dimension Z (which can semi-rigorously represent an altitude).

- One applies a motion displacement to the Cartesian coordinates (U, V, W) i.e. a translation and rotation in order to define regional coordinates (XS, YS, ZS) centered at point O.

- On the ZS axis one chooses an inversion pole P (OP = 2R) and applies the inversion (P, $4R^2$) followed by a symmetry relative to the plane (XS, 0, YS). Hence one obtains the regional inverse coordinates (X, Y, Z).

If one passes from an inverse space V(0, P) to another inverse space V'(0', P'), with the following notation:

ω = Rotation matrix of the 1st system to the second: $\omega(ij) = \alpha_{ij}$

$\overrightarrow{P'P}$ = Vector relating the inverse poles

$$\overrightarrow{P'P} = \begin{Bmatrix} \alpha \\ \beta \\ \gamma \end{Bmatrix} \text{ in } V \; ; \quad \overrightarrow{P'P} = \begin{Bmatrix} \alpha' \\ \beta' \\ \gamma' \end{Bmatrix} \text{ in } V'$$

$$(P'P)^2 = a^2 \; ; \quad OP = 2R \; ; \quad O'P' = 2R'$$

then one has:

$$D \cdot X' = \left(\frac{R'}{R}\right)^2 \left[\alpha_{11}X + \alpha_{12}Y + \alpha_{13}Z \right] + \frac{R'^2}{4R^2} \alpha'(X^2 + Y^2 + Z^2)$$

$$\vdots$$

$$D \cdot Z' = \left(\frac{R'}{R}\right)^2 \left[\alpha_{31}X + \alpha_{32}Y + \alpha_{33}Z \right] + \frac{R'^2}{4R^2} \gamma'(X^2 + Y^2 + Z^2) \qquad (24)$$

$$D = 1 + \frac{a^2}{16R^4}(X^2 + Y^2 + Z^2) + \frac{1}{2R^2}(\alpha X + \beta Y + \gamma Z) \; .$$

These formulae have the inversion poles P (for X, Y; Z) and P' (for X', Y', Z') for their respective origins. Normally they must be brought back to points O and O' by the following translations:

$$Z \quad \text{replaced by } (2R - Z)$$
$$Z' \quad \text{replaced by } (2R' - Z')$$

which gives more complicated expressions of type (5). In the particular case of the sphere (R = R', $\overrightarrow{P'P}$ interconnected with rotation ω) one arrives at Eqs. (22), (23).

7. REGIONAL INVERSES OF THE ELLIPSOID (E)

In the proximity of point O of the ellipsoid, of geographical coordinates (λ_0, ϕ_0), one can define a regional stereographic representation, having the point O for its centre (cf. Section 4), but the exterior region does not have a simply definable image. For many reasons it seems preferable to define a regional inversion of centre O, and pole at point P, on the normal from O to the ellipsoid. The straight line PO cuts the axis of revolution at I_1, eccentric to the origin I, quantified by

$$\zeta = \overrightarrow{II_1} = Ne^2 \sin \phi_0 \quad \text{(Fig. 2)}.$$

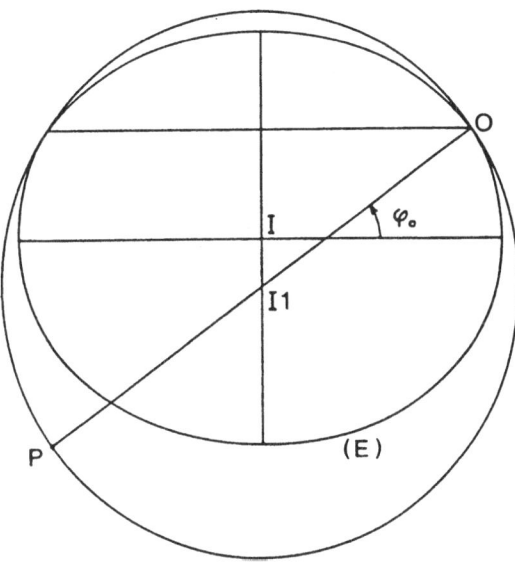

Fig. 2 Regional inverses of the ellipsoid (E)

We have adopted the principle of making P the antipole of 0, on the bitangential sphere of radius $N_0 = a/\sqrt{1 - e^2 \sin^2 \phi_0}$, but this solution is perhaps not optimal in certain cases.

In the inversion $(P, 4N^2)$ followed by the symmetry with reference to the tangent plane at 0 on the ellipsoid, the image of (E) is found entirely in the part of the coordinate system under the plane (0, XS, YS, ZS) with contact all along the parallel of latitude ϕ_0. The final coordinates are called (X, Y, Z).

The residual depression of the horizon in the image space is positive towards south or the north, and of the form

$$a \frac{e^2}{2} \cos^2 \phi_0 (\phi - \phi_0)^2 :$$

zero at the north pole and a maximum when 0 is equatorial, and being 650 m for $/\phi - \phi_0/ = 10°$.

The formulae giving the correspondance between the altitude H and the variable Z must be corrected to take account of this depression.

8. THREE-DIMENSIONAL CORRECTIONS

Referring to Fig. 1 for the notation = the formulae for the sphere are valid in the general case because they only use the radius R of the inversion sphere, the spatial coordinates (X, Y; Z) and in particular the expression of the scale factor K. They are completely soluble using the three-dimensional inversion formulae.

$$\text{Scale:} \quad K = \frac{(Pq_1)^2}{4R^2} = 1 - \frac{Z}{R} + \frac{X^2 + Y^2 + Z^2}{4R^2}$$

$$\text{Inverse:} \quad E = \frac{1}{K} = \frac{4R^2}{(PQ)^2} = 1 + \frac{ZS}{R} + \frac{XS^2 + YS^2 + ZS^2}{4R^2} .$$

(25)

The real length $\overline{AB}$ of a spatial segment

If $\overline{ab}$ is the image of AB in the inverse representation, one has the relationship

$$AB = ab/\sqrt{K(a)K(b)} .$$

(26)

Correction to a direction

In the image space it is necessary to introduce the point J, symmetric of pole P with reference to 0 $(\vec{OJ} = -\vec{OP})$.

To reproduce the straight line AB, it is necessary in projection to place the point b', shifted along the line Jb, in the following manner:

$$\frac{\vec{Jb'}}{\vec{Jb}} = \frac{1}{1 - \alpha} , \quad \text{with } \alpha = (ab)^2/(4R^2 K(a)) .$$

(27)

In the projection onto OZ, this relationship introduces an important depression of the horizon. On the axes (OX, OY) it reduces to:

$$\frac{\vec{Ob'}}{\vec{Ob}} = \frac{1}{1 - \alpha}$$

(28)

which is none other than a generalization of expression (16).

Partition of the spatial segment AB

The image "g" of G, divisor of AB in the relationship (p, q), has for coordinates:

$$D \cdot \vec{Jg} = pK(a)\vec{Jb} + qK(b)\vec{Ja}$$

$$D = pK(a) + qK(b) - pq(ab)^2/4R^2 .$$

(29)

The horizontal expression is not identical to Eq. (18), because the spherical arc $\stackrel{\frown}{AB}$ and the rectilinear arc $\overline{AB}$ are not compatible. (They are however tangent at a and b in the horizontal projection.)

Azimuths

The correction of direction is generalized when "b" approaches infinity. One can thus calculate, in the image space and in proximity to "a", the fundamental vector transformations, the physical vertical, parallel to the polar axis - and as a result, evaluate the azimuth AZ and the Zenithal distance DZ of the segment $\overline{AB}$ from the analogous elements in the STEREO image space. These calculations, as well as the corresponding differential terms (d(AB), d(AZ), d(DZ)), make finite formulae for a length of segment ab.

The fact that we have defined a rigorously conformal transformation is naturally essential for the rigour and the generality of the results.

9. POSSIBLE USES

9.1 The three-dimensional stereographic representation

The representation that one arrives at is the simplest of the conformal two dimensional representations - and at the same time, has a precise meaning for the third dimension.

We can say however, that there is a small difference between the stereographic representation of the ellipsoid as it has been defined in Section 4 and the image of the ellipsoid as it has been determined by the coordinates (X, Y) of the inverse space. But these differences are inconceivably small with respect to the facilities offered by the third dimension: we really consider the problem of the straight line (AB) in space without projection onto any surface.

Equations (26) and (27) are remarkable in their simplicity and diversity of use, a practical limitation issues from the variation of the scale and the difficulty of finding a working relationship between the variable Z and the altitude H (an extent of 3000 km can be proposed for practical purposes).

9.2 The refined analysis of three-dimensional networks

The regional inverse reference system allows one to maintain the advantage of the three-dimensional mode whilst carefully distinguishing the third dimension from the other two.

One has to be able to put formulae (2) into the following form:

$$dX = X' - X = T_X \left[K - \frac{X^2}{2R^2}\right] - T_Y \left[\frac{XY}{2R^2}\right] - T_Z \frac{X}{R} \left(1 - \frac{Z}{2R}\right) + \delta X$$

$$dY = Y' - Y = -T_X \left(\frac{XY}{2R^2}\right) + T_Y \left(K - \frac{Y^2}{2R^2}\right) - T_Z \frac{Y}{R} \left(1 - \frac{Z}{2R}\right) + \delta Y \qquad (30)$$

$$dZ = Z' - Z = T_X \frac{X}{R} \left(1 - \frac{Z}{2R}\right) + T_Y \frac{Y}{R} \left(1 - \frac{Z}{2R}\right) + T_Z (2 - K)(1 - \frac{2Z}{R(2 - K)}) + \delta Z$$

$$\delta X = aX(1 - \frac{Z}{R}) - rY + q(Z + \frac{X^2 - Y^2 - Z^2}{2R})$$

$$\delta Y = rX + aY \left(1 - \frac{Z}{R}\right) - p(Z + \frac{Y^2 - X^2 - Z^2}{2R}) \qquad (31)$$

$$\delta Z = (-qX + pY)(1 - \frac{Z}{R}) + a(Z + \frac{X^2 + Y^2 - Z^2}{2R}) \ .$$

These formulae allow the use of the fundamental parameters: translation (T_X, T_Y, T_Z), scale (a), and rotation (p, q, r) with the rigorous coefficients. These elements are expressed on the regional axis, and with reference to the origin O (which perhaps can be taken on the ellipsoid (E)), the Z axis being the normal from O to E including the inversion pole P, situated at the anti-pole position of O on the bitangent sphere at that point (see Fig. 2). One will note that the rotations (p, q) about the horizontal axis are different from the rotation r about the vertical axis from O.

These formulae can be completed without difficulty for the planimetric variations by the complementary terms which add to (δX, δY):

$$\left| \begin{array}{l} \delta X' = bX + cY + eX^2 + fXY + gY^2 \ldots \\ \delta Y' = cX - bY + e'X^2 + f'XY + g'Y^2 \ldots \end{array} \right. \qquad (32)$$

permitting a more comprehensive analysis of the two networks, incorporating in particular the linear deformation[4]):

$$v = \sqrt{b^2 + c^2}$$

Note:

The formulae (30) and (31) are rigorously equivalent to (2), if one removes the principal part of the translations ($\sqrt{TX^2 + TY^2 + TZ^2} < 50^m$) - and if one takes account of the fact that the term ($Z^2/2R^2$)dZ (quite negligible) has been neglected.

9.3 The possible reference systems for studying the local gravity field

We have already evoked time and time again the possible use that can be made of the local inverse coordinates X, Y, Z obtained by an inversion (and a reflection) from the cartesian points

(U, V, W). If T is the potential harmonic perturbation in the original space, then $T' = T/\sqrt{K}$ is a potential harmonic image in the inverse space (X, Y, Z)[5].

An analysis of the problem, still in progress, tends to eliminate the representation of T by the Fourier double series in the plane 0, X, Y or by the use of BESSEL functions - in favour of the use of translated functions rapidly decreasing in the plane Z = 0, their effect over the whole space would be simple to calculate = the possibility to create grid squares on the plane, with binary sub-divisions, gives a real advantage for the process, since one can then greatly shorten the calculations by precalculating the number of influencial terms between the grid points at level 0.

10. CONCLUSION

It is not impossible to imagine a global system of regional inverse representations organized around an ellipsoid or a mean sphere, with centres positioned according to a system of origins which should be precisely given. They can also serve to support the exchange of information in three dimensions, in concurrence with the actual geographical coordinate system, and the altitude (λ, ϕ, H). The multiplicity of origins is perhaps an inconvenience, the possible advantage in large calculations being the absence of trigonometrical lines. The materialization of the X, Y coordinates so created, by stereographic maps can be an important advantage.

One will note that the system can be used at any point where the curvature of level lines is not far from being spherical (on the Earth, at a particular height, or on the planets) - and that it naturally incorporates all the local systems that one can create.

For example, a local CERN system could be defined, which covers an area of 100 km^2 around a central point, and an altitude range between 0 and 1000 m. The system consists of (X, Y, Z) coordinates that can be considered as follows:

- X, Y are the stereographic map coordinates of an ellipsoid, with a precision which is better than
  0m,000 at altitude 0
  0m,050 for X = 0, Y = ±100 km, Z = 1 km
- Z represents the altitude:
  at level 0   , better than 2m,45 at X = 0; Y = ±100 km
  at level 1 km, better than 0m,16 at X = Y = 0
                better than 2m,57 at X = 0; Y = ±100 km

The spatial scale coefficient K, related to the origin, is a maximum when Z = 0, at the limit of the zone (K = 1,000062), and a minimum when Z = 1 km at X = Y = 0 (K = 0.999843).

The corrections of the direction for a distance ab = 10 km, correspond to a shift of the observed point b:

- Radially, towards the exterior, a distance of 15 mm,4 at 25 km from the centre,
- Vertically (downwards) a distance of the order of 7 m,80 (which approximately corresponds to the classical depression of the horizon).

REFERENCES

1) H.M. Dufour, The whole geodesy without ellipsoid. B.G. No. 88 (1968), p. 125 à 143.

2) H.M. Dufour, Géodésie tridimensionnelle et systèmes laplaciens de référence, RETRIG. Publication No. 14. München (1982).

3) T. Vincenty, Methods of adjusting space systems data and terrestrial measurements. B.G. No. 56,3 (1982), p. 231-241.

4) A. Fotiou and E. Livieratos, A strain interpretation in the comparison of network coordinate differences due to datum change. B.G: No. 58, 4 (1984(, p. 527 à 539.

5) H.M. Dufour, Utilisation possible de l'inversion tridimensionnelle dans les problèmes de Geophysique - 5th International Symposium - Geodesy and Physics of the Earth - Magdebourg (1984).

ON CONTINUOUS THEORY FOR GEODETIC NETWORKS

Kai Borre

Aalborg University, Fibigerstræde 11, DK-9220 Aalborg Øst, Denmark.

Abstract
Geodetic levelling networks and networks with distance measurements are analysed in order to elucidate their special features. This analysis naturally leads one to look for continuous analogons of these networks. By means of tools known from the method of finite elements we derive the corresponding Green's functions for various boundary value problems. The Green's functions act as a formal covariance function for the corresponding least squares problem and provide information on the general error behaviour in the networks. The theory is illustrated with examples relevant to the CERN survey problems.

1. INTRODUCTION

During the last decade the theoretical analysis of geodetic networks has advanced tremendously. Not least, the geodetic literature about optimization problems of various kinds has left an overwhelming impression.

For a better and more comprehensive understanding of the special features of geodetic networks, I am still of the opinion that we need more basic knowledge of the real matter, before we can solve most optimisation problems in a professional manner.

The number of mathematical theories is tremendous, but the problem is to introduce into geodesy only the relevant ones or, in the words of Robert M. Pirsig[1], "The purpose of scientific method is to select a single truth from among many hypothetical truths. That, more than anything else, is what science is about. But historically science has done exactly the opposite".

The mathematical analysis of geodetic networks is mostly based on matrix algebra, theory of elliptic difference and differential equations, Fourier analysis, statistical estimation theory, theory of electrical networks, elasticity theory and above all the finite element method which has very much in common with the theory of geodetic networks.

As the main aim of this paper is to emphasize the interplay between discrete and continuous models I shall now examine this topic. The original problem can be traced back to ideas published by G. Förster in 1931. He wanted to establish a numerical model for combining neighbouring networks with some common boundary points. (In fact, F.R. Helmert considered that problem already in 1893.) Förster proposed a solution which combines the two networks by means of a function of place which turns out mainly to be a conformal mapping. In later publications he uses concepts like "quadratic sum of residuals per unit area or the strength", homogeneity and isotropy, free networks, rotation of networks, scale-factor of net deformation, conformal mapping, etc. It is obvious that Förster was inspired by the theory of elastically deformable solids.

The finite element method combined with the ideas of Förster led to the publication by T. Krarup and myself[2] of the paper "Foundation of a Theory of Elasticity for Geodetic Networks".

The mathematical abstract of the method is the following:

The usual adjustment problem of a single triangle is set up. Next, the original observations are transformed into so-called pseudo-observations. In levelling they represent the slope of terrain in the direction of the coordinate axes and the closing error of this triangle. In the case of distance and azimuth networks they are linear functions of the elements of the metric tensor. Henceforth we calculate the weighted square sum of residuals. The kernel of this symmetric form is analogous to what in elasticity theory is called the stiffness matrix.

Now we look upon the network as a whole consisting of such elementary triangles. Here the crucial point appears, namely how we prolongate the discrete scalar and vector fields to continuous functions. We solve this problem by following simple principles from the method of finite elements. The weighted square sum of the residuals can now be extended to an approximating Riemann-Stieltjes' sum over the entire network. The search for a minimum of this sum leads to a variational problem the solution of which is a boundary value problem of the Neumann type.

This boundary value problem is the continuous analogue to the normal equations. The inverse of the partial differential operator (including the boundary conditions) is an integral operator, the kernel of which is the Green's function of the operator concerned. The Green's function acts analogously to the usual covariance matrix of the original discrete adjustment problem.

So the method yields one Green's function which is an approximating substitute for an n-dimensional covariance matrix. Instead of trying to grasp an n-dimensional covariance matrix we are furnished with a single covariance function which depends on certain characteristic parameters for the network. Such a simplification of course has its costs, e.g. we lose information about local phenomena of the network.

Let us illustrate the whole procedure by the following simple but characteristic example:

Along a line of length L we have observed the length of n sub-intervals.
As formal observation we introduce the logarithm of the length of the sub-
intervals rather than the lengths themselves:

$$\ln(x^0_{i+1}+u_{i+1}-x^0_i-u_i) = \ell_{i,i+1}+r_{i,i+1} \tag{1}$$

where x^0_i denotes the preliminary value of the abscissa of subpoint i, $\ell_{i,i+1}$
is the observational value, $r_{i,i+1}$ the residual and u_i the correction to the
preliminary abscissa x^0_i. (1) is expanded in a truncated Taylor series of order 1:

$$\ln(x^0_{i+1}-x^0_i)-\frac{u_i}{x^0_{i+1}-x^0_i}+\frac{u_{i+1}}{x^0_{i+1}-x^0_i} = \ell_{i,i+1}+r_{i,i+1}. \tag{2}$$

We introduce the following abbreviations

$$\epsilon_i = x^0_{i+1} - x^0_i$$
$$f_i = \ell_{i,i+1}-\ln(x^0_{i+1}-x^0_i)$$

and prescribe

$$u_{i+1}-u_i = \epsilon_i u'.$$

Then (2) reads

$$-f_i + u' = r_{i,i+1}. \tag{3}$$

The weighted square sum of residuals for the whole line L is then

$$\Delta E = \sum_{i=1}^{n} \frac{p_i}{\epsilon_i}(u'-f_i)^2. \tag{4}$$

At this point it is appropriate to make a remark. We are interested in
making a transcription of our algebraic problem into the language of differential operators. So we are discussing what is going to happen as we bring
our sub-points closer and closer together. We are obviously interested in the
limit to which our construction tends as the number of sub-points increases
to infinity by letting ϵ_i decrease to zero. In order to save the valuable
feature of a metrical space in relation to the study of continuous operators

we have to modify the usual concept of weight p_i by dividing it by ϵ_i exactly as shown in (4). This is an important modification!

Next we look for the minimum for E. Assuming $p_i/\epsilon_i=1$, which leads to a simple result and is no serious restriction, we get the result

$$-u'' = f'. \tag{5}$$

For a so-called free network we have the boundary conditions

$$u'(0) = u'(L) = 0. \tag{6}$$

The generalized Green's function for this problem is

$$G(u,v) = \begin{cases} \dfrac{u^2+v^2}{2L} + \dfrac{L}{3} - u & \text{for } v \le u \\ \dfrac{u^2+v^2}{2L} + \dfrac{L}{3} - v & \text{for } v \ge u. \end{cases} \tag{7}$$

This Green's function acts as a formal covariance function for the adjustment problem. An example is the variance of the difference of abscissae ϕ_{pq} of any two points p and q as follows:

$$\phi_{pq} = x_p - x_q, \qquad p < q$$

$$\sigma^2(\tilde{\phi}) = G(p,p) + G(q,q) - G(p,q) - G(q,p)$$

$$= (\frac{p^2}{L} + \frac{L}{3} - p) + (\frac{p^2}{L} + \frac{L}{3} - p) - (\frac{p^2+q^2}{2L} + \frac{L}{3} - p)$$

$$- (\frac{p^2+q^2}{2L} + \frac{L}{3} - q)$$

$$= q-p \tag{8}$$

i.e. the variance of a difference of length is proportional to itself. This result is in good agreement with the result known from the discrete case.

2. SYSTEMATIC ANALYSIS OF CERN-LIKE NETWORKS

What are the special features of the CERN survey problems?

The detail network is (nearly) 1-dimensional, but along the circumference of a circle, and mainly distance observations are performed. So we shall concentrate on a thorough analysis of the error behaviour of that type of network.

In the following investigations we make certain assumptions in order to obtain explicit results. All observations are taken with variance σ^2, and they are supposed to be uncorrelated, i.e. the a priori covariance matrix is $\sigma^2 I$. Furthermore, observations are only performed between neighbouring points which are equally spaced.

In an appendix we include a systematic investigation of 1-dimensional networks. Remember that levelling and distance observations give rise to similar normal equation matrices.

2.1 Distance observations along a straight line

For reasons of reference we shall briefly repeat the characteristics about distance observations along a straight line.
The normals for a so-called free network are, with the above-mentioned assumptions, proportional to

$$
N = \begin{bmatrix}
1 & -1 & & & & \\
-1 & 2 & -1 & & & 0 \\
& & \cdot & \cdot & \cdot & \\
& 0 & & -1 & 2 & -1 \\
& & & & -1 & 1
\end{bmatrix}
\tag{9}
$$

cf. $N_3 = R_n(2,-1,-1)$ in the appendix. Obviously, N is singular, so for its inverse we use the uniquely determined pseudo-inverse N^+. The elements of N^+ are given as:

$$
\frac{1}{2n}
\begin{bmatrix}
i^2 - i + \dfrac{(n+1)(2n+1)}{3} + j(j-1-2n) & \text{for } i \leq j \\[4mm]
i^2 - (2n+1)i + \dfrac{(n+1)(2n+1)}{3} + j(j-1) & \text{for } i \geq j.
\end{bmatrix}
\tag{10}
$$

The variance of a difference of distance between points P and Q, say with ordinates p and q, p < q, is

$$
\sigma^2{}_{PQ} = N^+_{PP} + N^+_{QQ} - 2N^+_{PQ} = \frac{1}{2n}\left[p^2 - p + \frac{(n+1)(2n+1)}{3} + p(p-1-2n) \right]
$$
$$
+ \frac{1}{2n}\left[q^2 - q + \frac{(n+1)(2n+1)}{3} + q(q-1-2n) \right] - \frac{1}{n}\left[p^2 - p + \frac{(n+1)(2n+1)}{3} \right.
$$
$$
\left. + q(q-1-2n) \right] = q-p.
\tag{11}
$$

We see, as we also expected, that the variance of a distance difference grows in proportion to the difference itself.

The continuous analogon to the problem just treated is the following boundary value problem:

$$
-u''(\xi) = \delta
$$

with
$$
u'(0) = u'(L) = 0.
\tag{12}
$$

The (generalized) Green's function for the problem (12) is

$$G(x,\xi) = \begin{cases} \dfrac{x^2+\xi^2}{2L} + \dfrac{L}{3} - x & \text{for } \xi < x \\[2mm] \dfrac{x^2+\xi^2}{2L} + \dfrac{L}{3} - \xi & \text{for } \xi > x, \end{cases} \tag{13}$$

cf. the appendix.

For the same variance we now have

$$\sigma^2_{PQ} = (\frac{p^2+p^2}{2L}+\frac{L}{3}-p)+(\frac{q^2+q^2}{2L}+\frac{L}{3}-q)-2(\frac{p^2+q^2}{2L}+\frac{L}{3}-q)$$

$$= q-p. \tag{14}$$

This result is exactly the same as that one obtained in the discrete case (11). The approximation error is zero.

2.2 Distance observations along the circumference of a circle

In this case the normals are modified slightly to

$$N = \begin{bmatrix} 2 & -1 & 0 & \ldots & & \ldots & -1 \\ -1 & 2 & -1 & 0 & \ldots & & \ldots & 0 \\ 0 & -1 & 2 & -1 & 0 & \ldots & 0 \\ & & \ddots & -1 & 2 & -1 \\ -1 & 0 & \ldots & & \ldots & 0 & -1 & 2 \end{bmatrix} . \tag{15}$$

But this modification has great consequences. By the way, in the case of observations along a circle the boundary conditions are of no interest, just because of the absence of a boundary!

This matrix - a circulant matrix - is also singular, so we look for its pseudo-inverse. It is explicitly given by

$$N^+_{ij}=\frac{1}{2n}\left[\frac{n^2-1}{6}-n(i-j)+(i-j)^2\right] \text{ for } i \geq j \tag{16}$$

and is also symmetric.

The variance of distance difference along the circumference is

$$\sigma^2_{PQ}= \frac{1}{2n}\left[\frac{n^2-1}{6}+\frac{n^2-1}{6}-2(\frac{n^2-1}{6}-n(p-q)+(p-q)^2)\right]$$

$$= (p-q)-\frac{(p-q)^2}{n}. \tag{17}$$

In comparing (14) and (17) we note that the variance on the circle has decreased as compared to the "free-line" case. This was also to be expected. The variance has a maximum, when the points P and Q are situated diametrically opposite!

The continuous formulation of the problem is

$$-u''(\phi) = \delta$$
$$u'(0) = u'(2\pi) \tag{18}$$

where the solution now has to be periodic with the period 2π.

We obtain the Green's function formally as the solution B.3 in the appendix with x and ξ substituted by the angles ϕ and ψ, and $L = 2\pi r$, where r denotes the radius of the circle:

$$G(\phi,\psi) = \begin{cases} \dfrac{\phi^2+\psi^2}{4\pi r} + \dfrac{2\pi r}{3} - \phi & \text{for } \psi \le \phi \\[2mm] \dfrac{\phi^2+\psi^2}{4\pi r} + \dfrac{2\pi r}{3} - \psi & \text{for } \psi \ge \phi. \end{cases}$$

Next we obtain

$$\sigma^2_{PQ} = (\psi-\phi)\sigma^2. \tag{19}$$

Seemingly, r has been eliminated, but observe the following fact. The distance PQ along the circumference of the circle with radius r is, of course, r times the corresponding one on the unit circle. If we want to preserve the same number of network points in the sector PQ along the circle with radius r as along the unit circle, there will be r times the distance on the unit circle between the points, and consequently the variance increases (possibly proportional to r, but this depends on the specific circumstances of the observational method).

There exists an alternative method for obtaining N^+, viz. through the spectral decomposition of (15).

We shall report the results in detail, as it may be of interest to geodesists. Those interested in the procedure used in deriving the results are advised to consult Polozhii[3].

The eigenvalues λ_j of N in (15) are given as

$$\lambda_j = 4 \sin^2 \frac{(j-1)\pi}{n}, \quad j = 1,2,3,\ldots,n. \tag{20}$$

Immediately, we observe that $0 \le \lambda_j \le 4$, but also it turns out to be essential to distinguish between whether n is odd or even. In both cases we have $\lambda_1 = 0$, while for n even

$$\lambda_{\frac{n}{2}+1} = 4.$$

In order to describe the set of eigenvectors in an easy manner we first **reorder** the eigenvalues determined by (20) according to magnitude. Then the corresponding eigenvectors are as follows - in the right succession, and with $\alpha = \frac{2(j-1)\pi}{n}$:

n odd

$$\Phi = \sqrt{\frac{2}{n}}\begin{bmatrix} \frac{1}{\sqrt{2}}\cos\alpha & \sin\alpha & \cos 2\alpha & \sin 2\alpha,\ldots,\cos\frac{(n-1)\alpha}{2} & \sin\frac{(n-1)\alpha}{2} \\ \frac{1}{\sqrt{2}}\cos 2\alpha & \sin 2\alpha & \cos 4\alpha & \sin 4\alpha,\ldots,\cos\frac{2(n-1)\alpha}{2} & \sin\frac{2(n-1)\alpha}{2} \\ \frac{1}{\sqrt{2}}\cos 3\alpha & \sin 3\alpha & \cos 6\alpha & \sin 6\alpha,\ldots,\cos\frac{3(n-1)\alpha}{2} & \sin\frac{3(n-1)\alpha}{2} \\ \ldots & & & & & \ldots \\ \frac{1}{\sqrt{2}}\cos n\alpha & \sin n\alpha & \cos 2n\alpha & \sin 2n\alpha,\ldots,\cos\frac{n(n-1)\alpha}{2} & \sin\frac{n(n-1)\alpha}{2} \end{bmatrix}. \quad (21)$$

n even

$$\Phi = \sqrt{\frac{2}{n}}\begin{bmatrix} \frac{1}{\sqrt{2}}\cos\alpha & \sin\alpha & \cos 2\alpha & \sin 2\alpha,\ldots,\cos\frac{(n-2)\alpha}{2} & \sin\frac{(n-2)\alpha}{2} & \frac{1}{\sqrt{2}} \\ \frac{1}{\sqrt{2}}\cos 2\alpha & \sin 2\alpha & \cos 4\alpha & \sin 4\alpha,\ldots,\cos\frac{2(n-2)\alpha}{2} & \sin\frac{2(n-2)\alpha}{2} & \frac{1}{\sqrt{2}} \\ \frac{1}{\sqrt{2}}\cos 3\alpha & \sin 3\alpha & \cos 6\alpha & \sin 6\alpha,\ldots,\cos\frac{3(n-2)\alpha}{2} & \sin\frac{3(n-2)\alpha}{2} & \frac{1}{\sqrt{2}} \\ \ldots & & & & & & \ldots \\ \frac{1}{\sqrt{2}}\cos n\alpha & \sin n\alpha & \cos 2n\alpha & \sin 2n\alpha,\ldots,\cos\frac{n(n-2)\alpha}{2} & \sin\frac{n(n-2)\alpha}{2} & \frac{1}{\sqrt{2}} \end{bmatrix}. \quad (22)$$

Examples

n=3 $N^+ = \Phi\,\Lambda^+\,\Phi^T$

where Λ^+ is a diagonal matrix with the inverse eigenvalues - in the reordered succession. If $\lambda_k = 0$, then $\lambda_k^+ = 0$.

$$\Phi = \sqrt{\frac{2}{3}}\begin{bmatrix} \frac{1}{\sqrt{2}} & \cos\frac{2\pi}{3} & \sin\frac{2\pi}{3} \\ \frac{1}{\sqrt{2}} & \cos\frac{4\pi}{3} & \sin\frac{4\pi}{3} \\ \frac{1}{\sqrt{2}} & \cos\frac{6\pi}{3} & \sin\frac{6\pi}{3} \end{bmatrix} = \sqrt{\frac{2}{3}}\begin{bmatrix} \frac{1}{\sqrt{2}} & -\frac{1}{2} & \frac{\sqrt{3}}{2} \\ \frac{1}{\sqrt{2}} & -\frac{1}{2} & -\frac{\sqrt{3}}{2} \\ \frac{1}{\sqrt{2}} & 1 & 0 \end{bmatrix}$$

$$\Lambda^+ = \text{diag}(0, \tfrac{1}{3}, \tfrac{1}{3})$$

$$N^+ = \frac{1}{9} \begin{bmatrix} 2 & -1 & -1 \\ -1 & 2 & -1 \\ -1 & -1 & 2 \end{bmatrix},$$

n=4

$$\Phi = \frac{1}{\sqrt{2}} \begin{bmatrix} \frac{1}{\sqrt{2}} & 0 & 1 & \frac{-1}{\sqrt{2}} \\ \frac{1}{\sqrt{2}} & -1 & 0 & \frac{1}{\sqrt{2}} \\ \frac{1}{\sqrt{2}} & 0 & -1 & \frac{-1}{\sqrt{2}} \\ \frac{1}{\sqrt{2}} & 1 & 0 & \frac{1}{\sqrt{2}} \end{bmatrix} \quad , \quad \Lambda^+ = \text{diag}(0,\frac{1}{2},\frac{1}{2},\frac{1}{4})$$

$$N^+ = \frac{1}{16} \begin{bmatrix} 5 & -1 & -3 & -1 \\ -1 & 5 & -1 & -3 \\ -3 & -1 & 5 & -1 \\ -1 & -3 & -1 & 5 \end{bmatrix}.$$

2.3 Distance observations in an annulus

Now one can argue that the CERN networks are not of the simple line nature discussed so far. In order to deal with this objection we shall report some results covering this case.

The fundamental figure consists of a quadrilateral with measured diagonals. This case has just been investigated by Meissl[4] who also includes an additional observation, but this does not change the results qualitatively.

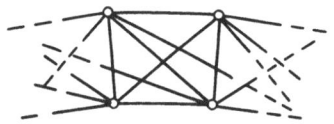

The variance in the tangential direction is asymptotically proportional to $\frac{n}{12}$ and in the normal direction asymptotically proportional to $\frac{n^3}{720}$, where n denotes the number of quadrilaterals around the circle. The reader is referred to Meissl[4] for further details.

In the continuous case we shall derive the Green's function for an annulus. The problem has, with Neumann boundary conditions, the following formulation:

$$-\Delta u(z) = g(z) \qquad \text{in } \Omega \tag{23}$$

$$\frac{\partial u(z)}{\partial r} = 0 \qquad \text{on } \omega . \qquad (24)$$

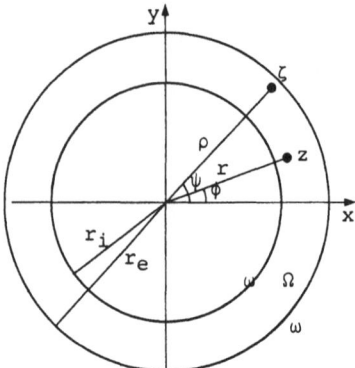

The corresponding Green's function $G(z,\zeta)$ shall satisfy the following conditions, cf. Meschkowski[5]:

(i)   For a fixed $\zeta \epsilon \Omega$ and $z \epsilon \Omega + \omega$ G must be harmonic for any $z \neq \zeta$,

(ii)   The function $G + \ell n |z - \zeta|$ must be harmonic in the neighbourhood of $z = \zeta$,

(iii)   At the boundary $z \epsilon \omega$, $\zeta \epsilon \Omega$, $\frac{\partial G(z,\zeta)}{\partial r_z} =$ constant for a fixed z.

The index z indicates that the differentiation has to be performed with respect to the variable z.

Furthermore, at the boundary G has to be normalized through the condition

$$\int_\omega G(z,\zeta) d\omega = 0 . \qquad (25)$$

Then in $\Omega + \omega$ the harmonic function $u(z)$ may be obtained in the following manner

$$u(\zeta) = \frac{1}{2\pi} \int_{r_i}^{r_e} \int_0^{2\pi} G(z,\zeta) \frac{\partial g(z)}{\partial r} d\Omega . \qquad (26)$$

By standard methods we obtain for G

$$G(r,\phi;\rho,\psi) = \frac{r_i \ell n r - r_e \ell n \rho}{r_i + r_e} - \frac{r_i^2 \ell n r_i - r_e^2 \ell n r_e}{(r_i + r_e)^2}$$

$$+ \sum_{n=1}^\infty \frac{\left[\left(\frac{r_i}{r}\right)^n + \left(\frac{r}{r_i}\right)^n\right]\left[\left(\frac{\rho}{r_e}\right)^n + \left(\frac{r_e}{\rho}\right)^n\right] \cos n(\phi - \psi)}{n\left[\left(\frac{r_e}{r_i}\right)^n - \left(\frac{r_i}{r_e}\right)^n\right]}, \qquad \text{for } r < \rho. \quad (27)$$

3. CONCLUSION

By a simple example it was demonstrated how we can proceed from a discrete network to a continuous analogous one. The Green's function for the continuous case acts as a formal covariance function which describes the propagation of random errors. In the appendix the influence of various boundary conditions is studied in more detail.

As specific examples relevant to the CERN networks, we investigated circulant matrices and the pertaining error propagation. In nearly all cases we are faced with error functions which are described by means of various powers of n. Only in the case of the annulus do these functions change character to be functions involving the logarithm of n. This feature reflects the general behaviour. 1-dimensional networks are described by powers of the distance between points, 2-dimensional networks with the logarithm of the distance between points, and 3-dimensional networks with the inverse of the distance between points.

The theory is described in a more elaborate publication which will be published in the near future. The title is: Untersuchung geodätischer Netze mittels partieller Differentialoperatoren.

APPENDIX

COVARIANCE MATRICES/FUNCTIONS FOR VARIOUS 1-D GEODETIC PROBLEMS

A.   DISCRETE CASES

Consider a straight line with n equidistant points. Any height or distance difference between neighbouring points is measured with variance σ^2. Furthermore, all observations are considered to be uncorrelated.
    Thus the weight matrix is $\sigma^2 I$. Using the method of least squares we ob-

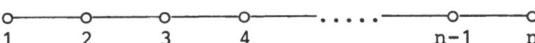

tain the following coefficient matrix of the normal equations (adjustment by variation of parameters).

(i)   Control is given at points 1 and n:

$$
\underset{n,n}{N_1} =
\begin{bmatrix}
2 & -1 & & & & & \\
-1 & 2 & -1 & & & 0 & \\
& -1 & 2 & -1 & & & \\
& & & \cdot\ \cdot\ \cdot & & & \\
& & & & -1 & 2 & -1 \\
0 & & & & & -1 & 2
\end{bmatrix} .
$$

(ii) Control is given only at point 1:

$$
\underset{n,n}{N_2} =
\begin{bmatrix}
2 & -1 & & & & & \\
-1 & 2 & -1 & & & 0 & \\
& -1 & 2 & -1 & & & \\
& & & \cdot\ \cdot\ \cdot & & & \\
& & & & -1 & 2 & -1 \\
0 & & & & & -1 & 1
\end{bmatrix} .
$$

(iii) No control available at all. Free network:

$$
\underset{n,n}{N_3} =
\begin{bmatrix}
1 & -1 & & & & & \\
-1 & 2 & -1 & & & 0 & \\
& -1 & 2 & -1 & & & \\
& & & \cdot\ \cdot\ \cdot & & & \\
& & & & -1 & 2 & -1 \\
0 & & & & & -1 & 1
\end{bmatrix} .
$$

In the sequel we shall use a notation (apart from the sign) introduced by D.E. Rutherford[6]:

$$R_n(x,a,b) = \begin{bmatrix} x+b & -1 & & & & \\ -1 & x & -1 & & & 0 \\ & -1 & x & -1 & & \\ & & \cdot & \cdot & \cdot & \\ & & & -1 & x & -1 \\ 0 & & & & -1 & x+a \end{bmatrix}.$$

The subscript n denotes the dimension of the tri-diagonal symmetric matrix. Hence

$$N_1 = R_n(2,0,0)$$
$$N_2 = R_n(2,-1,0)$$
$$N_3 = R_n(2,-1,-1).$$

Next we give a useful summary of characteristics of these matrices and their inverses.

(i) $R_n(2,0,0)$
_____

The eigenvalues are $\lambda_i(R_n(2,0,0)) = 4 \sin^2 \frac{i\pi}{2(n+1)}$, $i = 1,2,\ldots,n$, and the normalized eigenvectors are

$$\psi_i(R_n(2,0,0)) = \sqrt{\frac{2}{n+1}} \sin \frac{ij\pi}{n+1}, \quad i,j = 1,2,\ldots,n.$$

The matrix composed of the eigenvectors is not positive definite.

Let cond(A) denote the spectral condition number of a symmetric, real matrix A, defined by

$$\text{cond}(A) = \frac{\lambda_{max}(A)}{\lambda_{min}(A)}$$

then

$$\text{cond}(R_n(2,0,0)) \sim \frac{4(n+1)^2}{\pi^2}.$$

Further,

$$\det(R_n(2,0,0)) = n+1.$$

Finally

$$
R_n^{-1}(2,0,0) = \frac{1}{n+1}
\begin{bmatrix}
n & n-1 & n-2 & \cdots & 1 \\
n-1 & 2(n-1) & 2(n-2) & \cdots & 2 \\
n-2 & 2(n-2) & 3(n-2) & \cdots & 3 \\
\vdots & \vdots & \vdots & & \vdots \\
1 & 2 & 3 & \cdots & n
\end{bmatrix}
=
\begin{cases}
\dfrac{i(n+1-j)}{n+1} & \text{for } i \leq j \\[2ex]
\dfrac{j(n+1-i)}{n+1} & \text{for } i \geq j
\end{cases}
$$

(A.1)

(ii) $R_n(2,-1,0)$
___

$$
\det(R_n(2,-1,0)) = 1
$$

$$
R_n^{-1}(2,-1,0) =
\begin{bmatrix}
1 & 1 & \cdots & 1 \\
1 & 2 & \cdots & 2 \\
1 & 2 & & \\
\vdots & \vdots & & \\
1 & 2 & \cdots & n
\end{bmatrix}
= (\min(i,j)).
$$

(A.2)

$R_n^{-1}(2,-1,0)$ is a so-called Stieltjes matrix. Such a matrix has very nice features, especially in connection with SOR-methods. The optimal convergence rate can be predicted (see Varga[7]).

The largest eigenvalues of $R_n^{-1}(2,-1,0)$ are very well-conditioned and the smallest very ill-conditioned. For n=12 the first few λ's are of order unity, while the last three are of order 10^{-7}. With increasing values of n the smallest eigenvalues become progressively worse conditioned. As stated, $\det(R_n(2,-1,0))= 1$, but if the (1,n)-element is changed to $(1+\varepsilon)$ the determinant becomes $1 \pm (n-1)!\varepsilon$. If $\varepsilon = 10^{-10}$ and n=20 the determinant is changed from 1 to $(1-19!10^{-10}) = -1.2 \cdot 10^{7}$. Now, as the determinant equals the product of its eigenvalues, at least one eigenvalue of the perturbed matrix must be very different from that of the original matrix (see Wilkinson[8]).

(iii) $R_n(2,-1,-1)$
___

The eigenvalues are $\lambda_i(R_n(2,-1,-1)) = 4 \sin^2 \dfrac{(i-1)\pi}{2n}$, $i = 1,2,\ldots,n$ and the normalized eigenvectors

$$
\psi_i(R_n(2,-1,-1)) =
\begin{cases}
\dfrac{1}{\sqrt{n}} & i = 1 \\[2ex]
\sqrt{\dfrac{2}{n}} \cos\dfrac{(i-1)(2j-1)\pi}{2n} & i = 2,3,\ldots,n \\
& j = 1,2,\ldots,n
\end{cases}
$$

$$\text{cond}(R_n(2,-1,-1)) = \infty$$

$$\det(R_n(2,-1,-1)) = 0.$$

For the inverse we use the pseudo-inverse given by

$$R_n^+(2,-1,-1) = \frac{1}{2n}\left[\frac{1}{3}(n-1)(2n+5-6j)+(j-1)(j-2)+i(i-1)\right] \quad \text{for } i \leq j \quad\quad (A.3)$$

$$\text{else symmetric.}$$

B. CONTINUOUS CASES

(i) The continuous analogon is given through the problem:

$$-u''(\xi) = \delta$$

with boundary conditions $u(0) = u(L) = 0$. The corresponding Green's function is

$$G(x,\xi) = \begin{cases} \xi - \dfrac{x\xi}{L} & \text{for } \xi \leq x \\[2ex] x - \dfrac{x\xi}{L} & \text{for } \xi \geq x \end{cases} \quad\quad (B.1)$$

cf. Lanczos[9], Problem 203.

(ii) The problem is $\quad\quad -u''(\xi) = \delta$

with boundary conditions $u(0) = u'(L) = 0$. Solution

$$G(x,\xi) = \begin{cases} \xi & \text{for } \xi \leq x \\[1ex] x & \text{for } \xi \geq x \end{cases} \quad\quad (B.2)$$

cf. Lanczos[9], Problem 202.

(iii) The problem is $\quad\quad -u''(\xi) = \delta$

with boundary conditions $u'(0) = u'(L) = 0$. Solution

$$G(x,\xi) = \begin{cases} \dfrac{x^2+\xi^2}{2L} + \dfrac{L}{3} - x & \text{for } \xi \leq x \\[2ex] \dfrac{x^2+\xi^2}{2L} + \dfrac{L}{3} - \xi & \text{for } \xi \geq x \end{cases} \quad\quad (B.3)$$

cf. Lanczos[9], Problem 228.

C. MEAN VARIANCE OF ALL POINTS ALONG THE LINE

We define

$$\sigma^2_{mean} = \frac{1}{n} \, tr \, (R^{-1}(2,.,.)) \sigma^2,$$

cf. e.g. Meissl[4], eq. (6). Consequently

(i) According to (A.1) we get

$$\sigma^2_{mean} = \left\{ \frac{1}{n} \, \frac{1}{n+1} \sum_{i=1}^{n} i(n-1+i) \right\} \sigma^2 = \left(\frac{n}{6} + \frac{1}{3} \right) \sigma^2.$$

(ii) According to (A.2) we get

$$\sigma^2_{mean} = \left\{ \frac{1}{n} \sum_{i=1}^{n} i \right\} \sigma^2 = \left(\frac{n}{2} + \frac{1}{2} \right) \sigma^2.$$

(iii) According to (A.3) we get

$$\sigma^2_{mean} = \left\{ \frac{1}{n} \, \frac{1}{2n} \sum_{i=1}^{n} \left[\frac{1}{3}(n-1)(2n+5-6i) + (i-1)(i-2) + i(i-1) \right] \right\} \sigma^2$$

$$= \left(\frac{n}{6} - \frac{1}{6n} \right) \sigma^2.$$

Notice that we obtain minimum variance in the case of the free line!

Next, in the continuous case we define

$$\sigma^2_{mean} = \left(\frac{1}{L} \int_0^L G(x,x) \, dx \right) \sigma^2.$$

So

(i) According to (B.1) we get

$$\sigma^2_{mean} = \left\{ \frac{1}{L} \int_0^L \left(x - \frac{x^2}{2} \right) dx \right\} \sigma^2 = \frac{L}{6} \, \sigma^2.$$

(ii) According to (B.2) we get

$$\sigma^2_{mean} = \left\{ \frac{1}{L} \int_0^L x \, dx \right\} \sigma^2 = \frac{L}{2} \, \sigma^2.$$

(iii) According to (B.3) we get

$$\sigma^2_{mean} = \left\{ \frac{1}{L} \int_0^L \left(\frac{x^2}{L} - x + \frac{L}{3} \right) dx \right\} \sigma^2 = \frac{L}{6} \sigma^2 .$$

Notice that in the continuous case we miss the constant terms. In general the fixed-free-case (ii) is the poorest as far as the overall determination of points is concerned.

D.   VARIANCE OF POINT DIFFERENCES. APPROXIMATION ERRORS

We shall consider the variance of a point difference $H_Q - H_P$ between any two points P and Q with nodal numbers p and q, p < q. The discrete formulations are considered first:

(i) $\sigma^2_{PQ} = \left\{ R_n^{-1} (2,0,0)_{QQ} + R_n^{-1} (2,0,0)_{PP} - 2R_n^{-1} (2,0,0)_{PQ} \right\} \sigma^2$

$= \left\{ (q-p) - \frac{(q-p)^2}{n+1} \right\} \sigma^2 ,$ cf. eq. (A.1)

(ii) $\sigma^2_{PQ} = R_n^{-1} \left\{ (2,-1,0)_{QQ} + R_n^{-1} (2,-1,0)_{PP} - 2R_n^{-1} (2,-1,0)_{PQ} \right\} \sigma^2$

$= (q-p) \sigma^2 ,$ cf. eq. (A.2)

(iii) $\sigma^2_{PQ} = \left\{ R_n^{-1} (2,-1,-1)_{QQ} + R_n^{-1} (2,-1,-1)_{PP} - 2R_n^{-1} (2,-1,-1)_{PQ} \right\} \sigma^2$

$= (q-p) \sigma^2 ,$ cf. eq. (A.3)

The corresponding expressions for the continuous formulations follow from

$$\sigma^2_{PQ} = \left\{ G(q,q) + G(p,p) - 2 G(p,q) \right\} \sigma^2$$

(i) $\sigma^2_{PQ} = \left\{ (q-p) - \frac{(q-p)^2}{L} \right\} \sigma^2 ,$ cf. eq. (B.1)

(ii) $\sigma^2_{PQ} = (q-p) \sigma^2 ,$ cf. eq. (B.2)

(iii) $\sigma^2_{PQ} = (q-p) \sigma^2 ,$ cf. eq. (B.3).

There is an astonishingly good agreement between the exact discrete formulation and the corresponding results of the continuous approach. The latter is often easier to perform than the discrete one, thus we have demonstrated how the continuous approach is superior to the discrete analysis which is also the more difficult one. On the other hand, we normally lose the constant terms in the continuous solution.

* * *

REFERENCES

1) Pirsig, R.M., Zen and the art of motorcycle maintenance, the Bodley Head, London, Sydney, Toronto, (1974), p. 116.

2) Borre, K. & T. Krarup, Foundation of a theory of elasticity for geodetic networks, 7th Nordic Geodetic Meeting, 6-10 May, København, (1974).

3) Polozhii, G.N., The method of summary representation for numerical solution of problems of mathematical physics, Pergamon Press, (1965), p. 250.

4) Meissl, P., Über zufällige Fehler in regelmäßigen gestreckten Ketten, Zeitschrift für Vermessungswesen, (1969), p. 14-26.

5) Meschkowski, H., Hilbertsche Räume mit Kernfunktionen, Springer, Berlin, Göttingen, Heidelberg, (1962), p. 83.

6) Rutherford, D.E., Some continuant determinants arising in physics and chemistry I, II. Proceedings of the Royal Society of Edinburgh. Vol. LXII, p. 229-236, (1947) and Vol. LXIII, p. 232-241, (1952).

7) Varga, R.S., Matrix iterative analysis, Prentice-Hall, (1962), p. 85.

8) Wilkinson, J.H., The algebraic eigenvalue problem, Clarendon Press, Oxford, (1965), p. 92.

9) Lanczos, C., Linear differential operators, D. Van Nostrand, London, (1961).